The Evolution Wars

The Evolution Wars

A Guide to the Debates

Michael Ruse

Foreword by Edward O. Wilson

ABC-CLIO

Santa Barbara, California
Denver, Colorado
Oxford, England

Library of Congress Cataloging-in-Publication Data
Ruse, Michael.
The evolution wars : a guide to the debates / Michael Ruse.
    p.   cm.
Includes bibliographical references and index.
   ISBN 1-57607-185-5 (hardcover : alk. paper)
   1. Evolution (Biology)—History.   I. Title.
QH361.R874   2000
576.8'09—dc21   00-009037
                                                 CIP

06   05   04   03   02   01   00   10 9 8 7 6 5 4 3 2 1

ABC-CLIO, Inc.
130 Cremona Drive, P.O. Box 1911
Santa Barbara, California 93116-1911

This book is printed on acid-free paper ∞.
Manufactured in the United States of America.

For my friends and fellow teachers
Ronald Brooks
Brian Calvert
Jay Newman

Contents

Foreword

Edward O. Wilson

The history of biology has been ruled by two very powerful and fruitful ideas. The first is that all living structures and processes, from heredity and physiology to ecosystem cycles and the working of the brain, are organically based and obedient to the laws of physics and chemistry. This absolute physicalism of biology, after a tumultuous history in the nineteenth and early twentieth centuries, triumphed over the élan vital and other mystic forces of previous explanation and is today nowhere seriously challenged. Almost all the disputes in molecular and cellular biology, consequently, came down to data clashes and competing models that nevertheless fall within the secure framework of a universally materialist worldview. This is the reason that in mastering the principles of functional biology, thus how cells and organisms work, you need learn in addition very little history and virtually no philosophy.

The second, complementary ruling idea is that all of biological phenomena have arisen by the evolution of the aforementioned organic machinery. Further, the overwhelmingly most important creative force behind this historical process is natural selection. Ever since the idea of evolution by natural selection exploded forth from Charles Darwin's *Origin of Species* in 1859, it has proceeded not only to change biology but also to challenge—and dangerously so—those of humanity's most cherished self-images engendered by religion, philosophy, and ideology. And because evolution deals with the outer reaches of the space-time scale, encompassing particularities of whole populations as they change over generations, its study is technically more difficult than ordinary functional biology. Old and new controversies still roil the deeper waters of evolutionary scholarship. To understand these disputes, and the status of evolutionary thought generally, it is very useful to know both the fundamentals of the subject and something of its history and philosophy.

It is therefore important to employ a seasoned guide who is an expert on the substance of the key controversies as well as their historical origins. Michael Ruse, although at times a participant himself, has kept the equilibrium of a good war correspondent whose dispatches are consistently reliable even in the midst of noisy bombardments. I trust him and enjoy reading his work even though he has the irritating habit of frequently disagreeing with me in print.

Let me put my endorsement another way. Suppose I were told that all my memory of the evolution controversies, from Darwin's time forward, were to be erased an hour hence, and, before this calamity (there have been times I would have thought it a blessing) I were allowed to choose a book to begin my reeducation. I would select, and therefore here recommend, for clarity and good humor as well as substance, *The Evolution Wars.*

Acknowledgments

My hope of course is that this book can be read with interest and profit by anyone, but if I were pressed to name a specific target audience, I would say that the reader I had most in mind is the person who does not know a great deal about evolutionary thinking but who has heard enough to want to find out more. In other words, I have in mind students of all ages. I have myself been a student and teacher all of my life, for thirty-five years now as a professor at the University of Guelph in Canada. For this reason I wanted particularly to dedicate this book to three men who have taught alongside me, not merely offering me friendship and encouragement but also models of how our job should be done. Brian Calvert and Jay Newman are fellow members of the Philosophy Department, and Ron Brooks—with whom for many years now I have team-taught a course on the philosophy of biology—is a fellow member of the Zoology Department. I truly have been their student.

I am greatly indebted to the editors at ABC-CLIO both for suggesting the idea of this book and for their friendly help and advice during its writing and production: Kristi Ward, Kevin Downing, and Connie Oehring. My secretary, Linda Jenkins, and my research assistants, Alan Belk and Aryne Sheppard, as always did those boring jobs that I should have done myself but did not. Once again my former student and present colleague in the Zoology Department, Dennis Lynn, took time and effort to go carefully over my Glossary, preventing my making several silly mistakes. And finally my family, my wife, Lizzie, and our children, Emily, Oliver, and Edward, showed their usual enthusiastic support for matters Darwinian and for those who write on such subjects.

Prologue

L ondon, the capital of England, is one of the great cities of the
world. Within its boundaries, on either side of the River
Thames, which winds and curls its way down to the sea, live over 10
million people. The home of the Queen and of the British Parliament,
London teems with government and with business, with factories and
with shops, with hotels and restaurants and theaters, with sports are-
nas and with prisons, and with much, much more. It is an old town,
existing long before the Romans came conquering nearly two thousand years ago. And every part has its history: the Tower with its stories of traitors and of executions and (thanks to the little princes) of
foul murder; the great railway stations, which saw so many young
men leave between 1914 and 1918, bound for France, never to return; Highgate Hill, where poor and dispirited Dick Whittington
heard the bells imploring him to return, "thrice Lord Mayor of London"; and today, a special favorite of mine, on the South Bank of the
Thames, a replica of the open-air Globe Theatre of William Shakespeare, which first saw the production of so many of his great plays.

A particular part of London's glory is its churches, especially
those wonderful buildings designed by Sir Christopher Wren and
erected in the frenzy of building that followed the Great Fire of London in 1666. Supreme among these is the Cathedral, Saint Paul's,
which stood defiant as London burned all around in the early months
of World War II during the German bomber blitz. Here was the scene
of some of the grandest state occasions: the burial of Sir Winston
Churchill and the ill-fated marriage of Prince Charles and Lady
Diana. Now go west, out of this vicinity, along the Strand, home of
some of the greatest hotels and dining places. Go past Trafalgar
Square, celebrating Admiral Lord Nelson's victory over the French at
Trafalgar, and turn left down Whitehall. Pass the Life Guards and the
entrance to Downing Street, home of the British Prime Minister, and
come to the end, before another great church. Smaller than Saint
Paul's, older and more sacred, Westminster Abbey is the true heart of
the city. It is here that the British monarchs are crowned; it is here

that you will find monuments to artists and poets and writers and musicians; and it is here that some of the truly great are buried.

Go down the nave and you will come to the grave of the incomparable Isaac Newton, the man who discovered the law of gravitational attraction and who thus could explain why Copernicus was right and why it is that the planets circle endlessly around the sun, never pausing, never stopping, always spinning and moving. It is true that Einstein's theories and discoveries in the twentieth century dislodged Newton's work from the pedestal on which it once stood, but no one now or in the past has ever failed to celebrate Newton's genius and his achievements. But move now to the side, where you will find another grave, that of Newton's countryman, living and dying nearly 200 years later. Charles Robert Darwin was laid to rest nearly a century and a quarter ago, yet his bones surely do not rest easily, even today. Like none other, he had and has his defenders: passionate defenders. Like none other, he had and has his critics: passionate critics.

Charles Darwin himself is controversial. There are those, scientists particularly, who see in Darwin the ideal researcher—dedicated, persistent, innovative, comprehensive, working patiently and professionally toward his ends, troubled by illness yet not distracted. A man for whom the truth is the only value appropriate for a scientist, himself willing to give and to sacrifice all to this end. At the same time, this Darwin is a man of personal generosity, offering friendship and support to all, close acquaintance and stranger alike. When his lieutenant Thomas Henry Huxley fell sick, it was Darwin who at once passed the hat, making a typically generous personal donation. This was the man he was. There are others, however, who see a different Darwin. These people, frequently trained professionally in history and related subjects such as cultural studies, see a man who is a classic upper-middle-class Victorian with the prejudices of that class: racist, sexist, chauvinist, capitalist. Their Darwin has a second-rate mind; he was one who stumbled upon ideas that were truly beyond his grasp; and he was a man who quite probably stole most of his discoveries anyway. Rather than a man of genuine warmth and generosity, these critics see a user who concealed a heart of ice behind a facade of congeniality. They see one who used his illness to avoid responsibility, and they find many failures stemming from Darwin's personal inadequacies.

Controversial though Darwin himself may be, this is nothing to the work he produced. At the center is his major book, *On the Origin of Species.* His supporters and enthusiasts regard this work as a paragon of scientific excellence, a model of how to do good science—clear, thorough, balanced, suggestive, innovative. They think Darwin anticipated problems and—the mark of really important science—left work to do for generations to come. Darwin's detractors, however, see a mishmash of ideas and suggestions and hypotheses and half thoughts—half-baked thoughts!—that were strung together without order or reason, not just in the *Origin* but also in a series of secondary writings of genuine Victorian length and tedium. And these were just the first editions. By the time Darwin had written and rewritten his works in the face of criticism, one was left with material that showed as many disparate pieces as a crazy quilt, and with about as much organization. Only those with their own personal agendas to satisfy could find in Darwin that of real worth and value.

Finally, there is Darwin's legacy. His supporters today—neo-, ultra-, or just plain Darwinians—think that he left us one of the most important theories humankind has yet discovered. After the *Origin,* our thinking about the world and about ourselves could never again be the same. Darwin's was a revolution that equaled that of Copernicus. Indeed, one might even say that in the secular realm, Darwin's ideas and influence equal those of Jesus Christ in the spiritual realm—equal and perhaps supersede. Never before or again can there be a body of work of this significance. But his detractors think just about the opposite, although they appreciate the dangers of Darwinism. They argue that Darwin's ideas are overblown, unsubstantiated, and little more than ideology—secular religion—masquerading as disinterested description and explanation. They think that Darwinians are deluded, arrogant, and mischievously influential, especially on the young. Destroying the legacy of Charles Darwin must be the aim and obligation of every right-thinking person.

Even politicians have entered into the act. In the past century, several U.S. states have banned the teaching of Darwin's ideas, and to this day we find boards of education warning teachers and students against the dangers of accepting his theories. Of course, you might respond, one should always keep an open mind about anything one is told, especially in science. Was it not the great philosopher Karl Popper who warned us that nothing in science is permanent, that every idea may and someday probably will fall to the ground? However, which high school teacher feels the need to caution about the Copernican revolution, telling the class that it may be necessary to revise and revamp, perhaps one future day going back to an earth-centered static universe? None, obviously! But in the case of Darwin, students are told to beware and to take heed. Perhaps one is going to be seduced from the true faith by vile heresies and misrepresentations.

Now, obviously, when people fall out like this, there is something interesting going on. There is no smoke without fire—although what is burning is perhaps another matter. Indeed, trying to find the answer to this puzzle is one of the reasons why I have written this book. I want to introduce you to Charles Darwin and to his ideas, to the people who came before him and the people who came after. I want to see what it is that makes people so passionate, either for or against. And achieving this aim is the reason why this book is structured as it is. I shall go more or less historically, from the past to the present, and each chapter will be introduced by a clash between people or groups: hence my title, *The Evolution Wars*. But although I love a good fight as much as anyone, truly I want to dig out, going behind the arguments and the polemics. I want to see why it is that people disagree and what is at stake and whether there was or is or ever could be a solution to what so divides the antagonists. A word of caution: I am not a social worker or psychiatrist, so frankly I do not care whether a resolution is ever reached or whether anyone feels happier when I have finished. I am a teacher, so I do care very much whether you understand a lot more when I am finished.

Which brings me to my final point, and then we can begin in earnest. You have a right to know where I stand. I think Darwin was a great scientist, and I think his ideas were truly important. Although I think he was often wrong—and I shall be telling you much more about his mistakes—I believe that essentially Darwin got it right. This mattered back then, and it matters right now. Darwin told us things of importance about the world and about ourselves. I think Darwin's ideas impinge on other areas of human inquiry and interest. Most importantly they rub up against religion, the Christian religion in particular. And those who say that religion and science can never be in conflict are deluding themselves. Science and religion can be at war, they have been at war, and Darwinism is right in the thick of it. But science and religion can work together; that is the other side to the story. And Darwinism is in the thick of this too. Those who say that religion and science must always be in conflict are likewise deluding themselves.

But I have said that I am a teacher, and I take that responsibility seriously in two ways. First, I am not here to convert you one way or the other. It is my job to give you the information, the tools, and then to let you work on things yourself. I can fault you on your knowledge of the facts, but when it comes to the interpretations, you are on your own. To be honest, I am indifferent as to whether you end up agreeing with me or disagreeing with me. I always tell my students that before I assign their marks, I do not look at their final sentences, in which they give their conclusions. I do care about the arguments they use to get to their conclusions, and I feel the same way about you. Agree or disagree with me as you wish, but show me that I should take you seriously.

And this brings up my second responsibility as a teacher. If, when you have finished, you do not care to argue with me, then I have let you down. Above all, Darwin, Darwinism, the Darwinian legacy, is absolutely fascinating. It is the story of terrific people and terrific ideas. These are important issues, and they matter—to me and to you. I am not going to trivialize, and I am not going to glamorize. You are not about to get the Disney version of Darwin. But I shall be very disappointed if you do not think that this topic is something that makes learning worthwhile. We may be grubby little primates on a grubby little planet, but every now and then we rise above ourselves. We escape the tawdry humdrum of everyday life and make sense of the Christian claim that through our intellect we are made in the image of God. Thinking on the questions raised in this book is one of those times.

Further Reading

The standard history of evolutionary thought is Peter Bowler's *Evolution: The History of an Idea,* 2d ed. (Berkeley: University of California Press, 1989). This is very comprehensive and fair, although to be honest a little bit of a textbook and reads like one. Very different is Robert J. Richards's *The Meaning of Evolution: The Morphological Construction and Ideological Reconstruction of Darwin's Theory* (Chicago: University of Chicago Press, 1992). Short, opinionated, brusque with the views of others, it is fun to read and legitimated by its author's very deep learning and understanding of his subject. My own *Monad to Man: The Concept of Progress in Evolutionary Biology* (Cambridge: Harvard University Press, 1996), is very long and detailed. Only graduate students working on their theses have to read it through from beginning to end; others should read the short introduction and then dip into it as it interests them. It is written in a kind of modular form so you can easily move around from point to another. You will find that there is a lot of detail about the personalities and ideas of many of the people mentioned in this book.

My *Mystery of Mysteries: Is Evolution a Social Construction?* (Cambridge: Harvard University Press, 1998) covers some of the same ground and is much easier going. It is the best short overall introduction to the history of evolutionary thought around at the moment. Also let me recommend the *Dictionary of Scientific Biography* (New York: Scribner, 1970, and supplementary volumes later).

There are many excellent articles on the major figures in the history of evolutionary thought and useful guides to further reading. For a recent guide, take a look at my review article: "The Darwin industry: a guide," *Victorian Studies* 39, no. 2 (1996): 217–235.

The Evolution Wars

1

Monad to Man:
The Birth of the Idea

18 30. Georges Cuvier was angry. And when Cuvier, the most powerful scientist in France, was angry, he was really livid. Pompous too. And very dangerous. He knew more than anyone else, and he set the standards and judged the results (Coleman 1964). You crossed him at your peril. For thirty years now he had been listening to this stupid, unfounded, dangerous nonsense from his fellow scientists. First there had been Jean Baptiste de Lamarck, and now when finally Lamarck had died and peace was in the offering, Etienne Geoffroy Saint Hillare—an old friend and a man who should have known much better—had taken up the cudgels and was promulgating the same detritus of the intellectual world, pseudoscience if ever there was such a thing. Action had to be taken. No longer could this be a civilized debate between savants of the same stature and learning. Things had to go public (Appel 1987).

No better forum could be found than the chief learned scientific society of France, the Académie des Sciences, of which both Cuvier and Geoffroy were members, and where indeed Cuvier was a Permanent Secretary, one of the chief positions of power and authority. Yet as so often happens when things explode after many years of provocation, the ostensive topic of debate was very minor and arcane. In October 1829, two unknown naturalists, Pierre-Stanislas Meyranx and a Monsieur Laurencet—a man so obscure that no one today knows his first name!—had submitted a memoir to the Académie on the subject of molluscs (a well-known group of marine invertebrates, that is, animals without backbones). They argued that there are significant similarities between the molluscs—they took the cuttlefish as a typical example—and the vertebrates,

Georges Cuvier

that is, animals with backbones. At least they argued—for nothing in this world is simple and straightforward—that if you bend a vertebrate backward in a bow, so that its head is virtually sticking up its butt, then you can see similarities. Geoffroy (as a member of the Académie) was asked to make a report on their claim, and his response came in very positively. Rubbing salt into open sores, he quoted (without identifying either source or author) an old paper of Cuvier's that denied forcefully that there could be any similarities between vertebrates and invertebrates. Now, claimed Geoffroy, we see that this kind of zoology is outdated and unneeded.

Incandescent with rage—so much so that the unfortunate authors of the memoir wrote earnestly to Cuvier, denying that their work had any implications whatsoever or that they intended in any way to contradict "the admirable work that you have written and that we regard as the best guide in this matter" (Appel 1987, 147)—Cuvier held forth before the Académie, with charts and tables showing that similarities are absent and that only the truly deluded could think otherwise. At which point, realizing that the best form of defense is attack and that Cuvier had forgotten far more about the invertebrates than he could ever learn, Geoffroy switched topics, arguing now that real similarities across species could best be discerned within the vertebrates (rather than across the vertebrate/invertebrate line). Now his point of argument was focused on the bones in the ears of humans and cats, which although different in size, shape, and number were (according to Geoffroy) essentially similar. Again Cuvier responded, and again his arguments were mixed with scorn and derision. Define your terms, he thundered at Geoffroy. "If our colleague had made a clear and precise response to my requests, that would be a fine point of departure for our discussion." Unfortunately, all he does is introduce one airy-fairy philosophical construction after another. All words and no substance. "It is to say the same thing in other terms, and in much more vague, much more obscure terms" (p. 150).

And so the debate went back and forth, with Geoffroy bobbing and weaving, always changing ground. Chasing him round the ring was Cuvier, flailing away, every now and then landing a good hard punch but never able to strike his opponent on the chin and end the contest. Finally, the fight petered out, with the contestants threatening their opponents with long series of justificatory memoirs. But not before the audience had had a wonderfully good time. Including the aged poet Johann Wolfgang von Goethe, who exclaimed to a

Etienne Geoffroy Saint Hilaire

friend, "The volcano has come to an eruption, everything is in flames"—an event that he saw as being "of the highest importance for science" (p. 1).

But, even accounting for poetic license, could this really be so? An event "of the highest importance for science"? Are we truly talking about the same things: the similarities between a cuttlefish and a vertebrate bent backward until it resembled nothing so much as a participant in a prerevolutionary Cuban sex show? The bones in the ears of humans and of cats? Who cares? Or rather, since some obviously did care, why should we care? To answer these questions, we must go back a hundred years and start our story: then we shall see why it was that two distinguished French scientists did hammer it out in the spring of 1830, to the joy of onlookers then and of historians ever since.

Defining Evolution

But before we do this, there is one thing we must do. We must not fall into the same trap that Cuvier accused Geoffroy of falling into. We must be careful to define our terms. At least, we must be careful to define one particular term. I realize that at this point you will probably start to groan and fear that I have forgotten already what I said at the end of the Prologue about my duty to be interesting and informative. You will find that I am a professional philosopher, and you will remember that someone once told you that the trouble with philosophers is that they are obsessed with language. They get hold of an important problem, start defining and redefining the pertinent terms, turning them upside down and inside out, and then they end up by announcing triumphantly that there was no genuine problem to begin with!

I cannot deny that there is some truth to this. But terms and language are important, and unless one does take care one can waste an awful lot of time. I expect many of us have gotten into heated arguments about the existence of God, only to find at the end that we are arguing completely at cross purposes. The atheist is denying a God who looks a little bit like Santa Claus in a bed sheet, sitting on a cloud surrounded by angels with wings. The Christian is asserting a God who is the ground of our being or some such thing. The Christian would be appalled to learn that he or she is supposedly defending the odd entity that the atheist is denying. The atheist has never really thought seriously about the being that the Christian is affirming.

So, without further apology, let me turn to the term that is going to be at the heart of this book: *evolution*. And let me tell you that, traditionally, there are three things to which the term *evolution* applies (Ruse 1984). First, there is what we might call the very *fact* of evolution. By this is meant the idea that all organisms—you and I, cats and dogs, cabbages and kings, living and dead—are the end result of a long process of development, from forms vastly different. Usually it is thought that the original forms were very simple and today's forms are rather complex—some of them at least—and that everybody and everything is related in some form through descent. We shall see, however, that there are variations on this. Usually it is also thought that if you go back far enough then you pass from the living to the merely material—chemicals and so forth. In other words, the organic (that is to say the living) came from the inorganic (that is to say the nonliving). We shall have to go into this. And usually evolution is said to be "natural," in the sense that the processes (more on these in a moment) that fuel evolution are simply regular laws of nature—there is no need for divine or any other kinds of interventions. Again this is a matter that will get a lot more attention.

Second there is the *path* or paths of evolution, known technically as *phylogeny* (phylogenies). Here we are dealing with the tracks that evolution takes through time. When did life first occur on earth? When did multicellular organisms evolve from simpler forms? Was the Cambrian explosion one of a kind, or are there many such events? Did the birds come from the dinosaurs or simply from ordinary kinds of reptiles? When did the dinosaurs vanish, and was this associated with any grand terrestrial events? What do we know of human origins? Did humans get up on their legs and then the brains explode in size, or was it the other way around? In many respects, it is this aspect of the idea that most people think of when they think of evolution. "Missing links" is a favorite refrain of the critics of evolutionism, meaning that there are gaps in the fossil record (so the critics claim) where there should be transitions between one major kind (like land mammals) and another major kind (like sea mammals, such as whales). As we shall learn, in various ways the finding and establishing of paths stand somewhat aside from much else in the evolutionary enterprise. How, why, and what this all means will be a matter of some considerable interest.

Third and finally we have the question of the *causes* or *mechanisms* or *theory* of evolution. What makes the whole process go and work? What drives evolution? What is its motive force? In physics, this was Newton's great achievement. He did not discover that the planets go around the sun. This was the job of Copernicus. He did not map the heavens accurately. Tycho Brahe did this. He did not find the planetary motions. Kepler's job. Nor did he work out what happens down here on earth. Galileo. But he did find the law of inverse gravitational attraction and show how everything follows from this—orbiting planets and soaring cannon balls. For this reason alone, we venerate Newton and his genius. Likewise we have such questions in evolutionary biology. Is there a biological equivalent to the force of gravitational attraction and, if so, does it work in the same way? Is there indeed one prime cause, or are there many such forces that collectively make for the overall mechanism? And is the whole thing theoretical, and if so in what sense?

This division of evolution into three is somewhat artificial. Obviously you cannot have a path of evolution or a cause without the fact of evolution. And certainly any thoughts that you have about causes are going to be very much influenced by the paths that you think that evolution took. For instance if you thought—what nobody in fact has ever thought—that trilobites (a form of marine invertebrate that went extinct over three hundred million years ago) gave birth in one step to elephants, you would have a very different theory of evolution from thinking that the trilobite-elephant link (even if it existed) took 500 million years with many, many intermediates. Indeed, if your fact of evolution includes the origin of life itself, then you are probably going to be thinking differently causally than if you think that the question of ultimate origins lies outside your ken. But, for all of the artificialities, it is useful to make a three-part division—fact, path, cause—and it will help us to structure our discussions in this book. Let us use it but not be ruled by it.

Now we are ready to start into our story, so let us go back to the eighteenth century.

Erasmus Darwin

The eighteenth century is called the Age of the Enlightenment, the time when the discoveries in science were consolidated and extended and when in the arts and in literature people started to turn from the past and look to the future. It is the time when we find such great writers and critics as Voltaire; philosophers such as David Hume and (a little later) Immanuel Kant; and the beginnings of social science in the hands of such men as the Scottish political economist, Adam Smith. Physics had had its great revolutions in the two centuries previously. Chemistry was to have its revolution toward the end of the century, thanks particularly to Antoine Lavoisier—whose reward was to be the loss of his head under the guillotine. Biology was still looking forward (Roger 1997). But the way was being prepared, thanks

Erasmus Darwin

especially to the labors of two men. On the one hand, there was the Frenchman Georges Louis Leclerc, Comte de Buffon, author of the multivolumed *Histoire Naturelle* (from 1849 on), a discursive series of books that covered nature from one end to the next. Then on the other hand, there was the Swedish naturalist Carolus Linnaeus (Karl von Linné), whose ever-expanding *Systema Naturae* (first version 1735) introduced the modern system of organic classification, wherein every animal and plant can be fitted into its own unique place in the order of things.

Although neither was entirely successful in holding the dike, essentially both men had static pictures of nature. They had pictures that were, if not directly biblically based, then at least were views of life that might be called "Creationist," in the sense that God had created animals and plants basically in the forms that we see today, subject perhaps to a certain amount of variation, particularly of a degenerative kind. But the Age of the Enlightenment was above all an age of change, both as people saw the world around them and as the leading thinkers of the time saw the course of history. It is true that Christianity is itself a historical religion. One starts back with the Creation in Genesis and works through the Old Testament until the Incarnation, in the form of Jesus Christ. Then one moves forward until some time in the future when God judges us all, for good or for ill. But although our actions are certainly relevant, it is not a history over which we have much control. Indeed, ultimately, our greatest gains "count for naught" and we are dependent on God's grace for our salvation. With the development of science, however, and the advances of literature and philosophy and political economy and more, people began to develop the confidence that not only is there change, but this can be permanent and brought about by us, through our own efforts. Moreover, whatever the naysayers may have claimed to the contrary, this was thought to be change for the good. Progressive change, in short. Such a philosophy, if one may so call it, was bound to have an effect on thinking about the organic world, and now as we shall see it truly did (Ruse 1996).

Carolus Linnaeus

Erasmus Darwin, the grandfather of Charles, was a physician in the British Midlands in the second half of the eighteenth century (McNeil 1987). Famed for his skill—his diagnostic abilities were formidable—Darwin several times refused the earnest en-treaties of poor oft-times mad King George the Third to come south and take on the role of court physician. He was happy in his station in life and particularly in his place in the country, which was just then experiencing the first wave of the Industrial Revolution. Around him enterprising engineers were putting to use the powers of coal and steam in the running of those machines that were to produce finished goods at a rate far more rapid than could ever be achieved by hand. The Midlands and the North of England were the sites of the action, and Darwin was in the thick of it, mixing with industrialists, scien-tists, engineers, and others, and himself contributing knowledge and advice drawn from his medical studies and experience, not to men-tion his general grasp of things scientific. A particular interest was the world of agriculture, something that had to experience no less of a revolution than industry, as people moved from the land to the cities, and as population numbers exploded, and hence as there was need to produce far more food with far less remaining available labor.

Erasmus Darwin was a man big in every sense of the word. His appetites were gargantuan. He loved his food so much that it was nec-essary to cut a semicircle in his table so that he could get close to the action. Preparing for one of his visits required considerable fore-thought and expense. Expensive dishes—preferably many of them—were expected and appreciated. But Darwin gave as he received. He was a wonderful conversationalist and a much-loved friend, valued for his sensible advice. Yet, for all that he was fat, was missing his front teeth, and (with or without them) stammered badly, there was a ro-mantic side to Dr. Darwin. Sexually, he was a man of some consider-able action. Three children with a first wife, two with a mistress dur-ing a kind of interregnum, and then seven more when, nearing fifty, he married the widow of one of his patients. This last he did in the face of several younger suitors. Intellectually also there was a lighter side to Darwin. In his day, he was one of England's better known and appreciated poets, as well as a writer of prose on many and varied subjects.

One of Darwin's closest friends was the potter Josiah Wedg-wood, he who was responsible for the development of the British china trade—cups and saucers, plates and dishes, as well as vases and other objects of great beauty. In this prerailway age, the chief mode of transportation—especially safe and careful transportation—was by water. Supplementing the sea and the rivers, the eighteenth century was a time of great canal building: something that required an intimate knowledge of geology, especially when there were questions of boring tunnels through mountains. Wedgwood was a major figure in this work, and Darwin was in the midst of this activity, looking and searching and thinking and exclaiming. "I have lately travel'd two days journey into the bowels of the earth, with three most able philosophers, and have seen the Goddess of Minerals naked, as she lay in her inmost bowers" (King-Hele 1981, 43).

Erasmus Darwin was absolutely fascinated by discoveries such as these. Looking back two centuries later, there is no "smoking gun" that proves definitively just what it was that tipped him toward evolutionism or (as, in those days, he would have called it) transmutationism. Most probably it was the marine remains (shells and fossil fish) found hidden away in that mountain, in the middle of England, where he journeyed with his companions. Certainly, soon thereafter Darwin adopted *E conchis omnia* (Everything from shells) as his personal motto, and to celebrate he had the phrase painted on the door of his own carriage. He did not rush into print, however. Setting a pattern that was to be followed by his grandson Charles, Erasmus Darwin took some 20 years before he felt ready to announce his thinking to the outside world.

His ideas were first written about explicitly in his major medical treatise *Zoonomia,* although one could hardly say that the treatment there was particularly systematic. Darwin made little or no attempt to disentangle the various threads of his thinking. Claims about the *fact* of evolution were mingled with ideas about the *paths* of evolution, and then threaded through the whole discussion were all sorts of hypotheses and speculation about the *causes* of evolution. Quite often he would start a paragraph talking about paths and then end up talking about causes. Or he would start off talking about causes and end up arguing for the general fact. He may have been an innovative thinker; he was no great systematist (Darwin 1794–1796, Vol. 1, 500–505).

Trying our best to disentangle his thinking, we find that probably there were two direct arguments that Erasmus Darwin put forward for the fact of evolution. First of all, he was much impressed by the analogy that he presumed between individual development and group development. If we can transform the individual—"from the feminine boy to the bearded man, and from the infant girl to the lactescent woman"—then why should we not transmute the group? Second, he thought very significant the similarities that he saw holding between the parts of the members of quite different species. These similarities, which today we call "homologies," were taken—as, in-

Man

Humerus

Radius

Ulna

Carpals

Metacarpals

Phalanges

1 2 3 4 5

Dog

1 2 3 4 5

Whale

1 2 3 4 5

Bird

1 2 3

The homology between the forelimbs of vertebrates

deed, they are taken today—to be evidence of common ancestry. Although, as I have just said, it is almost certain that it was fossil discoveries that made Darwin an evolutionist in the first place, he did not really bring in the fossils as a major piece of information in favor of the fact of evolution. They are mentioned but not as an important plank in the evidential foundation.

Today, we would surely want to use the fossils as evidence of pathways. Erasmus Darwin made no move in this direction either, although in fairness he had virtually none of the evidence that today makes the fossil record so important a source of information. As we shall see in a moment, he had an overall vision of the path of evolution, but as far as the specifics are concerned, he said little. In his opinion, the best source of information for actual pathways lay in the natures of living organisms. Take the presumed transition from sea to land. Erasmus Darwin touched on the peculiarities of animals like whales, seals, and frogs. He seemed to think that animals of this kind are somehow representative of those transitional forms that must have existed when life made its move from the sea to the land. Since we have such hybrid types today, it is reasonable to assume that they existed in the past, and these types today give us some clue as to their former nature.

What interested Erasmus Darwin more was the question of causes. He collected and offered all sorts of jumbled anecdotal bits and pieces of information. As you might expect, given that Darwin was living in a particularly important agricultural part of England, many of his suggestions were based on the folklore of animal and plant breeders. Indeed, Darwin spoke explicitly of "the great changes introduced into various animals by artificial and accidental cultivation." He was a strong supporter of the idea that characteristics acquired by an organism in one generation can be passed straight to members of the next generation. He instanced the docking of dogs' tails. Darwin believed that this practice eventually results in the birth of animals without tails at all, and therefore without any need of docking. This inheritance of acquired characteristics is today known as "Lamarckism" after the great French evolutionist of that name, although it should be noted that Lamarck's writings came at least a decade after Darwin put pen to paper. (Actually, the inheritance of acquired characteristics is an idea much older than either Erasmus Darwin or Lamarck, although indeed it was Lamarck who made much of the mechanism as a force for evolutionary change.)

Naturally, as a physician, Erasmus Darwin was much interested in the nature of the mind and in the ways in which mental attributes can affect and be affected by physical causes. The popular psychological theory of his day—the brainchild of the eighteenth-century thinker David Hartley—was known as "associationism." In line with the general associationist position, Erasmus Darwin thought that habits and experiences could lead to new beliefs, and that these beliefs could be passed straight on thanks to reproduction. Hence, people's mental attributes could be a result of things having happened in the past to members of earlier generations. From this, there was an easy analogical slide to the physical world: "I would apply this ingenious idea to the generation or the production of the embryon or new animal which partakes so much of the form and propensities of the parent"(p. 480). Also, most interestingly, there was an anticipation of an idea that was promoted by grandson Charles. Erasmus Darwin thought that it was entirely possible that the body throws off small parts; these are carried around, presumably by the blood; and finally they are gathered in and transmitted via the sex organs. This supposedly gave a physiological backing to the already mentioned Lamarckism. The blacksmith's arms get stronger and stronger through use. These newly developed arms cast off modified particles that go down to the sex organs. And so the children of the blacksmith are born with strong arms as part of their biological heritage.

Truly, though, for Erasmus Darwin it was the big picture that counted. He found the nuts and bolts of evolutionism to be rather boring. Later in life, he was much given to poetic expression of his evolutionary vision:

Organic Life beneath the shoreless waves
Was born and nurs'd in Ocean's pearly caves;
First forms minute, unseen by spheric glass,
Move on the mud, or pierce the watery mass;
These, as successive generations bloom,
New powers acquire, and larger limbs assume;
Whence countless groups of vegetation spring,
And breathing realms of fin, and feet, and wing.

Thus the tall Oak, the giant of the wood,
Which bears Britannia's thunders on the flood;
The Whale, unmeasured monster of the main,
The lordly Lion, monarch of the plain,
The Eagle soaring in the realms of air,
Whose eye undazzled drinks the solar glare,
Imperious man, who rules the bestial crowd,
Of language, reason, and reflection proud,
With brow erect who scorns this earthy sod,
And styles himself the image of his God;
Arose from rudiments of form and sense,
An embryon point, or microscopic ens!
(Darwin 1803, 1, 295–314)

Understanding the Past

We are going to be looking at a lot of evolutionists before we have finished, so I do not want to linger too long over any one. Fortunately, the overall ideas of Erasmus Darwin are not too hard to follow. We start at the bottom with the most primitive form, what was then often called the "monad," and we work our way up to the most complex and best form, what was then (unself-consciously) known as the "man." From butterfly (monarch) to king (monarch), as he expressed himself on another occasion. From that which is totally without value to that which we value above all else. A progressive rise up the chain of life. Yet, straightforward though this vision may be, I do want to make a couple of points before we move on.

The first is a general point but applied specifically to Erasmus Darwin. It is about the way in which we should treat figures in the past. There is a temptation to go too far in one way or the other, to see too many virtues or too many faults. Either we see the historical figure as a pure genius, with no flaws, and as having anticipated just about everything. Or we see him or her as a real fool, who found his or her way in the history books by chance or default or even fraud. Erasmus Darwin is a good case in point. On the one hand, he surely did come up with evolution as fact long before a lot of other people. He was right on there. Moreover, he did pick up on some good points. Fossils are important. The similarities between the bone structures of very different organisms are puzzling at the least, and surely suggestive of some hidden links. And embryology? Well, we do develop from primitive beginnings, so why should not the same be true of life itself?

On the other hand, if ever anyone was credulously open to absurd arguments it was Erasmus Darwin. There was no systematic treatment with things properly quantified—the very things that, by the end of the eighteenth century, one took for granted in the physical sciences. Again and again the reader would get something far more suited for Ripley's *Believe It or Not* than for anything with pretensions to being serious science. One prominent anecdote told by Dr. Darwin was of a man who had fathered a dark-eyed daughter in a family of otherwise very fair children. How had this come about? Darwin tells that when the man's wife was pregnant, he (the father) had become totally enamored sexually of the dark-eyed daughter of one of his tenant farmers. Yet, although the man offered the girl money for sex, she would have nothing to do with him. The obsession remained, however, and "the form of this girl dwelt much in his mind for some weeks, and that the next child, which was the dark-eyed young lady above mentioned, was exceedingly like, in both features and colour, to the young woman who refused his addresses" (Darwin 1794–1796, 523–524).

There was much more in this vein. For instance, we are told about the "the phalli, which were hung round the necks of the Roman ladies, or worn in their hair, might have effect in producing a greater proportion of male children" (p. 524). At times, even those who liked Darwin's work showed a tone of regret about the level at which he was writing. "If Dr. Darwin had indulged less in theory and enlarged the number of his facts our satisfaction would have been complete" (McNeil 1987, 174, quoting an anonymous writer in the *Monthly Review 1800*). The simple fact of the matter is that by the end of the eighteenth century, the notion that artificial penises hanging at the ends of chains supposedly affected the sex of future children was just not taken seriously by people who cared about serious science.

What am I trying to tell you? Basically, that at a certain level there was something rather ambiguous or questionable about both the quality and the status of the evolutionary speculations of Dr. Erasmus Darwin. Of course, we today would think this; but the point I am making is that even in the eyes of his contemporaries the ideas of Darwin were somewhat dubious or suspect. Which raises another question. If Darwin was indeed writing at such a loose or unsubstantiated level, why was he driven to do so? He was no fool, nor was he an unsophisticated thinker about technical issues. I told you that he truly had a great and justified reputation as a physician. Why then did he write as he did about evolution, and why was it that others at the time responded favorably to his ideas?

The answer has been given already. Darwin and his followers were absolutely obsessed with the new philosophy of the day: the philosophy or ideology of progress. For Darwin and his supporters, the Industrial Revolution—which was now going ahead at full steam, to use an apt metaphor—was the best thing that had ever happened to rural, sleepy, church-dominated England. What was needed, therefore, was a complete change of worldview. A worldview making central the success of machines and of the men of purpose who devised and drove them. That is to say, a worldview making central the achievements and aims of Darwin himself and of his industrialist friends. Evolution for Darwin, and for his supporters, was very much part and parcel of this philosophy or vision. Darwin (as we saw just above) did not see evolution as a slow, meandering process going nowhere. Rather, he saw it as an upwardly directed, progressive process reflecting the social progress that Darwin thought was now highly desirable. In fact, Darwin himself drew the connection, saying that evolution "is analogous to the improving excellence observable in every part of the creation; such as in the progressive increase of the wisdom and happiness of its inhabitants" (Darwin 1794–1796, 509).

All in all, therefore, the evolutionism of Dr. Darwin was the industrialist's philosophy of action made flesh—or embedded in the rocks! One goes from "an embryon point, or microscopic ens!" to "imperious man, who rules the bestial crowd." At work here is a full-blown circular argument, or perhaps more charitably one might say a feedback argument. You start with the idea of progress, the philosophy of the British industrialist. You read this into nature. And then you read it right back to confirm your philosophy. "All nature exists in a state of perpetual improvement . . . the world may still be said to be in its infancy, and continue to improve FOR EVER and EVER" (Darwin 1801, 2, 318).

Is this the philosophy of a man who has turned his back against religion? In a sense, this has to be true. Erasmus Darwin was certainly putting himself in opposition to conventional Christianity. For the Christian, the overall history of the world is one of miraculous creation, of subsequent sin and fall, and of the need for redemption that comes through, and only through, God's grace. Christ's great sacrifice on the cross and his miraculous rising from the dead wash away the sins of us all. For the Christian, therefore, Providence is the key to understanding history and the future. We humans can do nothing, save only with God's help and love. Darwin, as a progressionist, was arguing strongly that we humans are capable of improving our lot ourselves. So, in this sense, quite apart from the fact that as an evolutionist he had no place for the creation story of Genesis, Darwin was putting himself against traditional religion.

However, one should not at once conclude that Darwin was an atheist, or even an agnostic in the sense of having any doubts about God's existence. Darwin was no Christian, but like many intellectuals of his age (including many of the early American presidents), Darwin believed in a God who was an unmoved mover. He believed in a God who has put things in motion and who then stands back and watches how things work out through the agency of unbroken law. To use the technical language of scholars, Darwin was a

The progressive history of life (from Richard Owen's Palaeontology, 1860)

deist, as opposed to a theist, traditionally a Christian, a Jew, and a Muslim. A deist sees the greatest mark of God's power and forethought in the working out of unbroken law, as opposed to the theist who sees God's power in direct intervention, that is, in miracles.

Using a modern metaphor, what one might say is that Darwin's god—the god of the deist—has preprogrammed the world so that he did not have to intervene further. Evolution, therefore, can be seen as the greatest triumph of God. It is the strongest proof of his existence. It is certainly not something that disproves the need for or existence of a Creator or Designer. In Darwin's own words, "What a magnificent idea of the infinite power of *The Great Architect! The Cause of Causes! Parent of Parents! Ens Entium!*" (Darwin 1794–1796, 509)

I am not now sure that you would want to say that Darwin's evolutionism was a religious theory, nor even am I quite sure what that might mean. But this is the first moment at which you should start to realize that the science-religion relationship—the relationship in the context of evolution—is more complex than you might have thought. Those people (and there are many) who seem to think that evolutionists become atheists in the morning and then think up their theories in the afternoon, as a kind of bad joke, could not be more mistaken. Certainly, Erasmus Darwin—the man who can first claim unambiguously the label of "evolutionist"—became an evolutionist as much because of his religious beliefs as despite them. And that is a good point on which to move forward.

Jean Baptiste de Lamarck

As it happens, forward and sideways, for we cross over the Channel to France. Had things been normal, there is no telling what effect Erasmus Darwin might have had. But things were not normal. At the end of the century came the French Revolution, that bloody explosion that destroyed the Old Regime and absolutely terrified the rest of Europe, especially Britain. At once, all radical progressivist ideas came under heavy attack, being seen (with some considerable justification) as one of the major factors that brought on the events in France. Erasmus Darwin, enthusiast for the American and then the French Revolutions (until the latter got out of hand), ardent progressionist, came under particularly bitter attack from the conservatives. Devastating was a brilliant and cruel parody of one of his major poems—where he extolled the love of the plants, his detractors extolled the love of the triangles! The world laughed at him, his reputation sagged, and his evolutionism was crushed beneath the reaction.

But in France, for all of the revolution, evolution proved a more hardy plant. The key figure was Jean Baptiste Pierre Antoine de Monet, chevalier de Lamarck, son of minor nobility, who on being invalided out of the army became a botanist under the patronage of Buffon in the Jardin du Roi (Burkhardt 1995). Scientists did well in the revolution: they represented the kind of forward-looking attitude the leaders cherished (Lavoisier, the obvious exception, lost his head because he was also a tax collector). Although Lamarck found it politic to change his name from the hitherto more aristocratic de la Mark, he found himself in the newly reconstituted Jardin, now called the Museum d'Histoire Naturelle, in charge of (what he himself was to name) the invertebrates.

Jean Baptiste de Lamarck

It is often said that the really revolutionary scientists tend to be young—mathematicians are all washed up by the time they are thirty. You need to have the vitality to move into new fields and not yet to have acquired the vested interests to stay with the old. I am not sure how true this really is—the man who recently cracked Fermat's Last Theorem was 40—but Lamarck is certainly an exception to the rule. Although by century end he was 56, it was not until then that he swung from a lifetime's commitment to a static world picture and became an evolutionist. The particular trigger apparently was those invertebrates over which he had just assumed control. There were many fossil specimens for which Lamarck could find no living counterparts, yet since most were marine he could think of no competitors so strong and violent as to make them go extinct without trace. Hence, Lamarck came to the conclusion that they must have changed into other forms, or rather have given birth to other forms, without leaving descendants like themselves. This insight, if we may so call it, was enough to spur Lamarck to further speculation, and before long he was a full-blown evolutionist, a position he articulated fully in his major work, *Philosophie zoologique* (1809).

Lamarck believed unequivocally in the fact of evolution. Complex forms come from older simpler forms. Moreover, for Lamarck there can be no question but that evolution encompasses the production of life from nonlife. He endorsed venerable ideas of "spontaneous generation," believing that heat and lightning (just at that time, electricity was a very trendy phenomenon thanks to the experiments of Franklin and others) and other natural causes would stir up mud and other substances. From this, supposedly, would emerge primitive life-forms—worms and mites and the like. Indeed, not only did Lamarck believe in spontaneous generation, but he thought that it is going on all of the time, in the past and down to and including the present.

It is when we come to the path of evolution that Lamarck starts to get confusing and really quite interesting. Today, thanks to Charles Darwin, we tend to think of life's history in terms of the metaphor of a tree—the tree of life. Primitive forms are down by the roots, and then (going up the trunk with time) we have branching out into the major life-forms, with today's organisms (including us) up at the top, facing up to the sun. In diagrams given in the *Philosophie zoologique,* Lamarck rather gives the idea that this is his position also. But it was not. I have spoken of the main motive force for Erasmus Darwin as being that of progress—the progress, as we have seen, of a British industrialist who thinks that through human effort things will get better and happiness and so forth will be distributed and maximized. Lamarck likewise was a progressionist, but of a distinctively French variety. For him, from a near feudal country with no major industry, the improvement of progress tended to be intellectual: improvement in the arts and literature and science and philosophy. To this he tied a very old idea—it goes back to the Greek philosopher Aristotle—that all organisms can be put in a line from the simplest to the most complex. This idea, the great Chain of Being or the *scala natura* as it is called in Latin, which was very popular in medieval times, was based on the idea that God would have left no gaps. It would have been incompatible with His Goodness and Greatness that, had it been possible to create intermediate forms, He would have failed to have done so (Lovejoy 1936).

Before Lamarck, the Chain was completely static. It was a way of laying out the living world. It followed, supposedly, from the creative nature of God and as such had no implications about origins. Lamarck fused it with his progressivism—things getting better all of the time—and his evolutionism emerged. One point of immediate interest therefore is that Lamarck had a somewhat ambiguous relationship with the fossil record. It was fossils that made him into an evolutionist. You might therefore think that he was then going to use the fossils to trace out the path of evolution through time, from simpler to more complex. Indeed, he was read that way by later commentators, notably by Charles Lyell the Scottish geologist (who thus gave a distorted picture of Lamarck to a whole generation, including Charles Darwin). But in fact Lamarck (like Erasmus Darwin) was basically uninterested in the record as evidence of the path of evolution. The path—from monad to man—was given to him through the Chain, and no more was needed.

Somewhat connected to all of this is the fact that, appearances to the contrary, Lamarck's evolution was not essentially treelike. Rather

A medieval rendering of the Chain of Being

it was a series of climbs up the Chain: a staircase no longer but now an escalator. Organisms hopped on at the beginning, thanks to spontaneous generation, and then kept going right up to the top, penultimately as orangutans and then as humans. Thus, rather than a tree, we have parallel upward progressions, as life keeps starting over and over again. For someone who believes in a tree, extinction is forever. The dinosaurs will not reappear. Their branch has come to an end. For Lamarck, however, extinction is always a matter of time. If tigers were wiped out, then it might be a while before they reappeared, but they would—when the next escalator reaches the appropriate point. It is as simple as this.

But then how do you account for the tree-diagrams given in Lamarck's book? Here we need to turn to the third arm of the evolutionary picture: we need to look at causes. Notoriously, Lamarck believed in the inheritance of acquired characteristics—the giraffe's neck is long because ancestral giraffes with short necks stretched and stretched and stretched to reach the leaves high up in tree. Now their descendants are born with necks suited to the job. Although, as we have seen, this mechanism is to be found in Erasmus Darwin—it is indeed part of a cluster of very old ideas, although not previously used in a full-blooded evolutionary context (remember how Jacob tricked Laban by altering his sheep and goats before they were born)—the mechanism is today known as Lamarckism. It is this that Lamarck thought makes for irregularities in the chain of being—some organisms get deflected off the main path—and it was this that Lamarck was trying to show in his pseudotree diagram.

But if Lamarckism is the minor mechanism, what are the main mechanisms? Here things get a little fuzzy, mainly because Lamarck's thinking was itself a little fuzzy! He thought in a mechanical materialist fashion, that there are bodily fluids (the *sentiment interieur*) that flow through organisms, carving out new paths and constantly complexifying things. Hence, we get a constant movement up the Chain, brought about by purely mechanical causal factors. Yet one might well ask why it is that organisms stay on the path that leads upward to human beings. Here we get no answer from Lamarck, but the impression one has is that in some sense this upward passage is foreordained. In other words, to use the language of the philosophers, Lamarck's is a "teleological" system, meaning that the end point in some sense influences the activities before it is achieved. Whereas normal causation works from back to front, from past to present to future—the banging door (past) made the servant jump (present) and then she dropped the plate (future)—in teleology (or, as it is sometimes called, "final causes" or "purposeful" or "end-directed" situations), the future somehow reaches back to affect the present.

Now normally, there is nothing terribly mysterious about any of this teleological thinking: what we have are human beings or God thinking about the future, and based on these thoughts (which although referring to the future are in the present) we take action. But some people, Aristotle 2,500 years ago was one and at the beginning of this century the French philosopher Henri Bergson was another, have thought that life itself—even if it is not conscious—has a kind of forward-looking aspect to it. This is said to happen, not through thought but through something analogous to it—a kind of life force or "vital force" as it is often called. (Bergson called it an *élan vital* and his contemporary, the German embryologist Hans Driesch, spoke of an "entelechy." Supporters of such a position are known as "vitalists.") Although Lamarck denied strenuously that he was a vitalist—the opposite position is often known as "materialism," implying that there is nothing but material substance and forces—one has to say that there is a whiff of this about his thinking. Somehow everything fits together just too patly. I am not implying that Lamarck was a hypocrite or deceitful—claiming one thing and doing another—but rather that he was in respects more of a prisoner of his own past than he realized himself.

You might want to say that, given all of this, you do not really want to speak of Lamarck as an "evolutionist" at all. After all, evolution has been defined in terms of unbroken regular laws. But this seems to me to be too strict. What we have to recognize is that his evolutionism was not as unambiguously scientific as one might find in physics and chemistry. Which of course makes us all rather wonder if Lamarck, like Erasmus Darwin, had religious factors at play, driving him in his thinking. And the answer is that he certainly did. In fact, in respects his thinking was very much like that of Erasmus Darwin: Lamarck was no orthodox Christian, but he was a deist, seeing God as working through unbroken laws. It is just that the laws for Lamarck probably included something akin to vital forces. But the important point is that for Lamarck, as for Darwin, together with the vital influence of progress, it is true to say that he was an evolutionist far more because of his religious beliefs than despite them. A god who works through law, rather than through miracle, is a god who creates through evolution rather than in one creative spurt in six days at the beginning of time.

Lamarck had a pretty shaky reputation. People admired and respected him for his taxonomic skills, but he was altogether too given to wild hypotheses. He had some really daft ideas about meteorology, which he suckered the French government into supporting at great expense. Supposedly on one occasion he offered his *Philosophie zoologique* to the Emperor Napoleon, who spurned it with contempt. It turns out that this was less because the Emperor was a creationist than because he thought he was being offered yet more wild and inaccurate weather forecasts! For most people, Lamarck's evolutionism had altogether too much of the speculative about it. It was not that they were close minded, but that they had heard much of this kind of guff from Lamarck before.

No one felt more strongly on this subject than Cuvier. So, as I reintroduce him, let me start by stressing that he had every right and authority to feel this way. As a student of the life sciences, Georges Cuvier was head and shoulders above his contemporaries. His anatomical studies were simply outstanding—rightly he is known as the "father of comparative anatomy"—and then he turned to paleontology, taking what was a mess of fragmented ideas and hypotheses and leaving a full-blown scientific discipline. By any standard, this man was a really great scientist. He knew it and his contemporaries knew it. But he did not like evolution: he thought it unnecessary, he deemed it bad science, he found it philosophically offensive, he knew it was socially dangerous, and he found it threatening to him personally. Let us start to unpack these objections.

First, there is the question of the science. Cuvier appreciated that one had to speak to origins. Although he himself was rather inclined to think that the present state of knowledge did not make any real suggestions plausible or convincing, he was not faulting Lamarck for the very attempt to give an explanation. Cuvier's own geological explorations and his

A mastodon as reconstructed by Cuvier

work on the fossil record around Paris persuaded him that the earth is subject to violent periodic convulsions—what his English supporters were to deem "catastrophes"—and that life in some sense starts anew after each catastrophic event. He rather inclined to think that new life came in from elsewhere, invading now empty territory, but on this he did not say much. The point is that historical inquiry as such was certainly legitimate. In fact, judged as a historical record, Cuvier was inclined to accept the biblical account of the Flood as the last catastrophe. But, as a sophisticated French scientist, the last thing that Cuvier was going to do was to appeal to the Creation account of Genesis as the beginning and end of inquiry. This was not how one did science. More on this point in a moment.

Why not evolution, especially since it was Cuvier's paleontological inquiries that were first starting to show in a definitive fashion that the fossil record is roughly progressive, leading up from strange and unknown forms to fossil remains not so very different from beings living and breathing today? The record itself, however, was taken as speaking against evolution, especially because, even if progressive, it was not continuously so. One got all sorts of gaps, with abrupt transitions from one distinct form to another. There was no way that this could be the record of continuous change. Better by far to speak of extinctions and then of restockings. In any case, argued Cuvier, drawing on specimens brought back by Napoleon's savants from the ill-fated French incursions into Egypt, the mummified forms of cats and birds and other organisms—beings that lived literally thousands of years ago—are absolutely identical to forms living today. Where then is the evolution, the change, in all of this? If Lamarck be right, we should expect to see some change right before our eyes, and this we do not see.

In a way, though, all of this was surface for Cuvier. He had much deeper reasons for dislike of evolution—reasons that were part scientific, part philosophical. Cuvier, born in a border state between France and Germany, was educated in Germany and clearly felt the influence of the philosophy of the great German philosopher Immanuel Kant. This was reinforced by readings of Aristotle—something Cuvier was able to do when, with enforced leisure during the worst excesses of the revolution, he lived far from Paris in Normandy, tutoring the children of a noble family. Like Aristotle, Kant took a teleological view of living nature—in particular, like Aristotle, Kant thought that one must try to understand organisms in terms of ends or purposes and not just prior causes.

We are not now dealing with the wide sweep of history, so we are not now dealing with vital forces. We are rather dealing with the way in which an individual organism is put together and organized—and for Aristotle and Kant (and Cuvier following them) the secret is that all of the parts are to be understood as seeming as if designed to serve the ends of the organism's well-being. Something like the hand or the eye is not just a piece of an organism, but rather an intricately integrated composition, which serves the end—which has the purpose or function—of the organism's well being. We have teleology, because we are trying to understand the present hand or eye in terms of what we think they will do in the future. Obviously, no one is saying that the hand and the eye are actually caused by the future well-being. What if the organism died young?

Cuvier spoke of this teleological way of regarding organisms as the "conditions of existence"—these are the kinds of integrative principles that one must have if organisms are to live and work and function. Random collection of bits will not do. He thought (with some justification) that with this approach he had a very powerful way of analyzing organisms, since one knows that many parts must fit together harmoniously with the whole. Apparently proving what he was doing, he was fond of taking some isolated fossil bone and "deducing" the whole of the rest of the organism. A carnivorous tooth, for instance, would imply feet and claws designed for chasing and holding and killing—one could not have the hooves of a horse—as well as a stomach ready to digest huge chunks of raw meat—the digestive system of the cow would not do—and so on and so forth. One or two cases where he did this inference correctly, working out from fragmentary bone parts the nature of the whole organism (which was later discovered), convinced his fellows that his method was indeed as powerful as he claimed.

The conditions of existence did more than this. Translated into practice, which Cuvier called the "correlation of parts," it gave him a way of classifying organisms in a "natural" manner. If once you have a basic part in place—the backbone for instance—then you cannot have many of the features of an invertebrate—an exoskeleton for instance. Then, if the vertebrate is a meat eater, once you have got the carnivorous teeth in place, you cannot then have the features of a herbivore. With the carnivore, if once you have the features of the cat in place, you cannot then mix them up with the features of the dog. And so forth. Everything has it place—starting (Cuvier thought) with four great divisions, what he called *embranchements:*

vertebrates, molluscs (like clams), articulates (like insects), and radiates (like starfish).

You can see now why Cuvier had to be, absolutely and completely, against evolution. Moving from one form to another would smash to smithereens his beautiful static picture of the organic world. Nothing would be permanent and every inference would be open to doubt. And you can see now why Geoffroy's attack was so powerful and why it had to be resisted. Geoffroy, by endorsing the analogy between the vertebrate and the mollusc, was suggesting precisely that Cuvier's nice neat system was open to fundamental revision. Ultimately nothing remains the same. Everything is open to change. If you can go from one *embranchement* to another, or if you can find evidence that there are links between one *embranchement* and another, then the game is over. The way is open for evolution to come flooding in and spoil everything.

What about religion? If the evolutionists were all deists, might we infer that Cuvier was not—that in fact he was a theist, a Christian, and that this was part of his opposition to evolutionism? As a matter of fact, Cuvier was a Christian, and interestingly a Protestant—a legacy of that border state where he had been born. (It was not in fact incorporated into France until after he was born.) There is no question but that, as their deism influenced the evolutionists, so also Cuvier's theism influenced his antievolutionism. He did believe that God was the Creator and that He had intervened miraculously to place organisms here on earth. They could not have appeared naturally. But, as I have explained, Cuvier was anything but a literalist. It may have been legitimate to use the Bible as a historical record. It could never be a source or substitute for serious scientific research. Genesis should simply not be read that way. It is the story of our moral relationship to God. It is not a scientific text.

Finally, let me raise some social questions. Cuvier was a powerful scientist, but his was a power circumscribed and defined by his circumstances. He had been trained in Germany, and his training was less as a professional scientist and more as a civil servant—as a bureaucrat. He was big on deference, by him to his superiors, to him by his inferiors. Lamarck and Geoffroy galled precisely because they would not accept his status and thought of him as an equal. Cuvier knew that it was politic for him to serve the ends of his masters, first Napoleon and then the government after the Restoration. And he knew that, given the revolution, his masters were terrified of any social upheaval or of any philosophy that tended that way. Since evolution was so blatantly a tool of change and turmoil, he saw it as his task to oppose it as best he could whenever he could. If he was going to be really useful to the State, this was a place where he could show it. And so, Cuvier did.

But there was more than this, and here the personal factor comes in. As a Protestant, and by no means high-born, in Catholic France—Catholic France, which became increasingly conservative in the early decades of the nineteenth century—Cuvier had to tread carefully. Not only had he to show his personal worth to the state, but he had to be nonthreatening. Here, science was the perfect medium. At least, a science shorn of value and culture and ideology was the perfect medium. Cuvier could, as it were, say to his masters: "Look, give me power and status in science, and feel no threat because science is precisely that area of inquiry where there is no place for culture or value. The fact that I am a Protestant might be worrisome in a sensitive area like education or the like [in fact, Cuvier was put in charge of Protestant education in France], but in science uniquely my religion does not count. Trust me, for my personal ideological and religious commitments are irrelevant."

When people like Lamarck and Geoffroy came along, touting their philosophies and ideologies and religions dressed up as serious science, using their authorities as senior scientists, they threatened to wreck Cuvier's careful social strategy no less than they threatened to wreck Cuvier's careful scientific strategy. No wonder he was drawn into public dispute. And now we can see what hidden depths there were beneath a technical and dry debate about cuttlefish classification, and why Goethe was spot on when he explained to his friend: "I am speaking of the contest, of the highest importance for science, between Cuvier and Geoffroy Saint-Hilaire, which has come to open rupture in the Academy" (Appel 1987, 1).

Further Reading

There are several good books on the main characters in this chapter. The aeronautical engineer Desmond King-Hele is somewhat of an Erasmus Darwin buff and has written many books on and around his subject. The latest version is *Erasmus Darwin: A Life of Unequalled Achievement* (London: De La Mer, 1999). Richard Burkhardt has produced the standard biography of Lamarck, *The Spirit of System: Lamarck and Evolutionary Biology (with a New Foreword by the Author)* (Cambridge: Harvard University Press, 1995); and William Coleman wrote a really good scientific biography of Cuvier: *Georges Cuvier Zoologist: A Study in the History of Evolutionary Thought* (Cambridge: Harvard University Press, 1964). Somewhat more technical, but top-quality scholarship, is Toby Appel's account of the dispute between Geoffroy and Cuvier, *The Cuvier-Geoffroy Debate: French Biology in the Decades before Darwin* (New York: Oxford University Press, 1987). She is really sensitive to the science of the day and to the institutional background.

Unfortunately, in a book like the *Evolution Wars,* you do have to be awfully selective, else you just end with a massive encyclopedia that only recommends itself because it leaves no one unmentioned. There was a terrific amount of activity between the disputes at the beginning of the nineteenth century and the controversies that erupted once Charles Darwin had published the *Origin of Species.* A really great book dealing with some of this activity in England in the pre-*Origin* years is Adrian Desmond's *The Politics of Evolution: Morphology, Medicine and Reform in Radical London* (Chicago: University of Chicago Press, 1989). I should tell you that Desmond is an ardent "social constructivist," meaning that he thinks that there is no ultimate truth, that science does not progress in any absolute way, and that evolution is to a great extent less a description of objective reality and more a reflection of the culture of its day. In *Mystery of Mysteries* I argue strongly against this philosophy of history, but this is not at all to deny Desmond's brilliance as a historian and the deep understanding he brings to the history of evolution. Almost always, I learn more from those with whom I disagree than from those whose thinking parallels my own.

2

Mystery of Mysteries:
The Legacy of Charles Darwin

Charles Darwin and Alfred Russel Wallace really liked each other. This was just as well, for their names are forever linked as the two men who discovered the chief cause of evolutionary change. It would have been so easy for them to have quarreled: Darwin resenting Wallace, who came many years later but who yet drove Darwin into action; Wallace resenting Darwin because the older man had beaten him to the punch and then hogged all of the limelight. But although their followers and supporters have tried their best to divide the two, the friendship and respect lasted all of their lives. Wallace admired Darwin for the great scientist that he was; Darwin appreciated Wallace for his genius and his modesty and his firm convictions in the search for the truth.

This said, they rarely agreed about anything. They battled over their jointly parented child in a way that makes today's custody battles look like Quaker meetings. If Darwin had an idea, then Wallace opposed it. If Wallace had a thought, Darwin thought he must be wrong. You thought that cuttlefish classification was a boring topic. Try female bird coat color. Darwin had an elaborate theory to explain what is known as "sexual dimorphism": the differences between males and females in the same species. Of course, you have got to have some differences. If everyone had a penis you would be as badly off as if no one had a womb. But why do you have the big and visible differences? Why do human males have beards when the women are hairless—at least on their faces and chests and so forth. Why do men go bald, for that matter, and not women? Why are male walruses so much bigger than the females—so much bigger that sometimes the females get crushed to death during copulation? Why do stags have massive heads of

Charles Darwin

antlers and the females go around with little or nothing? And why, why, does the peacock have such a magnificent backside when the female has nothing—magnificent and yet kind of stupid, because who can escape with tail feathers like that when the predator comes calling?

Darwin tended to put the emphasis on the male. Stags have massive heads of antlers because they do combat with each other in the rutting season—winner takes all, and that is why there are the horns. Females hang around passively, not competing, and so they do not need or obtain such appendages. The same sort of thing is true of walruses, who fight like mad for possession of a harem of females. And in the case of the peacocks, it is basically the males' showing off that counts. It is true that the female chooses the male with the most magnificent display, but it is the male who is (literally) the center of attention.

Wallace felt very uncomfortable about this. He could not deny that the stags fight and so do the walruses and that such an attribute probably is the cause of the differences in those sorts of species. But he disliked intensely the claim that the peacock grew his feathers because the peahen was attracted to beautiful backsides. Rather, he suggested that Darwin had got things bottom backward, as one might say. It is not so much that the males are beautiful and showy as that the females are drab and inconspicuous. Sometimes, Wallace argued, being the center of attention is precisely what one does not want—and this sometimes occurs particularly when one is sitting on eggs, incubating them. Wallace thus claimed that sexual dimorphism is a function of female camouflage, protecting the females from predators, rather than male gaudiness with consequent female preference.

Why the quarrel? Was it just a matter of fact or facts? Well, in a sense it was, as obviously the cuttlefish classification was a matter of fact or facts. Wallace thought he had good evidence of the significance of such things as coloration in mimicry and camouflage, so it was natural to apply his findings to an important question such as dimorphism. But there was a lot more than just that. Today's feminists would at once suspect that prejudice and attitudes were involved—Darwin was excluding the active input of females, whereas Wallace was making this absolutely central. In fact, as we shall see, the feminist would not be so far wrong. Darwin was a bit of a male chauvinist, and Wallace was exceptional in his sensitivity to the significance of the female sex. Yet, there was something even more important, and without now giving away the game completely, let me point out to you that Darwin was claiming that the female peahen's aesthetic sense was very much like a human aesthetic sense. She chooses a feather display for the same reasons as we find it beautiful. And if a peahen's aesthetic sense is like a human's aesthetic sense, then a human aesthetic sense is like a peahen's aesthetic sense. And this was a matter that neither Darwin nor Wallace thought trivial.

The Making of a Scientist

Charles Robert Darwin was born to a life of upper-middle-class English privilege (Browne 1995; Desmond and Moore 1992). His father, Robert—oldest son of Erasmus Darwin—was a very successful physician and financier, and his mother was the daughter of Erasmus Darwin's old friend, Josiah Wedgwood. There was simply lots of cash in Darwin's background, and this was augmented when at the age of 30 he married his first cousin, Emma, another grandchild of Josiah Wedgwood. I make this point right at the beginning because it is an absolutely vital key to understanding Darwin's actions and much of his thinking. For instance, it is often said that Darwin never worked for a living, with an implication that he simply was not bright enough to obtain and hold down a proper university professorship. But this is to distort matters entirely. Darwin never worked for money (although he was good with his investments and canny in his dealings with publishers) because he never had to. Not for him were boring department meetings and officious administrators and whining students intent on mark grubbing. He could avoid all of that.

More significantly, because Darwin did so well out of Victorian society, one should not expect to find him a rebel in the sense of repudiating all of his background. Why should he? He was doing very nicely out of it, thank you! This is not in any sense to minimize Darwin's achievements but to point out that Darwin's achievements will most probably involve taking what he has been given and rearranging them into a new pattern. We should not look for Darwin to be the Christian God, making everything out of nothing. Rather, Darwin will be the sculptor or modeler who takes what he has and makes of it something new.

As a boy, Darwin was sent to one of England's famous private schools (misleadingly they are known as "public" schools, but they are anything but). Something of a square peg in a round hole—the main educational diet was Latin and Greek, a terrible bore and burden for the already science-sensitive Darwin—he went next, as had his father and grandfather before him, to the University of Edinburgh to train as a physician. Revolted by the operations and driven to madness by the tedium of the lectures—Darwin hated having to rise on dark Scottish winter mornings to listen to dry old men with incomprehensible accents lecture on dry old topics with incomprehensible significance—by the age of 19 he was back home and at a loose end. Desperate that young Charles not slouch into a life of indolent ease—one son was already going that way—Robert Darwin (himself an atheist) somewhat cynically pushed Charles to the path of an Anglican clergyman, a traditionally safe and respectable position for a young man of wealth and minimal career objectives. This meant getting a degree from an English university, and so, in 1828, Charles Darwin enrolled at Christ's College in the University of Cambridge.

It was a fortuitous move. Although there were then no formal courses in the sciences—Darwin did not get a science degree at Cambridge because there were none—this was just the time when a group of men was starting to take a serious interest in the natural sciences (including geology and biology). Anyone with a like concern, including an untutored undergraduate, was welcome to join in. For three years then, Darwin did formal courses—Latin, Greek, mathematics—and informal courses covering many aspects of the con-

A cartoon of Darwin as a student at Cambridge

temporary sciences. In 1831, when he graduated, came the big break. The Napoleonic wars now well behind, the Industrial Revolution was starting to get its second breath. Industry demands markets, and some of the biggest were in South America, long settled by Europeans and very wealthy. Ships were going out from the British ports—London, Liverpool, Glasgow—laden down with factory-made goods. There was a need for good naval charts, and so the British Navy was sending a ship down to the southern continent to map the coasts and shoals and waters. The captain of this ship, Robert Fitzroy of H.M.S. *Beagle,* was only 23 and—faced with a long and lonely trip, given that as captain he would be a person apart from the crew—was looking for a gentleman who could be his friend and traveling companion. It had to be someone outside the chain of command, personable and able to pay his own mess bills. Through a friend of a friend, Darwin got the call, he fit the ticket entirely, and so—for all that his father grumbled that he ought to be settling down and starting on the career as a clergyman—he spent the next five years (1831–1836) eventually going around the whole globe as what became, de facto, ship's naturalist, on H.M.S. *Beagle.*

Darwin did not become an evolutionist on the voyage, but it was the experiences and discoveries on the voyage that turned him into one shortly after he returned (spring 1837). Since religion is going to play a large role in our account, as a preliminary let me make a distinction that will help our understanding. Students of religion make a division between two kinds of inquiry: revealed religion or theology and natural religion or theology. Revealed religion is the religion of faith—it is what you get when you read your Bible or have direct insights from God or (especially if you are a Catholic) what the Church tells you to believe. So, for the Christian, revealed religion covers such things as Jesus' birth and death, the miracles and the resurrection, and that sort of thing. Natural religion or theology is the religion of reason—it is what you get when you try to get at God through pure thought. If someone says that a good proof for the existence of God is the fact that everything has a cause and so the world must have a cause—call this "God"—they are in the realm of natural religion. (This particular argument is known as the "cosmological" argument.)

There is a lot of debate between theologians as to the significance of the two branches of religion, and their relationships. Here, we need not bother with this. It is enough that they exist and that they will both prove pertinent in the Darwin story. Going back now to our hero, it is revealed religion that is first up front and relevant. When he

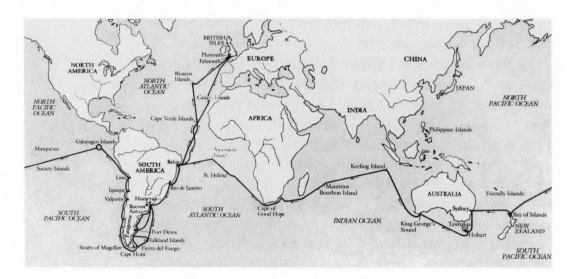

The voyage of the Beagle (1831–1836)

left England, by his own admission Darwin believed in the Bible pretty literally, and this extended to the origins of the earth and of organisms. But his views started to change as the *Beagle* worked its way around South America. It is clear that the major influence—the major influence always on Darwin—was a new book just appearing: *Principles of Geology* (1830–1833), in three volumes by the Scottish-born sometime lawyer Charles Lyell. (Darwin took the first volume with him, and the other volumes were sent out as they appeared.)

The full title to Lyell's work gives the clue to what it was about: *The Principles of Geology, being an Attempt to Explain the Earth's Surface by Reference to Causes now in Operation.* Lyell wanted to counter the catastrophic geology of Cuvier—a geology that had found much favor in Britain—by arguing that if one has enough time (indefinite time as far as he was concerned) then causes that we see around us today, governed by laws operating today, are quite enough to explain everything: seas, mountains, rivers, canyons, and all else. All one needs is time, and then rain and wind and earthquake and volcano and the rest can do the work. Above all else, one has no need of miracles, in the sense of divine interventions from above mixing things up and creating anew.

This forswearing of miracles and reliance on unbroken law will probably ring a bell and so it should! Could it be that Lyell had inclinations toward deism, away from conventional Christianity (which seemed to fit well with catastrophism)? The answer is that he did very much—his geological philosophy of "uniformitarianism" was deism in the stones, as it were. And this rang a bell or a chord in Darwin also. For all that he had had a conventional Christian (Anglican) education, intending to be a priest no less, in ways his formal belief sat lightly on him. His mother's family (the Wedgwoods) were practicing Unitarians—people who deny the Trinity, hence the divinity of Christ and the legitimacy of all of his miracles, and thus deists by another name. Before he was long into the voyage, it is clear that Darwin saw himself likewise moving toward deism (a position he was to hold almost the rest of his life), and we know already how that inclines one to views like evolution.

But Lyell had another part to play. Not only was he an enthusiast for unbroken law and causes—causes of a kind and intensity we see around us today—he had a particular theory that was intended to reinforce this uniformitarianism. The catastrophists tended to see the earth as directional, cooling from an original incandescent state down to the temperate state that it has today: this they saw as a background to the progressivism that Cuvier had found in the fossil record. As the world took on the form it has today, so its denizens took on the form they have today. Lyell to the contrary argued that there is no genuine direction to earth history. Yesterday was much like today. Today will be much like tomorrow.

The frontispiece of Lyell's Principles of Geology. *Lyell is using this picture to show that land sinks (hence the erosion of the pillars) and then rises (hence the pillars out of water), as confirmation of his theory of climate.*

Yet he could not deny some change: there is fluctuation. The fossil plants around Paris are definitely tropical, implying that the climate was warmer. So herein came Lyell's "grand theory of climate": he argued that temporary fluctuations of earth climate are a direct function of the distributions of land and sea around the globe. The Gulf Stream, that body of water that flows up across the Atlantic from the West Indies to Britain, makes for a much more temperate climate than the latitude would suggest. But, like all else, this will be temporary: the world is in a constant state of rising and falling. As rivers deposit silt at their bottoms, they press down the earth; then, like a gigantic water bed, another part of the earth rises upward. Thus the currents are altered and the local climates are changed. But overall, the general state is one of uniformity. Within limits, nothing changes. There is no direction to earth history.

Darwin bought into this theory all the way down (or up!). Much of his geologizing on the *Beagle* voyage was devoted to finding evidence that the earth is (and was) in a constant state of rising and falling. But what kind of evidence does count at a time like this? Fossils are helpful, of course, but even more so are the distributions of animals and plants around the globe: what is known as "biogeography" or "biogeographical distribution." Lyell was a bit vague about where he thought organisms come from—for all his deism, he was not keen on evolution because he thought it would downgrade the status of humankind—but he was fairly certain that they come into being on a regular basis and that by and large the new arrivals tend to be fairly similar to those most recently arrived. Thus, if (for instance) we find two groups of animals, very similar, divided by a natural barrier like a river or mountain, we can infer that the barrier is fairly recent. If, on the other hand, the animals are very different, we might infer that the barrier is ancient.

I explain this all in some detail, because here we are about to see one of the most important aspects of scientific creativity. Finding the answers is easy. It is asking the questions that is difficult, and important. Once you know where to look, you are on your way. It is finding the right direction that is what counts. Keyed by Lyell's climate theory, looking intently at biogeographical distributions, Darwin was well primed when the *Beagle* put in (in 1835) at the Galápagos Archipelago, a group of volcanic islands in the mid-Pacific. At first he saw nothing very peculiar, as he collected the birds on the various islands and goggled—as did everyone else—at the giant tortoises that live on the islands. Then, thanks to information furnished by the governor of the archipelago, Darwin realized that from island to island the inhabitants are different. Even on islands within calling distance one has different forms of bird and tortoise. This had to be significant, especially since on the South American mainland (which Darwin had just left), one sometimes found the same animal inhabiting the land from top to bottom, from steamy Brazilian jungle to snowy Patagonian desert.

Back home in England, John Gould, the leading ornithologist of the day, assured the young Darwin that his collections did indeed represent different species (Darwin was already making enough of a name for himself that the top people were happy to look at his specimens). For someone who was thinking in terms of unbroken law, for someone who was nevertheless trying to fit everything into a scheme that was designed by an understanding and good Creator, to someone who had read his grandfather's works and was well aware of Lamarck's ideas (Lyell conveniently gave a digest in the second volume of the *Principles,* intending to dissuade his readers of the attractions of evolutionism), there was only one answer to the problem. One simply had to argue that the birds and reptiles had come to the Galápagos, and then once there had changed in significant ways as they moved from island to island. Evolution had to be the key!

This was evolution as fact. But straightaway Darwin had his basic picture of the path of evolution. He was thinking of ancestors coming to the islands and then evolving as they moved around. This at once gives a treelike pattern to life's history. Not for Darwin was the upward parallelism of Lamarck—an aspect of the Frenchman's theory that, incidentally, Darwin quite missed. I have mentioned how, in Lyell's discussion of the French naturalist, he had mistakenly presented Lamarck's theory as a response to a progressive fossil record, that is, as a one-off phenomenon no doubt complicated by branching. Yet still there was the question of evolution as cause, and to show how he was far ahead of his grandfather in scientific sophistication—no evolution as pseudoscience for Charles Darwin—we find that the young naturalist now spent some 18 months searching systematically for an answer. His teachers at Cambridge had instilled in him the importance of causal thinking—after all, this was the achievement of the great Newton, and a biologist should aspire to no less (Ruse 1979).

As with earlier evolutionists, it is surely true to say that Darwin became an evolutionist because of his religious beliefs, rather than despite them. The same is true of his path to causal understanding. Darwin was ever a Lamarckian believing in the inheritance of acquired characteristics, but he knew that this alone could not be adequate. One needed some overall cause—a kind of force equivalent to a Newtonian power. But it could not be any kind of force. Cuvier may not have been an evolutionist, but his legacy hung over everything anyone thought about the organic world. In particular, one had to pay attention to function. This was a given, even for the evolutionist. Not that Darwin wanted to dispute this. By the time of the *Beagle*'s return, he was thinking of himself as a professional scientist, and as such he knew that one might modify and build on Cuvier's legacy, but one ignored it at one's peril. Moreover, his own personal Cambridge theological training had likewise convinced him of the significance of a functional—a teleological—approach to organisms. It is here that natural religion or theology starts to become important.

Actually, we have already encountered natural religion or theology at work in Darwin's thinking. When he worried about God's wisdom in creating separate species for each Galápagos island, he was appealing to the kind of Supreme Being that reason would dictate. But now natural religion was to become really important, a direct function of Darwin's having read at Cambridge the classic text on the subject: *Natural Theology* ([1802] 1819) by the Reverend Archdeacon William Paley (an Anglican clergyman). Paley gave the definitive version of the argument from design (for God's existence), also known as the teleological argument. He pointed out that in many respects the mammalian eye is just like a telescope—the lens, the way it focuses images, and so forth. But telescopes, argued Paley, have designers and creators. Hence the eye must have a designer and creator: the Great Optician in the Sky.

Darwin no longer accepted Paley's belief that this designer had to be a miraculous intervener—all was to happen through unbroken law—but he accepted entirely Paley's premise that the eye seems as if designed. And more generally, Darwin agreed entirely with the theologians that the definitive mark of the living is that organisms seem not have been put together randomly, but that they seem as if they were designed. The features that help organisms to thrive, to survive and reproduce—features that go under the heading of "adaptations"—were for Darwin, as they had been for Cuvier and the natural theologians, things that bore all of the marks of intentionality and forethought.

Archdeacon William Paley

The point is that any adequate evolutionary mechanism or cause had to be able to speak not just to change, but to change of a particular sort. It had to be able to speak to the evolution of adaptation, meaning it had to be able to show how designlike features come into being, even though—especially though—all was going to be done (by God as Darwin still thought) at remote control through regular laws of nature. The cause of evolution had to produce design. Darwin soon realized how this could be done in principle. He was perfectly stationed, living with his family in the heart of England, where the rural revolution was still in full swing. It had been necessary to produce such animals as cows and sheep and such plants as vegetables—especially the turnip, crucial for feeding overwintering animals—of far better quality than hitherto. Breeders had come to see that the secret lies in selective breeding: one chooses the animal or plant with the features that one most desires and one breeds from it, discarding all of the others. Fatter cows, shaggier sheep, fleshier turnips appear almost by magic, thanks to the selective skill of the professional breeder.

But how is this to occur in nature? Finally, after months of searching, at the end of September 1838, Darwin read a well-known political-economic tract by yet another English Anglican clergyman. This time it was the *Essay on a Principle of Population,* by the Reverend Thomas Robert Malthus, the sixth edition of which (the edition read by Darwin) had appeared a dozen years earlier, in 1826. Here we see in action the precise point made above about Darwin's rearranging parts that he had received from others. Malthus's work was conservative and appealed strongly to the segment of society from which Darwin arose. The *Essay* argued that state welfare schemes are pointless—worse than pointless—because population numbers have always a tendency to outstrip the supplies of food and space. There is bound to be a struggle for existence, which can only get worse if one feeds and coddles the poor and destitute. Better by far to let them suffer at the immediate level: then they will be persuaded to work and support themselves and to practice prudence and temperance and to restrict their family sizes.

A cartoonist's vision of what might happen if the struggle for existence is relaxed

This was music to the ears to people like the Wedgwoods, whose manufacturing enterprises depended on low taxes—no large, state welfare bills to pay—and lots of cheap and desperate labor. If it could all be wrapped up in the guise of God's stern unbending laws, so much the better. Charles Darwin, however, took Malthus's ideas, standing them on their head. He generalized from population pressures among humans to population pressures occurring throughout the animal and plant world, arguing that numbers will always have the potential to outstrip food and space. Consequently, there will always be an ongoing struggle for existence (and more importantly, struggle for reproduction).

But far from this having conservative do-nothing, go-nowhere effects, it is the motive force required to fuel a kind of selection: a lawbound natural kind of selection throughout the living world, which will lead to permanent and significant change. Only a few organisms will be able to get through—to survive and to reproduce—and those that do will tend on average to be different from those that do not. Those that do survive and reproduce (those that later Darwin was to call the "fitter") will do so precisely because they have features that the losers do not have. They will be faster, stronger, sexier, and so forth. In time, this will lead to a full-blown evolution, and moreover, it will be evolution in the direction of adaptive advantage. This new mechanism, that Darwin was to call "natural selection," has the effect precisely of producing the designlike effects of which Cuvier and the natural theologians had made so much.

The Long Delay

Having a bright idea is one thing. Having a full-blown theory that will convince other people, especially doubters and critics, is quite another. In science, no less than in the fast-food business, what counts is the sizzle not just the steak. Darwin realized fully that he was going to have to work to put things together into a fully finished form that would be presentable to others. In the end, it took him 20 years to do this, which is cautious by anybody's standards. If nothing else, it shows just how much science has changed over the past century and a half and how it has become a collaborative big business. No one today could sit on an important idea for 20 years. Jim Watson and Francis Crick discovered the double helical shape of the DNA molecule in 1953. Can you imagine if they had concealed their finding until 1973?! Of course, they could not have done so. Someone would have scooped them. In any case, today most scientists are funded by governments and big business, unlike Darwin, who was supported by the family fortune. If you want to keep the grants coming, you had better come up with a steady stream of results.

In fact, I do not think that Darwin suspected that the delay would be anything like as long as it eventually proved. Within a year or two he had put things together in theory form—he wrote a 35-page outline in 1842, and then a full version of 230 pages in 1844. But a number of factors intervened. One was that Darwin fell very sick from some unknown illness. It slowed him right down. From being a vibrant young man who had braved the elements on the *Beagle* and through South America, he became a near invalid, wracked with headaches and other ailments. He and his increasingly large family spent long periods at spas and other places of treatment as vainly he searched for relief. He became a recluse, totally dependent on his wife for every minor item of everyday life.

This was not a man to take on the scientific community with a daring and dangerous new hypothesis. Especially given that in 1844 there appeared an anonymously authored evolutionary tract: *Vestiges of the Natural History of Creation*. This caused a huge sensation, being wildly popular with the general public, especially women. Almost naturally, all of the Oxbridge science professors who were Darwin's teachers and mentors took a leading role in opposition. Adam Sedgwick, Cambridge Professor of Geology, evangelical Christian, and ardent catastrophist, led the attack with an 85-page critical review, followed by a 300-page Preface and a 500-page Afterword—all condemnatory of *Vestiges*—added to an inoffensive little 30-page essay on good conduct by undergraduates at university. Having suggested that the anonymous author had such low standards that it had to be a woman, Sedgwick drew back and denied that any member of the fair sex could have penned so vile a work.

In the same vein were the sentiments of David Brewster, Scottish man of science and biographer of Newton: "Prophetic of infidel times, and indicating the unsoundness of our general education, 'The Vestiges . . . ' has started into public favour with a fair chance of poisoning the fountains of science, and sapping the foundations of religion" (Brewster 1844, 471). He knew wherein lay the trouble: "The mould in which Providence has cast the female mind, does not present to us those rough phases of masculine strength which can sound depths, and grasp syllogisms, and cross-examine nature" (p. 503).

In the light of all of this, Darwin wisely decided to remain silent. He buried himself in a massive project of barnacle taxonomy, letting a few selected friends in on the great secret. His reputation grew, meanwhile, both as a scientist and as a general man of letters, thanks to a wonderful travel book that he produced from the diary kept on his long journey from England. Darwin was a man known, loved, and respected by the Victorian public, and so it was perhaps no great surprise that in the summer of 1858 a young naturalist and collector, then in the Malay Peninsula, should have sent to Darwin of all people a copy of an essay that he had penned just after recovering from a malarial attack. Shocked beyond belief, Darwin read this piece, by Alfred Russel Wallace, realizing that at last someone had hit on exactly the same ideas as he some 20 years earlier. Material was rushed into print, Darwin wrote frantically, and in the autumn of 1859, *On the Origin of Species by Means of Natural Selection, or the Preservation of Favoured Races in the Struggle for Life* finally saw the light of day.

Darwin's *Origin*

Darwin later referred to his work as "one long argument," and this it was. He knew he had a selling job to do. Darwin was never that much interested in the actual path of evolution, although with hindsight we can spot some fascinating speculations in the ostensibly nonevolutionary work on barnacles (published in the early years of the 1850s). But he had to persuade people of the fact of evolution, and he hoped also to convince them of his mechanism or cause for evolution. Running these two tasks together and influenced, I might add, by some of the leading methodologists of his age, Darwin reasoned in two quite distinct ways.

First, he tried to persuade people of evolution through selection by analogy. He thought that if he could introduce people to something they already knew and accepted—in this case the success of breeders in transforming animals and plants through artificial selection—then he might be able to persuade them of something they neither knew nor accepted—full-blown evolution through natural selection. To this end, Darwin trotted out all sorts of examples of the triumphs of animal breeders—with pigeons, with horses, with cows, with sheep, and much more—and then hinted heavily that this is no less than we might expect to find in nature. In a way, therefore, practical agriculture together with the work of those who breed for pleasure (pigeons, fighting cocks, bulldogs) was serving as the experimental evidence for the case that Darwin was building.

THE ORIGIN OF SPECIES

BY MEANS OF NATURAL SELECTION,

OR THE

PRESERVATION OF FAVOURED RACES IN THE STRUGGLE FOR LIFE.

By CHARLES DARWIN, M.A.,

FELLOW OF THE ROYAL, GEOLOGICAL, LINNÆAN, ETC., SOCIETIES;
AUTHOR OF 'JOURNAL OF RESEARCHES DURING H. M. S. BEAGLE'S VOYAGE ROUND THE WORLD.'

LONDON:
JOHN MURRAY, ALBEMARLE STREET.
1859.

The right of Translation is reserved.

The title page of The Origin of Species

A cartoonist's
thoughts on the
power of artificial
selection

When this was done, Darwin moved to the second phase of his agreement. First, the struggle leading to selection was introduced and discussed. To this end, Darwin turned to Malthus ([1826] 1914) and argued that as the political economist had claimed there is a struggle for existence in the human world, so likewise there is a struggle for existence in the organic world.

> A struggle for existence inevitably follows from the high rate at which all organic beings tend to increase. Every being, which during its natural lifetime produces several eggs or seeds, must suffer destruction during some period of its life, and during some season or occasional year, otherwise, on the principle of geometrical increase, its numbers would quickly become so inordinately great that no country could support the product. Hence, as more individuals are produced than can possibly survive, there must in every case be a struggle for existence, either one individual with another of the same species, or with the individuals of distinct species, or with the physical conditions of life. It is the doctrine of Malthus applied with manifold force to the whole animal and vegetable kingdoms; for in this case there can be no artificial increase of food, and no prudential restraint from marriage. (Darwin 1859, 63)

As Darwin himself recognized, strictly speaking, what he found in the organic world was not necessarily a struggle for selection, nor was it necessarily for existence. Rather, there is competition of a kind between organisms for space and food, and this competition centers more directly on reproduction than it does on existence. But either way, what one has is some kind of transference of a social idea from the human realm to the biological realm, where Darwin made of it a biological idea.

Then after the struggle, Darwin moved to say that, given that there is constant new variation in populations, one will get a natural form of the selection practiced by animal and plant breeders. This leads to ongoing change, but change of a particular kind: change in the direction of adaptive advantage.

> Let it be borne in mind in what an endless number of strange peculiarities our domestic productions, and, in a lesser degree, those under nature, vary; and how strong the hereditary tendency is. Under domestication, it may be truly said that the whole organization becomes in some degree plastic. Let it be borne in mind how infinitely complex and close-fitting are the mutual relations of all organic beings to each other and to their physical conditions of life. Can it, then, be thought improbable, seeing that variations useful to man have undoubtedly occurred, that other variations useful in some way to each being in the great and complex battle of life, should sometimes occur in the course of thousands of generations? If such do occur, can we doubt (remembering that many more individuals are born than can possibly survive) that individuals having any advantage, however slight, over others, would have the best chance of surviving and of procreating their kind? On the other hand, we may feel sure that any variation in the least degree injurious would be rigidly destroyed. This preservation of favourable variations and the rejection of injurious variations, I call Natural Selection. (pp. 80–81)

As a substitute for the term *natural selection,* later editions of the *Origin* introduced the term *survival of the fittest.* This was an invention of the English philosopher, social scientist, and biologist Herbert Spencer. The term was urged on Darwin by Wallace as less misleading than *natural selection.* But, whatever name the rose was given, do not think that natural selection was the only causal mechanism endorsed in the *Origin.* Darwin always endorsed secondary mechanisms of evolutionary change. We have seen the acceptance of Lamarckian acquired characteristics. More important—indeed, the most important of all of the alternative mechanisms—was *sexual selection,* a kind of secondary mechanism to natural selection. This corollary, as one might call it, tacked onto Darwin's earliest (private) writings on selection, centers less on the struggle for existence and reproduction and more on the struggle for mates. There is a differential reproduction leading to evolutionary change as features that help in the mating game get selected and refined. Sexual selection clearly came by analogy from the breeders' world, where one selects, on the one hand, for physical characteristics like fleshier meat and shaggier skins (the practical agricultural side, the natural selection equivalent) and, on the other hand, for the kinds of characteristics that organisms have to attract mates and repel rivals (the pleasurable fanciers' side, the sexual selection equivalent). Things brought about by sexual selection include characteristics used for intraspecific fighting, like the antlers of the stag, and characteristics used for sexual attraction, like the peacock's tail. (As you will have realized, it was this sexual selection at the center of the dispute between Darwin and Wallace. Later we will see more on this topic.)

Selection in itself does not explain how a group of organisms might split into two. Most particularly, how one might start with one species and then end up with two species. (This process is known as *speciation.*) The model that Darwin had always in mind was the speciation that occurred within the reptiles and the birds on the Galápagos Archipelago. He needed something that would positively induce selection to tear groups apart, and he thought he had found it in what he called his *principle of divergence.* The essential idea here is that by breaking up into smaller groups, organisms can better exploit their ecological circumstances. Two groups with somewhat different adaptations can do better than one. Big finches can eat big nuts and plants, and small finches can eat seeds or insects.

From this, Darwin was led immediately into his well-known description of life's history, where he drew an analogy with a magnificent tree.

> The affinities of all the beings of the same class have sometimes been represented by a great tree. I believe this simile largely speaks the truth. The green and budding twigs may represent existing species; and those produced during each former year may represent the long succession of extinct species. At each period of growth, all the growing twigs have tried to branch out on all sides, and to overtop and kill the surrounding twigs and branches, in the same manner as species and groups of species

have tried to overmaster other species in the great battle for life. The limbs divided into great branches, were themselves once, when the tree was small, budding twigs; and this connexion of the former and present buds by ramifying branches may well represent the classification of all extinct and living species in groups subordinate to groups. . . . As buds give rise by growth to fresh buds, and these, if vigorous, branch out and overtop on all sides many a feebler branch, so by generation I believe it has been with the great Tree of Life, which fill with its dead and broken branches the crust of the earth, and covers the surface with its ever branching and beautiful ramifications. (Darwin 1859, 129–130)

It hardly needs saying that if Darwin's mechanism of natural selection were to work, then he needed a constant supply of new variations coming into every population of organisms. Otherwise everything runs down very quickly to a sterile uniformity. In addition, these new variations must be heritable. If they are not, then however effective selection may be in one generation, it cannot pass on its results to the next. Here Darwin's genius rather deserted him. At best one got a compendium of speculations that would have done credit to his grandfather, and indeed many of the speculations—Lamarckism had a prominent role—were the same as those of the earlier evolutionist. As I have mentioned, Charles Darwin (not in the *Origin* but in later publications) floated ideas about the transmission of particles from the body to future generations, via the bloodstream and the sex organs. (This theory was known as *pangenesis.*)

After this somewhat unsatisfactory discussion, Darwin moved to a quick survey of some of the difficulties of his theory, for instance, the evolution of features that are highly adaptive or complex. Then he was able to turn to the second major part of the *Origin*. It was here that Darwin really came into his own as he surveyed the different branches of biology showing how evolution through natural selection throws light on so many different areas and conversely, in turn, is supported by each and every one of these areas. One area of major interest to Darwin was that of instinct and behavior. Like many biologists of his era, he was absolutely fascinated by the social insects, particularly the ants and the bees. He was concerned particularly to show how their social characteristics, just as much as anything else, were things that could be explained by natural selection. "No one will dispute that instincts are of the highest importance to each animal. Therefore I can see no difficulty, under changing condi-

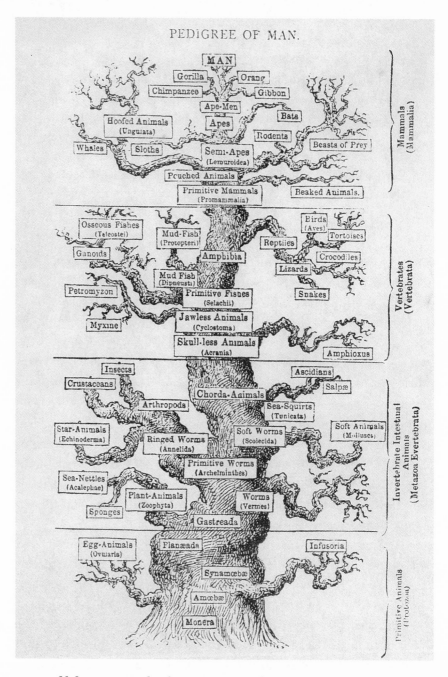

The tree of life (as drawn by Darwin's German supporter Ernst Haeckel)

PEDIGREE OF MAN.

tions of life, in natural selection accumulating slight modifications of instinct to any extent, in any useful direction" (p.243). As we shall see later, there was a lot more to the story than this, and perhaps Darwin was being overly optimistic in what he wrote. Indeed, he worried a great deal about how organisms can cooperate as tightly as they do in an ant's or bee's nest. However, as we shall see also, it was not until modern theories of heredity had been developed that the full story could be uncovered.

Geology and paleontology naturally got full treatment in The Origin of Species. On the one hand, Darwin was somewhat defensive. Like any evolutionist, he had to face the problem of the incompleteness of the fossil record. He had to show not only why he thought there would be few if any transitional forms but also why the fossil record starts so suddenly. The record does not go very gradually from the most primitive up to the most complex but starts off with a bang with really quite complex and sophisticated forms. (In fact, this is no longer quite true. In a later chapter, we will see that this problem has been remedied somewhat by new discoveries. Darwin, however, was driven to all sorts of speculations about how the early organisms would have lived where there are now seas, and how the weight of the land above them would have squashed their fossils to nothingness, and so forth.)

On the other hand, Darwin happily stressed the positive side to geology and paleontology. For all the problems, the fossil record does have a roughly progressive upward favor, which is what one expects given evolution. "The inhabitants in each successive period in the world's history have beaten their predecessors in the race for life, and are, in so far, higher in the scale of nature; and this may account for that vague yet ill-defined sentiment, felt by many paleontologists, that organisation has on the whole progressed" (p. 267). Moreover, Darwin was able to show that we find the more general and putative linking types of organisms lower down in the fossil record, and hence earlier. Conversely, the more specialized organisms come higher, and therefore later. This is just what one would expect if evolution were true. Darwin stressed also that once an organism has gone extinct it never reappears. Evolution through natural selection would lead one to expect this. On a theory of divine, miraculous, instantaneous creation, it is quite anomalous.

Moving on to geographical distribution, here (as you might expect) Darwin grew positively expansive. This was (and still is) always one of the really strong areas of biological inquiry supporting the evolutionist's case. For Darwin, given his Galápagos experience, it was an area of special importance, and naturally he made much of it. The Galápagos Archipelago itself gets a full treatment, and there is much discussion of oceanic islands in general. Then, following this, Darwin went quickly through a range of topics—classification, morphology, embryology, and rudimentary organs—showing how each and every one of these can be explained by evolution through natural selection, and conversely gives support to the mechanism. Embryology particularly got a detailed and vibrant discussion. Darwin was extremely pleased with his explanation of the fact that often organisms that are very different as adults have embryos that are very, very similar—humans and dogs, for instance. Darwin pointed out, using the analogy of artificial selection, that embryos have much the same selective environment and so are not ripped apart, whereas adults have different environments and so are driven apart by natural selection. In the world of animal breeders, no one cares much how the juveniles look. What really counts are the adults. "Fanciers select their horses, dogs, and pigeons, for breeding, when they are nearly grown up: they are indifferent whether the desired qualities and structures have been acquired earlier or later in life, if the full-grown animal possesses them" (p. 446).

And so the case was brought to completion. Truly, Darwin described what he had done as "one long argument" from beginning to end.

> From the war of nature, from famine and death, the most exalted object that we are capable of conceiving, namely, the production of the higher animals, directly follows. There is grandeur in this view of life, with its several powers, having been originally breathed into a few forms or into one; and that, whilst this planet has gone cycling on according to the fixed law of gravity, from so simple a beginning endless forms most beautiful and most wonderful have been, and are being, evolved. (p. 490)

Note incidentally the final word. You will often see it said that Darwin never used the word *evolution* in the *Origin of Species*—as if this tells you something significant, such as that Darwin did not really believe in and argue for evolution. This is misleading nonsense. The word *evolution* only came into the modern use—our use—around the time of Darwin, and he clearly had no strict objection to its use (Richards 1992). The more common language was that of "transmutation" or, Darwin's own preference, "descent with modification."

The argument of this second part of the *Origin* was as deliberate and as structured as was the argument of the first part. A good analogy can be drawn from criminal detection, where we have similar challenges to that faced by Darwin. Suppose you have a crime, let us say a murder. We have a suspect, but there were no eyewitnesses to give testimony. Let us suppose now that we find a similar crime, and there is good evidence that the suspect committed that crime. Now the detective would feel much more convinced of the suspect's guilt.

This is analogous to Darwin's use of the artificial selection analogy. No one saw evolution occur, but now we have something similar that produces a similar effect. The guilt may not be there, proven absolutely, but the detective/evolutionist feels that we are on the right track.

So what does the detective do next? He or she looks for circumstantial evidence. The search is on for clues. Lord Rake lies dead in the library, a dagger through his heart. The detective pins the guilt on the butler because of the bloodstain (the butler has a rare blood group that is found on the knife), because of the efficient way in which his lordship was killed (the butler was a commando in earlier life), because of the motive (the butler's daughter was seduced by Lord Rake), because of many, many more little bits of information and evidence. The clues point to the guilt and the guilt explains the clues. It is exactly the same for Darwin. The facts of instinct, the paleontological record, the distribution of organisms, morphology, embryology, systematics, all of these point to evolution through natural selection. Conversely, evolution through natural selection explains all of these facts of the biological world. Why the progressive fossil record, why the Galápagos distribution, why the homologies, and so forth. In Darwin's opinion, evolution through selection is proven "beyond reasonable doubt."

After the Origin

The year 1859 really was a watershed in the history of evolutionary thought. Before then, people knew about the idea of evolution (fact, that is) and would speculate about paths, and the Lamarckian inheritance of acquired characteristics was part of general lore (although of course most people did not think that it could cause more than minor effects and certainly did not think it would lead to the change of one species into another). But as a general belief, evolution was looked down upon by serious thinkers, and among professional scientists in particular it was regarded with scorn, and with not a little of the contempt and fear that had marked Cuvier's response and attitude.

If things were to change, something exceptional had to happen. And it did. Darwin was a person with stature: as a scientist and as a general figure in Victorian life. He could and did write extremely well—his travel book showed that and not a few remarked of the *Origin* that it had the same easy and inviting style. The reader is brought into the argument, never condescended to, and seduced by familiar examples and practices. Pigeon breeding—the classic working man's hobby (think of Andy Capp)—who could be scared of that? The main argument was convincing—after all, if not evolution, then how do you explain the Galápagos birds and tortoises, the homologies between the limbs of very different organisms, the facts of embryology? And on top of this, even if the old guard was never going to change, by 1859 Darwin had built up a group of younger scientists and supporters who would see that his ideas got full coverage and fair treatment.

It worked! At least, it worked in part. Virtually overnight people became evolutionists—evolution as fact that is. It was a little like the Hans Christian Anderson story of the Emperor's new clothes. Once the child had spoken—"But Daddy, he doesn't have any clothes on"—everyone said the same. Said they had known it all along! Once Darwin had spoken—"But evolution does occur"—everyone said the same. Said they had known it all along! I do not want to exaggerate. Of course, some of the established scientists and their friends never accepted evolution. Sedgwick went to his grave (in the 1870s as a very old man) denying and denouncing the vile doctrine. Mr. Gladstone, four times prime minister, classicist and churchman, never deviated from very old fashioned religious belief. But generally, evolution became the flavor of the decade.

How can one be so certain? The most compelling evidence is from surveys of magazines and newspapers and other such organs—especially religious publications, where one might expect to find opposition (Ellegård 1958). It is quite remarkable how quickly contributors accepted evolution and urged it on their readers. Obviously, liberal writers more quickly than conservative writers, but before long—certainly by 1865, and usually much earlier—evolution was the norm, the orthodoxy. What convinced me personally of the rapidity of the change was looking at the examination papers that students had to attempt at the universities. (In England, examinations are all printed up and copies kept on file.) In 1851, when Sedgwick was an examiner, one question read: "Reviewing the whole fossil evidence, show that it does not lead to a theory of natural development through a natural transmutation of species." But just a few years after the *Origin,* students were being told to assume "the truth of the hypothesis that the existing species of plants and animals have been derived by generation from others widely different" and to get on with discussing causes! When something is part of the standard undergraduate curriculum, you can be fairly sure that it is established truth.

Evolution as fact went over quickly and well. As I have said, Darwin was never really that interested in evolution as path. This was the job for the professional paleontologist, and he never really had aspirations in that direction. But what about evolution as cause—what about natural selection—Darwin's real pride and joy? Here, to be candid, he had a lot less success. No one wanted to deny it outright, but by and large people looked to other mechanisms. There were a number of reasons for this, and in the next chapter we shall be looking at what I think is the most significant of these. But for the moment, let me tell you that it soon became apparent that there were some fairly serious scientific problems with Darwin's theory—Darwinism, as it was usually called, and as we shall now call it, meaning not just evolution as fact but the theory that makes natural selection the chief and central mechanism of change. Let me mention two such problems.

First, there were problems with Darwin's thinking about heredity—that is, about the means by which new variations come into populations (the "raw stuff" of evolution, necessary for indefinite change) and even more about the ways in which variations are passed on from generation to generation. It is clear that one must have some such theory of heredity or—as it is known today—of "genetics." Suppose natural selection picks out some feature as especially valuable—say a new predator comes along, and those potential victims that are darker than others are better camouflaged against the background and hence tend to be the "fitter." It does not matter how dark a successful organism may be, if it does not pass this feature on to its offspring. Without some way of preserving and transmitting good characteristics, the clock is put back in each generation and selection goes nowhere.

Unfortunately, at this point, Darwin took a false step. You might think that in each generation characteristics blend in with each other, and certainly this seems to happen sometimes. A black man and a white woman have brown children. Or you might think that in each generation characteristics stay distinct and entire. Eye color, for instance, or sex for that matter. You either have boy features or you have girl features. Depending on the way you go—what you take to be the norm—you explain the other side as anomalous or temporary or some such thing. For instance, if you think that sex is the norm (you take the "particulate" side), then you explain skin color as a temporary manifestation and probably the underlying causes are unaffected. The same obviously if you take the other side ("blending"), thinking skin color the norm and eye color and sex to be explained away.

The point about the two positions is that the particulate side lends itself immediately to a selection position—no matter what happens on the surface, the essential causes remain unchanged and preserved through the generations, always ready to show their effects again. Blending does not so lend itself—however good selection may be, in a generation or two a good new feature gets diluted right down and out to invisibility. One drop of black paint in a gallon of white makes little difference.

Let me be fair to Darwin. None of this is very obvious. There is no clear surface reason why inheritance should be particulate rather than blending, or conversely. It is a judgment call based on the overall background information that you have. But a false judgment is a false judgment, and this is where Darwin faltered: he went the wrong way, thinking that blending is the norm. And the critics pounced, showing how selection simply could not do what was needed. As we shall learn in time, it was in fact not until the twentieth century that the problem was resolved. But that was little consolation to Darwin in the nineteenth century!

The same is true of Darwin's other big scientific problem: although here one has rather more sympathy for Darwin, for he was the victim of the erroneous thinking of others (Burchfield 1975). The physicists, ignorant of the warming effects of radioactive decay, argued strenuously that the earth must be much younger than Darwin needed for the slow processes of natural selection. In the *Origin,* Darwin rather suggested that time was almost infinitely available, and now the physicists cut him down to a hundred million years maximum. This was still huge by what people had believed even a few years earlier. No one in the nineteenth century believed in the 6,000 years since creation that the sixteenth-century Archbishop Ussher had calculated from the genealogies given in the Bible. But one suspects that the catastrophists were thinking in terms of hundreds of thousands of years, a few million at most. The physicists were certainly being generous by earlier standards, but this was not enough for Darwin, who spent years trying to speed things up in the face of criticism. Ultimately, he had to tough it out and hope that something would turn up—which it did, but not until 25 years after he had died. Then and only then was radioactive decay discovered, its warming effects appreciated, the span of earth-life lengthened, and natural selection given full rein. (Today, there is lots of time. The universe is believed to be about 15 to 20 billion years old, the earth is about 4 and a half billion years old, and life started about a billion years later. More on this in due time.)

A cartoon of Darwin showing how the status of humankind was the real problem and fear

Genetics and geological history were somewhat technical questions. The big popular question of course was our own species: *Homo sapiens.* It is here that Wallace comes back into our story. In the *Origin,* Darwin made it clear that we are part of the scenario, but he did not want to make too much of this. "Light will be thrown on the origin of man and his history" (Darwin 1859, 488). This silence was deliberate. Darwin wanted to get the main ideas on the table before people got diverted straight into questions of human origins. Darwin knew that, once he published, his theory would be swallowed up by the "monkey question": a prediction that not only proved true but that was reinforced by the almost simultaneous first arrival from Africa, in Victorian England, of the gorilla.

At first, Wallace had been as hard nosed about humans as Darwin—"hard nosed" in a comparative sense, for Darwin certainly thought that God was creating everything including us, if through natural laws—but in the 1860s, the junior evolutionist became enamored with spiritualism. Wallace started to believe that there are occult forces ruling the world and responsible for our evolution. Selection alone could not do the job, because we humans are fundamentally different from all other organisms. Hence, Wallace's reluctance to allow that the peahen might have the same standards of beauty as humans. There must be something nonmaterial, nonphysical, about human evolution.

Darwin was appalled at Wallace's apostasy. But he realized that there was a challenge here to be met. The younger man had come up with all sorts of human characteristics that he claimed could not have

Alfred Russel Wallace

been perfected by natural selection—human hairlessness, big brains, racial differences, and more. Darwin's response, really driving in the wedge between him and Wallace, was to make more and more of sexual selection. In *The Descent of Man,* published in 1871, Darwin argued that it is males competing and women choosing that makes us what we are. The bigger, stronger, brighter men got the pick of the women; the nicer, sexier, more sensitive women got the pick of the men (or picked by the men). My favorite example, if that is the right term, is Darwin's explanation of why Hottentot women have big backsides. Apparently they are lined up, and the warriors crouch down and squint along the line. She who protrudes farthest (*a tergo*) is she who is chosen by the bravest warrior.

Supposedly, all of this tells us not only why there are racial differences but also why there are sexual differences, and even why we humans are different from the brutes. The most beautiful women were desired by the strongest and most intelligent men, and the women in turn were happy to lie down for a good cause. If nothing else, by going with the flow they would then determine that their own sons would have precisely the features that make for male success in the struggle for reproduction. And if this were not enough, for good measure Darwin threw in a defense of capitalism! "In all civilized countries man accumulates property and bequeaths it to his children. So that the children in the same country do not by any means start fair in the race for success. But this is far from an unmixed evil; for without the accumulation of capital the arts could not progress; and it is chiefly thorough their power that the civilised races have extended, and are now everywhere extending, their range, so as to take the place of the lower races" (Darwin 1871, 1, 169).

In the face of this kind of argumentation, Wallace got somewhat short shrift. Apart from the fact that, by the 1860s and 1870s, appeal to spirit forces was simply not acceptable in forward-looking science, his innate assumption that all humans are likewise distinctive in their intelligence and other defining characteristics was much against the temper of the times. The "lower races" were certainly considered human and much above the apes—remember there had just been a bloody civil war in America over this very issue, with forward-looking liberals arguing that slavery is immoral precisely because all humans are in one family. But even the most liberal were generally not about to equate the Negroes and aborigines and Indians and native north Americans along with Europeans, or even southern Europeans and Slavs and Jews with Anglo-Saxons. Darwin's approach, which rested ultimately on competition and differences between peoples and with some coming out ahead of others, fit perfectly with what people already knew (or "knew"). Especially since Darwin made it very clear that the inhabitants of a small island off the coast of Europe are the apotheosis of human development.

It was no wonder the Victorians loved Charles Darwin and ended by burying him in Westminster Abbey. This was a man who spoke a language they could all understand. The coming of evolution was indeed a momentous event, but you should not think that it faced united opposition and hostility. In respects, it lent itself very nicely to the most standard and basic of societal beliefs and prejudices and was welcomed accordingly.

Further Reading

There are many books on Charles Darwin, but start with *On the Origin of Species* (London: John Murray, 1859) itself. It is remarkably readable for a "great book." Darwin kept revising and rerevising his work, and by the time he had finished it had rather lost its original, clean, spare form. So try to get hold of the first edition. You can tell if you have found it, because in the fourth chapter where Darwin introduces natural selection, it is only in later editions that he adds Spencer's alternative name of "survival of the fittest." Harvard University Press has produced a facsimile of the first edition, and the Penguin edition is also of the first. This latter has an excellent introduction by the historian John Burrow.

For Darwin himself, the best single volume biography is by Adrian Desmond and James Moore: *Darwin: The Life of a Tormented Evolutionist* (New York: Warner, 1992). This is compulsively readable and simply packed with information about Darwin and his friends and family and the society in which he lived. Be warned, however, that it is written from a Marxist perspective that sees England on the verge of revolution and Darwin as a key figure in precipitating potential trouble.

Darwin is portrayed as racked with guilt because, through his promotion of godless evolution, he was thereby betraying his own social class (hence the subtitle of the book). I think this is silly nonsense. I am not convinced that Britain was so very unstable. In any case, Darwin was well liked and secure in his position in society, and although his theory was truly revolutionary it never bothered him that he had it. Moreover, as I shall be telling you in the next chapter, he saw that his basic ideas rapidly became orthodoxy. This was a man who was buried in Westminster Abbey.

The best overall biography of Darwin is as yet incomplete. Janet Browne's *Charles Darwin: Voyaging* (New York: Knopf, 1995) still awaits its second and concluding volume. The first part is thorough, judicious, well-written with keen insight into Darwin's psyche, and very detailed and knowledgeable about the pertinent science. I really liked it and am looking forward to more. Also may I recommend my own general history of the whole Darwinian revolution: *The Darwinian Revolution: Science Red in Tooth and Claw*, 2d ed. (Chicago: University of Chicago Press, 1999). It is (said he immodestly) particularly strong on the religious and philosophical factors in the revolution and has just been reissued with a new afterword discussing findings and interpretations since it first appeared 20 years ago.

Modified Monkeys: Evolution as Religion

"I should like to ask Professor Huxley, who is sitting by me, and is about to tear me to pieces when I have sat down, as to his belief in being descended from an ape. Is it on his grandfather's or his grandmother's side that the ape ancestry comes in?" And then taking a graver tone, he asserted, in a solemn peroration, that Darwin's views were contrary to the revelation of God in the Scriptures. Professor Huxley was unwilling to respond: but he was called for and spoke with his usual incisiveness and with some scorn: "I am here only in the interests of science," he said, "and I have not heard anything which can prejudice the case of my august client." Then after showing how little competent the Bishop was to enter upon the discussion, he touched on the question of Creation. "You say that development drives out the Creator; but you assert that God made you: and yet you know that you yourself were originally a little piece of matter, no bigger than the end of this gold pencil-case." Lastly as to the descent from a monkey, he said: "I should feel it no shame to have risen from such an origin; but I should feel it a shame to have sprung from one who prostituted the gifts of culture and eloquence to the service of prejudice and of falsehood." (Huxley 1900, 1, 200–201)

A wonderful confrontation. This was the meeting of the British Association for the Advancement of Science, held in Oxford in 1860, the year after the *Origin* was published. Thomas Henry Huxley was clashing with Samuel Wilberforce (son of William Wilberforce, famous in England for leading the fight against slavery), a leader of the "high church" movement in the Anglican Church. This truly was a David and Goliath encounter, for Huxley was young and vigorous, a morphologist and paleontologist and now professor at the London School of Mines—a worthy home but with virtually no status what-

Bishop Samuel Wilberforce of Oxford

soever—and Wilberforce was old and established and important and occupying one of the most distinguished of bishoprics—at Oxford, no less, the city of the most venerable and powerful university in the realm. The defender of science, the "bulldog" who spoke for the new theory of evolution, battled the champion of the Church of England, speaking for all that was set and important and traditional. And as David slew Goliath, so Huxley's verbal slingshots left the bishop vanquished and speechless.

A wonderful confrontation and a wonderful story, told and retold by generations of evolutionists. I myself first became interested in Darwinism thanks to a graphic reinactment by my history master when I was a schoolboy some 40 years ago. I still remember his striding about the room, smashing fist into hand as he made Huxley's rhetorical points. (He was a terrific teacher!) But probably more a myth than true, I am afraid, although (as these things tend to be) a very revealing myth for all that. Let us go back and set the scene, following the story through to the end.

Victorian Britain

Start at the beginning of the century (Ruse 1979). Thanks to the French Revolution, compounded by the rise of Napoleon, Britain was in a conservative phase. But it was a country in tension, with the seeds of change germinated and sprouting. On the one hand, Britain—southern England particularly—was ruled, owned, and controlled by large, generally aristocratic landowners (identified with the Whig party), with the spaces in between belonging to smaller landlords, the squires or gentry (identified with the Tory party). Parliament consisted of an unelected house of peers (which included judges and senior bishops of the Anglican church) and an elected house of commons, although this latter was controlled by those who had power over the nomination and election of the memberships of parliament. In an age where the vote was open, tenants knew full well that the wishes of their landlords were paramount. The duke of Norfolk, for instance, was barred from taking his own seat in the House of Lords because he was a Roman Catholic. Nevertheless, through his holdings he controlled the occupancy of several seats in the House of Commons.

Laws tended to be very much in favor of those in power and naturally tended to reflect rural interests. Hunting was given full rein and poaching was heavily prosecuted. Most notorious of all the laws were the so-called Corn Laws, enacted at the end of the Napoleonic Wars. During the wars, thanks to the French navy, imports of corn (the term by which the English referred to wheat, not to the North

Thomas Henry Huxley

American maize) had been difficult or impossible; so landlords had done well as their land was used for every last cultivated patch. Now with supplies coming in and rents dropping, the government brought in laws that specified that imported corn would be subject to restrictive taxes unless locally grown corn reached a certain price. Thus rents were pushed back up again, to the delight of the landowners— which landowners incidentally included Darwin's teachers and mentors at Oxford and Cambridge, for they were all fellows (members) of colleges at those universities, and the colleges got their incomes from rents of very large rural holdings.

H. G. Wells, the novelist who trained under T. H. Huxley to be a schoolteacher

But things were starting to change—things had to change no matter what the authorities in power wanted. I have spoken several times already of the Industrial Revolution. This started in the second half of the eighteenth century, particularly in the North and the Midlands, close to major supplies of coal and water and minerals. This change brought great wealth to many people, including landowners, but the interests of the industrialists and the landowners tended not to be the same. The industrialists, for instance, wanted cheap corn so that they did not have to pay high wages. They did not care if the materials for bread were imported from around the world. And they resented very much that they tended to be outside the corridors of power—seats in the elected House of Parliament were not distributed equally. "Rotten boroughs" might have but a handful of voters—all controlled by some powerful interest—whereas a new city might have little or no representation at all.

In any case, not everyone could vote. Not women obviously. And by and large, not workers either. You had to be a property owner. And, not dissenters (Protestant, non-Anglicans) and certainly not Catholics. This latter was a grave injustice, particularly because of the Ireland question, where almost everyone (parts of the north excepted) was a Roman Catholic. At that time there was a united kingdom of Great Britain (England, Wales, and Scotland) and the whole of Ireland. This was a major factor. To give you some idea of how major, look at population numbers. Today, there are 60 million people in Britain, and 5 million in Ireland. Then, at the beginning of the nineteenth century, there were 10 million people in Britain and already 5 million in Ireland—poor, rural, uneducated, not overly fond of the British, and (in the opinion of those British) appallingly superstitious and priest ridden. Right or wrong, the point is that the United Kingdom had a major fault line running right down it: the Irish Sea.

An ironworks, the epitome of the Industrial Revolution

Most dramatic of all was the population explosion. For reasons that are still not fully understood, numbers started to climb at a higher rate, and it was not only an absolute climb but one away from the countryside toward the towns. Just to give you some figures: between 1831 and 1851, London jumped from 1,900,000 to 2,600,000 people, Manchester from 182,000 to 303,000, Leeds from 123,000 to 172,000, Birmingham from 144,000 to 233,000, and Glasgow from 202,000 to 345,000. Between 1801 and 1851, Bradford grew from 13,000 to 104,000. With growth like this things have to happen. You have to have food, and law and order, and sewage, and education, and much much more—especially, you have to have the entertainments and occupations of a large closely packed urban group rather than the ways and means of traditional small village groups. No longer can you leave charity to the wives of the squire and the vicar. No longer can a couple of old women in the village give out remedies for everything from childbirth to cancer. No longer can literacy be a privilege of the spoiled few. You have to have a more modern society.

Society and its institutions did start to respond—slowly and unwillingly but inexorably. Catholics were emancipated—this did not mean that they could necessarily vote but that their religion did not at once exclude them. Some of the worst "rotten boroughs" were abolished and parliamentary seats given to major new urban centers. At the same time there were moves to reform education. New universities, starting with London, were being formed—University College started by Radicals and then Kings College started in response by Anglicans. The old universities had reform thrust upon them. They were forced, for instance, to offer science degrees. The Church itself was made to distribute a little of its wealth a little more equitably, with provision for the unchurched, new, urban areas. And as the century went on, reform also came to places like elementary education—something that was always tense given conflicting religious interests. The military and the nursing and hospital and medical professions generally showed that they needed change, especially after the appalling conditions that were revealed during the Crimean war. The civil service also had to start thinking in terms of a meritocracy, as it became clear that connection without talent and industry simply was not enough to run a modern country.

So it went through the nineteenth century, and as the needs arose the men (and sometimes, as in the case of Florence Nightingale, the women) rose up also to tackle and meet the challenges. One such person was Thomas Henry Huxley, born of a mentally distressed schoolteacher, who triumphed over his own personal demons and who grew up to be the Cuvier of late Victorian Britain: the most important and influential scientist of his day (Desmond 1994, 1997).

The Professional Scientist

The contrast with Darwin is the most striking. Whereas the author of the *Origin* was born to upper-middle-class security, driven to work only as his ambition dictated, never once in his life ever having to worry about mortgage or school fees or that little extra cash for a house extension, Huxley had to make his own way from the beginning. Apprenticed to medical relatives, he started his rise through brilliant performances at Charing Cross Hospital. He joined H.M.S. *Rattlesnake* as ship's surgeon—significantly, whereas Darwin took his meals with the captain, Huxley ate with the midshipmen—and it was on his journeys through the South Seas that Huxley started to build his scientific reputation and career.

Daily, fishing up delicate marine invertebrates like jellyfish and sponges, Huxley dissected them, showing in wonderful detail their structure and morphology, and most especially the relationships between different forms. No one could be ignorant of or indifferent to teleology, the Cuvierian functional approach; but, from the beginning, Huxley was less interested in ends and workings and more interested in the very ways in which things are put together and how they are transformed from species to species. Hence, from the beginning, he was attracted to a biology that put an emphasis on similarities and isomorphisims: homologies. This of course was the approach of Geoffroy in opposition to Cuvier, but even more (especially by mid-century) it was the approach of a school of German biologists—*Naturphilosophen*—for whom homology was the defining mark of the living, far more than functionality (Gould 1977b). Thus we find that, from the beginning, Huxley stood outside the tradition in which Darwin was trained and in which he excelled. It is characteristic of Huxley that, having determined the significance of German thought, he immediately set about teaching himself the language. Darwin was never able to do this.

Returning to England, Huxley rose rapidly through the ranks of science. He became a Fellow of the Royal Society—Britain's premier society for distinguished scientists—and got himself good jobs in London institutions. These did not carry the prestige of an Oxford or Cambridge post, but Huxley saw them as stepping-stones to the control of science as he envisioned it. At the same time he continued to establish himself as a master of the science of living form—morphology—as well as beginning to turn his gaze backward toward paleontology. The most powerful and influential man in these fields in England in the 1850s was Richard Owen, for many years an employee of the Royal College of Surgeons. Almost naturally, the touchy older Owen and the pushy younger Huxley fell out, and this set up a lifetime's rivalry. But Huxley had a strong capacity for friendship, and so he forged links with other younger scientists, like him determined to take over British science and convert it into the kind of university-based, professional, government-supported enterprise that they saw as the needed component of a modern forward-looking society. Darwin was linked to these scientists, especially through the botanist Joseph Hooker, and so it was natural that when Darwinism needed a champion at the British Association in 1860, Huxley was there to play the role.

For a moment longer, however, let us leave evolution on one side. Huxley and his chums did take over British science. By the 1860s and 1870s, they controlled the Royal Society—the presidency, the secretaryship, and the like. They got plum university posts and saw

Richard Owen

that their students got the same; although Huxley himself never left London and turned down offers from Oxford and Cambridge. They set and marked the examinations. They influenced elementary and secondary education, seeing that science got a firm foothold. They invaded the civil service and insisted that there be a place for science—and for properly trained scientists. They started journals and supported the efforts of others in this direction. Early issues of *Nature* owed much to Huxley. They took over the museums and much, much more. A little dining club started by Huxley and friends—the X-Club—became the very center of the English scientific establishment.

But what sort of science did these men want for their society, or rather what sort of biological science did Huxley want? Here we start to get some very interesting answers, which have a surprising significance for our tale (Ruse 1996). There were two branches of science particularly that caught Huxley's attention and concern. One was physiology, the study of the workings of organisms: a field in which Huxley himself was not a great practitioner, for it puts a premium on experimental technique and expertise, not something in which he shone. But through his students, especially H. N. Martin in London (and later at Johns Hopkins in Baltimore) and Michael Foster in Cambridge, Huxley supported and encouraged the science. Moreover—and this is absolutely crucial when you are founding and building a professional science—he found cash for its practitioners, and students for its teachers, and jobs for its students. In particular, he persuaded the medical profession—desperate to start curing rather than killing patients, and no less desperate to exclude pre-tenders—that physiology was just the training required for would-be doctors. And, the message finding very receptive ears, physiology was off and running.

Morphology was the other area of science that found professional favor with Huxley. This was indeed his own field, and here it was the teaching world that was the object of attack and persuasion. Huxley argued that a modern society puts aside such useless subjects as Latin and Greek and takes up science, morphology in particular. He was forever trumpeting the moral virtues of individual empirical experience. There is an intentionally biblical echo to his most famous dictum: "Sit down before fact as a little child, be prepared to give up every preconceived notion,

Joseph Hooker

follow humbly wherever and to whatever abysses nature leads, or you shall learn nothing" (Huxley 1900, 1, 219). And, to further his ends, we find that Huxley and his associates not only taught full courses in morphology but that they started summer schools for teachers, where the message could be passed on, as well as encouraging all that they could to take up cudgels on behalf of the science. H. G. Wells, the novelist, is probably the most famous product of the Huxley system—his fascination with science shows right through his writings—but he was one of many.

Physiology and morphology. Where does evolution fit into all of this? Well, of one thing you can be absolutely certain. Thomas Henry Huxley was a fanatical evolutionist. This was not always the case. Early on he had been against the idea, and when he returned to England one of his first publications was an absolutely savage review of a later edition of the *Vestiges of the Natural History of Creation,* originally published in 1844. (Although he knew that this was not really true, Huxley rather pretended that Owen might have been the author and went at the task with extra zeal!) But then like Saint Paul, also a convert to a belief that hitherto he had rejected, Huxley swung round and became an absolute fanatic on the subject of evolution. He too had his Romans and his Ephesians and his Corinthians and he preached and wrote accordingly. A brilliant showman and rhetorician, he knew precisely both the weak points of the opposition and all the flashy persuasive ideas that support evolution. Shortly after the *Origin* was published, the first full skeletons of archeopteryx—the reptile-bird—were uncovered in Germany. Given the propaganda value of the find, Huxley brought the discovery to the public podium, complete with diagrams and illustrations (Huxley 1868a). (In those pre-slide days, it helped mightily that Huxley was a brilliant blackboard artist.)

Fossil finds bridging different kinds of organisms were important. But these bridging fossils—known as *missing links* (until they were no longer missing!)—were not the most important focus of evolutionary studies. This honor was taken by our own species, *Homo sapiens*. Realizing at once that the most pressing question was human evolution, Huxley hammered away incessantly on our likeness to the apes and to our simian ancestry. He wrote a little book on the subject (*Evidence as to Man's Place in Nature*), he lectured on the subject, he discoursed at length on the subject—applying special attention to the recently discovered Neanderthal remains—and all of the time, he constantly inflated the mythic memory of the encounter with the Bishop. Why indeed should not his grandmother and grandfather be lower ape forms? After all, it is as dignified to be modified monkey as modified mud. Huxley's opponents complained, sometimes bitterly, that he misrepresented them and painted them into far more conservative positions than they truly held—in fact, if you look at Bishop Wilberforce's written review of the *Origin* you will find that although it is critical it is anything but negative. Huxley knew full well, however, that to make an effective positive case you need to portray yourselves as fighting forces of reaction and prejudice and succeeding only against great odds and in spite of gross knavery and trickery.

But what about evolution as a science? Did Huxley think of it as a field like physiology or morphology, that could be developed as a field of professional endeavor? Now it is certainly the case that some people thought this, especially German biologists influenced by Darwin's great supporter and enthusiast in that land, Ernst Haeckel. He promoted evolutionism tirelessly and built around himself at Jena University and elsewhere a group who worked hard to make of the subject a professional discipline. In fact, for all the talk, it was not terribly Darwinian, for no one was much interested in natural selection, and the main emphasis was on that very part of the enterprise that Darwin himself had rather neglected, namely the tracing of paths or phylogenies. Haeckel himself was responsible for the notorious "biogenetic law," which states that ontogeny (the developmental path of the individual) "recapitulates" phylogeny (the developmental path of the group). Using this as a tool, and working with the ever-increasing fossil record, he and his followers strove mightily to map out the details of life's evolutionary history and to start the long and laborious task of filling in the details (Haeckel 1866, 1898; Bowler 1996).

Huxley took some interest in this work, and his younger followers and students—most notably the leading end-of-the-century morphologist E. Ray Lankester—worked hard at this German-inspired activity. But by and large this was not the use or role at all for which Huxley intended evolution. For all that he took proudly the label of "Darwin's bulldog," and for all that I am sure that Darwin himself

Ernst Haeckel (left) with a friend

wanted to see evolution as a thriving professional discipline like morphology and physiology, Huxley was essentially uninterested in promoting evolution in this wise. His lectures to his students, for instance, would be spread over two years and would take over a hundred and fifty classes, not to mention practica where one would be dissecting specimens. They were marvels of detail and instruction—Huxley was a brilliant teacher. Evolution would be lucky if it got half a lecture! Natural selection five minutes!

Amazing but absolutely true: "One day when I was talking to him, our conversation turned upon evolution. 'There is one thing about you I cannot understand,' I said, 'and I should like a word in explanation. For several months now I have been attending your course, and I have never heard you mention evolution, while in your public lectures everywhere you openly proclaim yourself an evolutionist'" (Huxley 1900, 2, 428). This was a question by a puzzled student to his great teacher.

Why the silence? There were negative and positive reasons. Negatively, as a scientist himself—as a morphologist, and later as a paleontologist—Huxley really had little need of natural selection or any other cause, and to be quite frank evolution itself (evolution as fact) was not that pressing. His specimens were all dead and on the dissecting table by the time he got to them, so natural selection did not do much for him. Here Huxley was in a very different situation from students working on questions to do with ecology and behavior. You might nevertheless think that evolution as such had to be important, because (in paleontology particularly) one is dealing with change through time. But although developmentalism was certainly crucial, evolution in the sense of a natural (that is, lawbound) connected succession, from one form to another, was not essential. In fact, following Cuvier both in time and commitment, much of the record as known in Huxley's day had been worked out by nonevolutionists (Bowler 1976). They saw change, but they saw change that was God driven, without connected succession. Rather, a series of miracles. Now, as we shall see, this was not for Huxley. However, as a scientist, the paleontological succession was all he needed. Even if one invoked embryological analogies thanks to the biogenetic law, well, this approach too had in essence been formulated, pre-Haeckel, by people violently opposed to evolution!

On top of these negative scientific factors, there were the negative social factors. Huxley just could not see how one could find cash support for evolutionary studies as a professional science. Evolution did not help the physician, and in schools it was certainly going to be regarded with suspicion. And it was an absolutely key part of the strategy of Huxley and friends, as they worked to establish power in Victorian society, that they seem even more honest and conventional and moral than anyone else. They were pushing things regarded with doubt and misgiving, so they themselves had to be purer than pure. Huxley knew and liked the novelist George Elliot, but since she lived unmarried with a man, he would not allow her to visit his wife and children at his home.

A place was found for evolution, however, and this starts to push us toward the positive reasons for Huxley's attitude. Museums welcomed evolution into their halls. As we shall see more fully in the next chapter, museums were developing in a major way as places of instruction and entertainment as the century drew to a close—the British Museum (Natural History) and the American Museum of Natural History and others—and evolution had a natural role to play here and could find its support. For what Thomas Henry Huxley wanted positively of evolution was a popular science, a kind of metaphysics, or secular religion if you like—one that could be used to challenge and substitute for the conventional religion of Christianity, which he saw embedded in society and standing in the way of those many reforms he and his fellows were attempting.

Hence, in a fashion, Huxley stood right in the tradition of Erasmus Darwin and Lamarck, except he himself had dropped the deism and evolution was now a respectable doctrine, no longer revolutionary, and being used to reform society rather than break it or overthrow it. Which brings up the question of progress. For the earlier evolutionists, what really counted was that evolution represented progress, in all of its various manifestations: progress against Christian providentialism and progress as a philosophy that represented everything for which the evolutionists stood. Now this was likewise important for Huxley and his friends and associates and students. They too wanted to promote progress, and they too looked to evolution as the ideal vehicle. But here they felt they had to go beyond the work of Charles Darwin. It is true that Charles Darwin himself believed in progress, but it was given only a limited role in the *Origin*. A gap had therefore to be filled, and fortunately there was at hand the man of the hour. For real faith—faith slopping right over into fanaticism—no one could hold the candle to Darwin's and Huxley's fellow Englishman and ardent evolutionist, Herbert Spencer. He lived and breathed and wrote—at very great length in one long volume after

Herbert Spencer

another—the subject of upward development, from the simple to the complex, from the blob to the human. Sometimes change takes a break—it achieves a point of "dynamic equilibrium"—but then something disrupts the balance and we are off upward again.

Spencer was truly Mr. Progress and so it was he far more than anyone else who became Mr. Evolution to his countrymen (Richards 1987; Ruse 1996). Even the way in which Spencer wrote of the topic, from the undifferentiated or what he called the "homogeneous" to the thoroughly mixed up or what he called the "heterogeneous" had a very Victorian ring to it. Progress was not just a biological or a social phenomenon: it was an all-encompassing world philosophy.

Now, we propose in the first place to show, that this law of organic progress is the law of all progress. Whether it be in the development of the Earth, in the development of Life upon its surface, in the development of Society, of Government, of Manufactures, of Commerce, of Language, Literature, Science, Art, this same evolution of the simple into the complex, through successive differentiations, hold throughout. From the earliest traceable cosmical changes down to the latest results of civilization, we shall find that the transformation of the homogeneous into the heterogeneous, is that in which Progress essentially consists. (Spencer 1857, 245)

Evolution therefore took on the role of a substitute religion for Christianity, and whereas Christians worshipped in churches, evolutionists worshipped in museums, where one found grand displays intending to illustrate and confirm the faith.

Huxley, a close friend of Spencer (they were both members of the X-Club and it was in fact Spencer who in the 1850s first started to persuaded Huxley that evolution as an idea makes good sense), bought entirely into the view of evolution as secular religion. He preached the gospel nonstop, from every public platform he could find—at learned societies, before groups of fellow savants, in workingmen's clubs—traveling far and wide, including a highly successful trip to North America. In every sense of the word, Huxley was the Saint Paul of the movement, although later in his life the press took to calling him "Pope Huxley." There was even a Judas Iscariot of the movement. St. George Mivart, Catholic convert and student of Huxley, was seduced by the Jesuits and became the most bitter critic of the Darwinian establishment: labeled "not quite a gentleman" for an intemperate attack on one of Darwin's sons, he was excluded from positions of power and comfort.

Social Darwinism

A good and full religion has a moral code, directives that it gives to its acolytes. "Love your neighbor as yourself." "Honor thy mother and thy father." "Do not lust after the wives of other men." Evolutionists took very seriously, as part of their system, this need for obligation. This led to the full development of what came to be known as Social Darwinism—a moral code based on evolution—although truly it would be better known as Social Spencerianism. The way in which the directives were obtained were fairly simple and direct. One ferrets out the nature of the evolutionary process—the mechanism or cause of evolution—and then one transfers it to the human realm (if this has not already been done), arguing that that which holds as a matter of fact among organisms holds as a matter of obligation among humans (Ruse 1986).

Take the case of Herbert Spencer. Several years before Darwin published—although some considerable time after Darwin made his own discoveries—Spencer (1852) recognized the significance of the struggle for existence for human population development. He saw clearly that natural urges to reproduce would bring on a differential survival and reproduction of organisms within and between populations, and that this could lead to permanent biological change. Always more interested in humans than in the rest of the organic world, Spencer at once drew the implications for our species. Take, to use his example, the different natures and behaviors of the Irish and the Scots. In true Victorian fashion, Spencer argued that even though the Irish have lots of children, because of their lazy, indolent ways they are going to fail in life's struggles. The far more frugal and hardworking Scots will succeed and thrive, as indeed they do. Change in human nature will ensue.

From this satisfying biological inference, Spencer made an easy transition to economics, arguing that just as biology favors an unrestricted struggle and consequent selective success, so also economically this is the way that one should go for success. In particular, one should promote policies based on extreme laissez-faire socioeconomics. States should stay away from the activities of people following their own self-interest. In no way should politicians try to regulate or otherwise control unrestricted competition. Spencer felt, with some considerable regret, that mid-Victorian Britain was far from the ideal libertarian society, but he thought that if it was to continue and to thrive and to succeed, then it should strive to maximize to the fullest extent its citizens' freedoms to pursue their own interests and ends. The state should be helping people to do what they want to do rather than acting as a deterrent and barrier.

> We must call those spurious philanthropists, who, to prevent present misery, would entail greater misery upon future generations. All defenders of a poor-law must, however, be classed among such. That rigorous necessity that, when allowed to act on them, becomes so sharp a spur to the lazy, and so strong a bridle to the random, these paupers' friends would repeal, because of the wailings it here and there produces. Blind to the fact, that under the natural order of things, society is constantly excreting its unhealthy, imbecile, slow, vacillating, faithless members, these unthinking, though well-meaning, men advocate an interference that not only stops the purifying process, but even increases the vitiation—absolutely encourages the multiplication of the reckless and incompetent by offering them an unfailing provision, and *discourages* the multiplication of the competent and provident by heightening the prospective difficulty of maintaining a family. (Spencer 1851, 323–324)

Spencer could sound positively brutal about those who would help the unfortunate within society: "If the unworthy are helped to increase, by shielding them from that mortality which their unworthiness would naturally entail, the effect is to produce, generation after generation, a greater unworthiness" (Spencer [1873] 1961, 313). And one can find similar sentiments in the writings of Spencer's followers. Listen, for instance, to the turn-of-the-century American sociologist William Graham Sumner, who makes the converse case:

> The facts of human life . . . are in many respects hard and stern. It is by strenuous exertion only that each one of us can sustain himself against the destructive forces and the ever recurring needs of life; and the higher the degree to which we seek to carry our development the greater is the proportionate cost of every step. For help in the struggle we can only look back to those in the previous generation who are responsible for our ex-

istence. In the competition of life the son of wise and prudent ancestors has immense advantages over the son of vicious and imprudent ones. The man who has capital possesses immeasurable advantages for the struggle of life over him who has none. The more we break down privileges of class, or industry, and establish liberty, the greater will be the inequalities and the more exclusively will the vicious bear the penalties. Poverty and misery will exist in society just so long as vice exists in human nature. (Sumner 1914, 30–31)

But there is much more to the story than this. Quite apart from the fact that Spencer had somewhat ambiguous feelings about natural selection—feelings shared by just about everyone else but Darwin— if anything Spencer's ethical theory was due chiefly to his background of Protestant nonconformism, which saw the Poor Laws and the like as keeping much of the population in a state of perpetual poverty and dependency. Spencer (rightly) saw establishment Christianity as serving the ends of the rich and powerful (represented by the Anglican church), who inherit their wealth and status and who have no fear of the threat of competition from the more gifted and industrious (Spencer 1904; Duncan 1908). Spencer's evolutionism certainly moved in to confirm and support his alternative, supposedly secular, Social Darwinian views, but there was no simple deduction of ethics from biology. It was as much a question of one branch of Christianity set against another branch as it was a question of science set against all of Christianity.

Confirming this claim that there were strong Christian elements lurking in even the most ferocious-sounding Social Darwinian systems, it is clear (from statements and from actions) that it was never the intent of Spencer or his followers to deny the importance of individual charity. Take two of Spencer's more notorious disciples. John D. Rockefeller spent the first part of his life building up the vast petroleum company Standard Oil and the second part of his life fighting the federal government as it tried to break up the monopoly he had established over so vital a national resource as fuel oil. From his childhood, Rockefeller had tithed to his church, and he gave seriously and deeply to charity. The University of Chicago would never have become the world institution that it is without Rockefeller munificence.

The same generosity is true of Andrew Carnegie, who came from Scotland and made his fortune by founding and building U.S. Steel. He always claimed that no man should die rich, and he gave huge amounts of money directed toward the founding of public libraries. Carnegie's charity was an immediate function of his reading of Spencer, a reading that stressed the positive rather than the negative side of laissez-faire. Carnegie (like other industrialists) was proud of what he had done, thinking it a credit to his own abilities rather than a black mark against the lesser abilities of others. That poor but gifted children might likewise have the opportunity to develop and use their talents, Carnegie wanted to found public places of instruction and learning where one might go to better oneself. A public library, seeded by Carnegie and then supported by the community, was a perfect outlet for his philanthropic drive (Bannister 1979; Russett 1976).

Alternatives to Laissez-Faire

It is interesting to note just how often the proponents of a Spencerian-inspired Social Darwinism had childhoods that were not only deeply and sincerely Christian but were from that branch of Protestantism indebted to Calvinism, in America particularly the Scottish version known as Presbyterianism. The great sixteenth-century religious reformer John Calvin had given prominent place to the doctrine of predestination: God has ordered things according to His stern, unbreakable laws, and thus the fates of all are decided before we are even born. Some are predestined to be saints, and others sinners. Many Victorians who thought this way embraced evolutionism with enthusiasm—a theory that stressed the force and power of law was music to their ears. Thus Dr. James McCosh, Scottish-born president of Princeton University and one of the most influential churchmen and educators in America in the second half of the nineteenth century, claimed that there are no accidents, all is foreordained:

> It is in the very constitution of things. It is one of the most marked characteristics of the state of the world in which our lot is cast. It is, in fact, the grand means by which the Governor of the world employs for the accomplishment of his specific purposes, and by which his providence is rendered a particular providence, reaching to the most minute incidents and embracing all events and every event. It is the special instrument employed by him to keep man dependent, and make him feel his dependence. (McCosh 1882, 164)

The belief that some are chosen by nature to be successes and some are doomed to failure, that not only are all humans not born equal but that this is a right and proper state of affairs, was to the likes of Rock-efeller and Carnegie as much a matter of theology as it was of scientifically based philosophy.

At first this was true also of Thomas Henry Huxley. He spoke of himself as a "scientific Calvinist," meaning that he thought that the stern laws of nature decided the fates of us all, determining some to succeed and others to fail. However, increasingly, as Huxley and his friends succeeded in their aims of changing and reforming Victorian Britain, he was drawn metaphysically toward a position where an individual's own free will and efforts are the true determinants of life (Huxley 1893). Socially, despite his continuing friendship with Herbert Spencer, he pulled away from laissez-faire. For the mature Huxley, ethical success lay not in a conformity with and acquiescence to nature's laws. It lay rather in fighting such laws and the evil consequences to which they lead. At the same time, Huxley saw the virtues of a functioning civil service and of intervention by the state into such things as education and medicine and the military and the like (Huxley [1871] 1893). (See also Jones 1980.)

One senses that for Huxley there was always a conflict within: his enthusiasm for naked evolutionism, which he always interpreted as based on a brutal struggle, battled with his innate decency and his conviction that it is our ultimate moral obligation to fight those vile personal attributes that come in a package deal as part of our biology. No such worries ever troubled the happy thinking of Alfred Russel Wallace. As a boy, he had been taken by one of his older brothers to hear the Scottish mill owner and early socialist Robert Owen (Wallace 1900). He always looked back to this moment as a real turning point and, for the rest of his very long life, Wallace was ever an ardent socialist (Wallace 1905; Marchant 1916). Against Darwin, he believed that selection can operate for the good of the group as well as for the individual, and he thought that evolutionary success would be something that promoted the harmonious whole over the selfish individual.

Similar sorts of views appealed to the exiled Russian Prince Petr Kropotkin. He claimed that there exists between all animals, including humans, a natural sense of sympathy, something that he called *mutual aid*. Kropotkin did differ from Wallace in having little or no time for the state whatsoever. One suspects that his anarchism owed as much to the fact that he hailed from czarist Russia, one of the nineteenth century's most repressive societies, as it did to anything in evolution. But one should not dismiss entirely the influence of the particular spin that Kropotkin put on the evolutionary process.

*Prince Petr
Kropotkin*

The terrible snow storms that sweep over the northern portion of Eurasia in the later part of the winter, and the glazed frost that often follows them; the frosts and the snow-storms that return every year in the second half of May, when the trees are already in full blossom and insect life swarms everywhere; the early frosts and, occasionally, the heavy snowfalls in July and August, which suddenly destroy myriads of insects, as well as the second broods of birds in the prairies; the torrential rains, due to the monsoons, which fall in more temperate regions in August and September—resulting in inundations on a scale that is only known in America and in Eastern Asia, and swamping, on the plateaus, areas as wide as European States; and finally, the heavy snowfalls, early in October, which eventually render a territory as large as France and Germany, absolutely impracticable for ruminants, and destroy them by the thousand—these were the conditions under which I saw animal life struggling in Northern Asia. They made me realize at an early date the overwhelming importance in Nature of what Darwin described as "the natural checks to overmultiplication," in comparison to the struggle between individuals of the same species for the means of subsistence. (Kropotkin [1902] 1955, vi–viii)

To survive, it was necessary to work together against the elements.

In the animal world we have seen that the vast majority of species live in societies and that they find in association the best arms for the struggle for life: understood, of course, in its wide Darwinian sense—not as a struggle for the sheer means of existence, but as a struggle against all natural conditions unfavourable to the species. The animal species, in which individual struggle has been reduced to its narrowest limits, and the practice of mutual aid has attained the greatest development, are invariably the most numerous, the most prosperous, and the most open to further progress. . . . The unsociable species, on the contrary, are doomed to decay. (p. 293)

By now you may be convinced that I must be exaggerating. In my eagerness to show to you a different, more friendly side to Social Darwinism, in my urge to counter belief in the extreme laissez-faire socioeconomic doctrine that so many people today associate with Herbert Spencer, I have to be ignoring much that is true and pertinent. Surely there was a side to Social Darwinism that stressed conflict and violence. Surely there was a side to Social Darwinism—perhaps more characteristic of continental thought, of German thought in particular—where violence and conflict and, ultimately, all-out warfare were seen to be the right and proper expressions of evolutionary principles. Indeed, can one not say that, in some ways, Social Darwinism was a major motivating force that led to World War I, not to mention the hateful systems that followed in its aftermath? I refer of course to Soviet communism and to German national socialism.

In fact, as with social and economic questions, matters rather mixed. Certainly one cannot and should not exonerate Social Darwinism from all responsibility for the monstrous happenings and philosophies of the century that has just passed. One does find people who argued strongly that war and violence are natural states of affairs and who happily expressed their sentiments in evolutionary or pseudoevolutionary language. One enthusiast claimed that war is "a phase in the life effort of the State towards complete self realization, a phase of the eternal nisus, the perpetual omnipresence strife of all beings towards self fulfilment" (Crook 1994, 137). Even though writing like this probably owes as much to the early-nineteenth-century German philosopher Hegel as it does to Charles Darwin, it was not a sentiment expressed by one person alone. Others put matters in similar language: "Man has always been a fighter and his passion to kill animals . . . and inferior races . . . is the same thing which perhaps in the dark past so effectively destroyed the missing link between the great fossil apes of the tertiary and the lowest men of the Neanderthal type. All these illustrate an instinct which we cannot eradicate or suppress, but can best only hope to sublimate" (pp. 143–144).

However, countering this kind of writing, one finds that there were many who argued that war and violence are, if anything, the antitheses of evolution, especially inasmuch as one thinks that the course of evolution can and must be progressive. Herbert Spencer himself spoke eloquently to this end. As always with Spencer, the

Christian training was never far from the surface—significant here was surely the fact that spicing the nonconformist elements in his intellectual broth was a large pinch of Quakerism—but there were other factors also that led him to deplore militarism. As one who was keen on free trade and open competition, Spencer had little or no tolerance for intersocietal rivalries. Quite properly he saw them as major barriers to such trade. Moreover, he deprecated strongly the arms races that began, at the end of the nineteenth century, to obsess and burden countries like Britain and Germany. He thought that expenditure on such things as ever bigger and more powerful battleships was an appalling waste of money and resources. Far better that these be spent on peaceful things. In these sentiments, Spencer was far from alone. In fact, some of the most important relief work done during World War I came at the hands of evolutionists. They thought that only by trying to ameliorate the appalling consequences of conflict could one have any possible hope of rescuing the desired upward progress of evolution from the degenerate state into which, sadly, it had fallen.

As always, evolutionism's relationship with people's actions and beliefs is ambiguous. This also proves true when we turn to look at the ideologies of the post–World War I period (Mitman 1992). You might think that communism—at least, the nineteenth-century communism of Karl Marx and Friedrich Engels—owed much not just to evolutionism in general but to Charles Darwin's *Origin of Species* in particular. We know that Marx spoke warmly of the *Origin* (Young 1985). At Marx's funeral, Engels went so far as to say that as Darwin had provided great insights about the biological world, so Marx had done likewise for the social world. But although the Soviet system ostensibly responded warmly to these sentiments—after the revolution there were always departments of Darwinism in Russian universities—the English materialistic science of Darwin was not the major influence on Soviet science. This honor was held by the Germanic idealistic philosophy of Hegel. The Soviet bible on scientific methodology was Engels's posthumously published *Dialectics of Nature,* which owes a great debt to *Naturphilosophie* and nothing at all to Darwinism and natural selection. Darwinism, it appears, was more something used to give people's thinking a veneer of intellectual respectability than something that profoundly altered the way that people thought. One amusing side note is that in America the compliment was returned. Early American communism owed more to evolutionism, to Herbert Spencer in particular, than it ever did to Karl Marx (Pittenger 1993). Socialists in the New World found the progressivism of the English evolutionist far more to their taste than the complex dialectic of the German thinkers!

National socialism likewise has a very ambiguous relationship with evolutionism. Darwinism—at least a bastardized form of Darwinism—found its way across the English Channel and ended up in Bismarck's newly unified Germany. Ernst Haeckel, a professor at Jena, preached nonstop "Darwinismus." It is true that, by the time that the German evolutionists had finished converting the Englishman's ideas to their own purposes, the doctrine bore little resemblance to anything to be found either in *The Origin of Species* or *The Descent of Man*. There was for instance a rather heavy bias toward the group as the major unit

Karl Marx

in evolution—something that Haeckel saw as nicely justifying the strong emphasis on the virtues of the state, a major theme in Bismarck's Germany. But genuinely Darwinian or not, there was much enthusiasm for an evolutionism applied to social and political issues. Moreover, this kind of thinking continued right through the time of the kaiser and resurfaced with the founding of the Third Reich (Gasman 1971). Even in that hotchpotch of half truths and lies that poured into *Mein Kampf,* one can find sentiments that seem on the surface to be strongly influenced by Social Darwinism. "He who wants to live must fight, and he who does not want to fight in this world where eternal struggle is the law of life has no right to exist" (Bullock 1991, 141).

However, it does not take much to see that there could have been no simple relationship between any philosophy based on evolutionary ideas and the ideology that was so important for the national socialists (Kelly 1981). Apart from anything else, evolutionism—Darwinism in particular—stresses the unity of humankind. The Victorians were quite happy to put themselves at the top of the evolutionary tree—others, including Slavs and Jews, came lower down. However, ultimately, we are all part of one family. A consequence like this was anathema to Hitler and his cronies. It is revealing that although Haeckel (like so many of his countrymen at the time) was anti-Semitic, his solution to the Jewish problem was one of assimilation rather than elimination. This was the very opposite of the policy endorsed and enacted by the Nazis. It is no surprise that celebrations in the Third Reich of the anniversary of Haeckel's birth were muted in the extreme. Truly, as scholars have shown, national socialism owed far more to the Volkish movements of the nineteenth century, and particularly to the so-called redemptive anti-Semitism of the group of Wagnerians at Bayreuth, than it did to anything to be found in the writings of the evolutionists (Friedlander 1997).

Women

Let me conclude this brief survey of the ways in which Social Darwinism was interpreted and molded and used by turning to a discussion of the nature and status of women. This was a matter of as much pressing interest at the end of the nineteenth century as it was at the end of the twentieth century. It is the popular view today that in many respects Social Darwinism was grossly sexist. "Darwin's theories were conditioned by the patriarchal culture in which they were elaborated. . . . The *Origin* provided a mechanism for converting culturally entrenched ideas of female hierarchy into permanent, biologically determined, sexual hierarchy" (Erskine 1995, 118). It is difficult not to feel sympathy for views like this. If one looks at the writings of Charles Darwin himself, particularly in *The Descent of Man,* there is much to justify the conclusion that Darwinian evolutionary theory was (and perhaps still is) little more than a thinly covered ideology intended expressly for the suppression and demeaning of the female sex. Darwin claimed that "man is more courageous, pugnacious and energetic than woman, and has a more inventive genius" (Darwin 1871, 2, 316). Women in compensation show "greater tenderness or less selfishness" (p. 326). In many respects, one could as easily be reading a novel by Charles Dickens or the most reactionary country vicar as a work of science.

Moreover, one finds that even those evolutionists who claimed to be favorable to the cause of women, notably Huxley, often behaved in ways that rather belied their good intentions. The general sentiment of leading evolutionists was that women simply do not have the intelligence and drive of men and that therefore they ought to be kept out of scientific societies and universities and the like. Lesser-known evolutionists shared these sentiments, albeit the real influence was often German idealism as much as anything written by Darwin or even Spencer.

In the animal and vegetable kingdoms we find this invariable law—rapidity of growth inversely proportionate to the degree of perfection at maturity. The higher the animal or plant in the scale of being, the more slowly does it reach its utmost capacity of development. Girls are physically and mentally more precocious than boys. The human female arrives sooner than the male at maturity, and furnishes one of the strongest arguments against the alleged equality of the sexes. The quicker appreciation of girls is the instinct, or intuitive faculty in operation; while the slower boy is an example of the latent reasoning power not yet developed. Compare them in after-life, when the boy has become a young man full of intelligence, and the girl has been educated into a young lady reading novels, working crochet, and going into hysterics at the sight of a mouse or a spider. (Allan 1869, cxcvii)

But, going the other way, once again what one finds is that just as Social Darwinians were divided on something like war and peace, so likewise they were divided on the woman question. For every Darwin or Huxley, there was someone on the other side. Take, one more time, Alfred Russel Wallace. So far was he from thinking that women are inferior to men that he came to the opinion that the only way in which the human race will achieve success and salvation is through putting all of our hopes in the hands of the females of the species. What we need is for young women to come to the fore and to choose only the better-quality young men. Thus, society and civilization will move upward. If we do not do this, then doom and destruction will be our fate.

> In such a reformed society the vicious man, the man of degraded taste or of feeble intellect, will have little chance of finding a wife, and his bad qualities will die out with himself. The most perfect and beautiful in body and mind will, on the other hand, be most sought and therefore be most likely to marry early, the less highly endowed later, and the least gifted in any way the latest of all, and this will be the case with both sexes. From this varying age of marriage, . . . there will result a more rapid increase of the former than of the latter, and this cause continuing at work for successive generations will at length bring the average man to be the equal of those who are among the more advanced of the race. (Wallace 1900, 2, 507)

To be honest, you may feel that ideas like this are about as naive as Wallace's already expressed faith in spiritualism. Wallace's daughters and their friends must have been very peculiar and distinctive young women if they were choosing only the better members of the opposite sex. But whether or not Wallace's ideas had any genuine connection with the real world, the simple fact of the matter is that he put forward these ideas in the name of evolution, in the name of Darwinism, even. And surely if anybody in the nineteenth century after Charles Darwin himself has the right to call himself a Darwinian, it was Alfred Russel Wallace. We can properly conclude that, as with other matters, the implications of Social Darwinism for the questions of sex and equality are by no means as straightforward and one-sided as critics today often claim.

Secular Religion

By now you will be starting to get the picture and realizing that secular religions tend to run into the same problems as regular spiritual religions. The problem with Christianity, the love commandment, is that it can mean as many things as there are Christians. Take capital punishment. Some Christians, like Pope John Paul the Second for instance, think that taking the life of another human as punishment is never compatible with loving him or her as oneself, no matter how vile that person may be. Others, the majority of church-going Americans I suspect, think that capital punishment is indeed allowable—mandatory—under the love commandment. How can one possibly show love of one's neighbors and their children if one does not support the ultimate punishment for ultimate sinners? The same is true of evolution. One can support just about anything in the name of Darwin.

Of course, this is an exaggeration of both Christianity and Darwinism. The two systems do put constraints on behavior, even if they allow much flexibility within these constraints. I cannot imagine either Christians or evolutionists positively welcoming wanton cruelty, at least not if they read their systems correctly. The point I would emphasize is the extent to which the two systems, spiritual and secular, have run so parallel in the past. Of course, when we come to reasons or foundations, there are and have been differences. For the Christian, presumably, ultimately all goes back to God and His will. One ought to obey the love commandment because this is what He wants of us. The evolutionist as evolutionist has no such recourse. Rather we find that, to a person, evolutionists turned to progress to justify the stands that they took. With Herbert Spencer, they would and did argue that evolution makes sense, it has meaning, for it is ever striving and moving upward. The reason why we ought to cherish evolution and its processes is that if we do not, if we let evolution be and perhaps even stop and reverse itself, then progress will end and perhaps decline. And this, by definition, cannot be a good thing. Hence, it is up to us to keep things going. Offering a public library where a poor but bright child can better him or herself is a way of making sure that society keeps up to the mark, perhaps even improving, rather than sinking back to a preenlightened state. The same is true of economic theory and of militarism and of sexual relations. Our various evolutionists prescribed and proscribed as they did because, in this way, they thought that progress could be kept moving right along upward.

I should say that by century's end, not everyone was entirely happy that a full and satisfying world-picture had been sketched

Phrenology explained (phrenology is the belief that you can read mental attributes from the shape of the skull)

out. Huxley himself before his death in 1895 began to have severe doubts about whether evolution is quite a beneficent as the neo-Spencerians preached. He himself was subject to quite wracking depressions and periods of guilt and worried more and more about whether real secular progress is ever possible or desirable. Moreover, the science on which this all depended did not flourish quite as people had hoped. If anything, increased knowledge of the fossil record and of embryology rather increased the problems than reduced their magnitude. The biogenetic law was seen to be a guide—and a rather misleading guide—at best. All too often, ontogeny and phylogeny take very different routes (Bowler 1996).

Increasingly, bright young biologists turned away from evolutionary problems and concentrated on other matters that seemed more tractable and of more immediate value, intellectually and practically. First cytology (the study of the cell) became important, and then heredity (or genetics, as it became known). Evolution, never very Darwinian, never very scientific, took on more and more of the guise of a secular religion, a world picture, than of a forward-reaching professional discipline. The *Origin of Species* lifted evolution up from the pseudoscience status of phrenology and astrology. But whether evolution became quite what Darwin had hoped and expected back in the fall of 1838, when he hit on natural selection, is another matter. My suspicion is that even when he died, in 1882, for all that he was honored for his achievements, Darwin must have been a little disappointed at the way in which things were turning out. But in the realm of ideas, as in real life, our children do not always grow up in quite the way that we had intended. They take on lives of their own. As was certainly the case with evolution.

Further Reading

I mentioned one of Adrian Desmond's books as additional reading for Chapter One. A good background source for Chapter Three is Desmond's massive biography of Thomas Henry Huxley: *Huxley: From Devil's Disciple to Evolution's High Priest* (Reading, Mass.: Addison-Wesley, 1997). As always with Desmond's writings, it is very strong on the characterization and the social factors within and without the science. Where it falls down I think is on the whole question of the professionalization of biology in the second half of the nineteenth century. My *Monad to Man* (Cambridge: Harvard University Press, 1996) gives a very different reading of Huxley, for there (and in this book) I see his attitude toward evolution as being very much at odds with the usual picture of an unqualified advocate. But do not think that, because Desmond and I differ in our interpretations, this means that everything is subjective and that you can simply believe what you like. The history of science has itself only been professionalized in the past 30 to 40 years. We are still digging out the pertinent archives and putting things together in coherent pictures. The dust has not yet settled on our labors. Specifically with respect to the history of post-Darwin evolutionism, although we are talking about persons and events of the nineteenth century, the history of this period is still being uncovered. You would expect alternative accounts at this comparatively early stage.

Social Darwinism has a vast literature. Thomas Henry Huxley was one of the most thoughtful writers on the evolution/ethics relationship. His classic essay "Evolution and ethics" was recently reprinted with two new introductions (one by a historian, James Paradis; the other by a biologist, George Williams): *Evolution and Ethics* (Princeton: Princeton University Press, 1989). This should not be missed. The best discussion of that very peculiar man Herbert Spencer can be found in Robert J. Richards's massive *Darwin and the Emergence of Evolutionary Theories of Mind and Behavior* (Chicago: University of Chicago Press, 1987). This book incidentally, which is even less easy to read at a sitting than *Monad to Man,* contains many fascinating details about evolution's history, backed by formidable scholarship. At the secondary level, two books I rather like are both by the American historian of ideas, Cynthia Eagle Russett. *Darwin in America: The Intellectual Response, 1865–1912* (San Francisco: Freeman, 1976) deals with the social and political issues around evolution in the new world. *Sexual Science: The Victorian Construction of Womanhood* (Cambridge: Harvard University Press, 1989) is a very good account of how biologists and others wrestled with the sex and gender issues in the light of evolution. The whole question of Darwinism and the link with national socialism is very controversial and tense. Two books by Daniel Gasman lay out the case for the prosecution: *The Scientific Origins of National Socialism: Social Darwinism in Ernst Haeckel and the German Monist League* (New York: Elsevier, 1971) and *Haeckel's Monism and the Birth of Fascist Ideology* (New York: P. Lang, 1998). Alfred Kelly speaks for the defense: *The Descent of Darwin: The Popularization of Darwinism in Germany, 1860–1914* (Chapel Hill: University of North Carolina Press, 1981). My feeling (as I hint in the chapter) is that simplistic connections are surely wrong but that something had to be responsible for that vile phenomenon and I am not sure that biology is entirely guilt-free.

The New World: Darwin in America

Darrow picked up the Bible and began to read: "'And the Lord God said unto the serpent, Because thou hast done this, thou art cursed above all cattle, and above every beast of the field; upon thy belly shalt thou go and dust shalt thou eat all the days of thy life.' Do you think that is why the serpent is compelled to crawl upon its belly?"

"I believe that."

"Have you any idea how the snake went before that time?"

"No sir."

"Do you know whether he walked on his tail or not?"

"No, sir, I have no way to know."

There was a howl of laughter from the crowd.

Suddenly Bryan's voice rose, screaming, hysterical: "The only purpose Mr. Darrow has is to slur at the Bible. . . . I want the world to know that this man, who does not believe in a God, is trying to use a court in Tennessee—"

"I object to your statement." Darrow was contemptuous. "I am examining you on your fool ideas that no intelligent Christian on earth believes."

Judge Raulston put an end to the argument by adjourning the court.

That night, at last, it rained. (Settle 1972, 108–109)

"The Scopes monkey trial"! This exchange comes directly from the transcript of a court case in Tennessee in 1925, when a young schoolteacher, John Thomas Scopes, was put on trial for having taught evolution to his class. Prosecuted by three-time presidential candidate William Jennings Bryan—an ardent evangelical Christian—Scopes was defended by well-known lawyer Clarence Darrow—a notorious agnostic and freethinker. In a country that loves a good court case—remember the O. J. Simpson trial—this was entertainment of the highest order. The lawyers put on a wonderful show, dueling openly before the whole American public. And how that public laughed! The best-known and most savagely funny reporter in the country, H. L. Mencken of the *Baltimore Sun,* wrote scathingly of a society that takes seriously "degraded nonsense which country preachers are ramming and hammering into yokel skulls." Which may have been true, but in the end Scopes was found guilty and fined $100.

Evolution in the New World! What has gone wrong? Let us put back the clock and see how ideas (and people) crossed the Atlantic and what happened when they did.

The Harvard Debate

Louis Agassiz was a striking, florid, self-confident man, who could charm dollars out of eager New Englanders like a conjurer with his hat and his rabbit (Lurie 1960). And a man who could spend those dollars just as rapidly. Born in Switzerland of a father who was a Protestant pastor in a Catholic canton, religion was ever a major factor in Agassiz's world picture. He left his home and country in midlife, crossing the Atlantic in 1846 to a professorship at Harvard. His first wife (from whom he was estranged) conveniently dying, he married again, this time into the Boston aristocracy. With the change, he moved also from the piety of his youth to the American Unitarianism of his new spouse. But on one thing Agassiz stood firm all his life: with all his being and with all of his formidable energy, he opposed the vile doctrine of transmutationism. He was against it in his youth and he was against it in his old age, and Agassiz being Agassiz, everybody knew this and the reasons for the opposition.

Louis Agassiz

So let me start by stressing that Agassiz was a great scientist. Toward the end of his life, his energies were given more to institution building, but his achievements were real and important. It was he who established the fact of ice ages in Europe—his coming from Switzerland, where he had first-hand experience of glaciation, played a major role in this achievement, but the triumph was Agassiz's nevertheless. And in the field of biological studies, specifically of ichthyology (fish) both in nature today and as represented in the fossil record, Agassiz was rightly recognized as the world leader. Indeed, his start in this had been fast and precocious, for as a young man he had visited the aged Cuvier in Paris. The French scientist had been so impressed that he had given to Agassiz his own notes on fish, that the visiting junior scholar might take them up and use them in his own studies.

This was clearly a defining moment for Agassiz, who felt for the rest of his life (with some justification) that he was carrying the mantle of Cuvier—the torch that had been passed on. He adopted and never relinquished Cuvier's fourfold *embranchement* division, he put his ice age thinking into a context of Cuvierian catastrophism (which was one reason why many people, Lyell and Darwin especially, had trouble with immediate acceptance), and one senses that the lifelong opposition to evolution was in major part very much a paying of debt and homage to Cuvier as Agassiz's mentor. There was an emotional bond struck at once between the two men, something that might have been based in part on their shared religious situations: Protestants in Catholic territory.

But Agassiz was always more than simply a reflection of Cuvier. He was educated in Germany, and at Munich had sat at the feet of two of the greatest of the *Naturphilosophen:* Schelling the philosopher and Oken the biologist (Agassiz 1885). This experience affected him greatly. Schelling was not only a great system builder but a charismatic lecturer, and Oken was little less and (a pattern inherited by Agassiz) was himself a great friend of the students: a real old-fashioned college prof who would drink beer and talk until the small hours of the night. Although the *embranchement* theory meant that Agassiz could never himself be a full-blown *Naturphilosoph,* he adopted many of its ideas, especially the belief that (within *embranchements,* the vertebrate branch particularly) one could see an upward rise from primitive to complex. "One single idea has presided over the development of the whole class, and that all the deviations lead back to a primary plan, so that even if the thread seem broken in the present creation, one can reunite it on reaching the domain of fossil ichthyology" (Agassiz 1885, 1, 241).

Moreover, Agassiz was very much into parallels between life history and embryological development, and indeed he saw a threefold parallelism that included living beings today. "One may consider it as henceforth proved that the embryo of the fish during its development, the class of fishes as it at present exists in its numerous families, and the type of fish in its planetary history, exhibit analogous phases through which one may follow the same creative thought like a guiding thread in the study of the connection between organized beings" (pp. 1, 369–370). With this perspective, Cuvier notwithstanding, one might question why Agassiz would be so strongly against evolution. Especially since he was forced to admit that, just before the appearance of any new form or type in the record, the fossils start to forecast what will come. They are "prophetic types." "It seems to me even that the fishes which preceded the appearance of reptiles in the plan of creation were higher in certain characters than those which succeeded them; and it is a strange fact that these ancient fishes have something analogous with reptiles, which had not then made their appearance" (pp. 1, 393).

But to ask questions about evolution is to misunderstand the philosophy within which Agassiz was reared. By the time of Haeckel, *Naturphilosophie* had matured into a philosophy that could bear an evolutionary interpretation, but the early beginning-of-the-century version—the version of Schelling and Goethe and Hegel and others—although deeply developmental was no less deeply idealistic. As shown by the passage quoted above from Agassiz, "the same creative thought like a guiding thread in the study of the connection between organized beings," it was the idea that counted, not the reality. No one believed that one could actually move from one species to another—

Kant's teleology (which is what Cuvier inherited) precluded the transition between types. But it was not needed anyway. German idealism was just that: idealism. Real evolutionism would have messed things up, and so Agassiz never thought or could think in terms of evolution. And when he went to America, to Boston, the people he mixed with tended less to be scientists of his own stature and more the intellectuals, the philosophers and poets, men like Emerson and Longfellow, who were themselves much taken with German philosophy ("transcendentalism"). So here was reinforcement for the beliefs.

Why did Agassiz want the funds that he sought? Five thousand dollars from this donor, 10,000 dollars from that legislature? His American dream was to build at Harvard a magnificent museum, one that could house collections drawn from the world over and where researchers could study and advance our understanding of the world of animals (Winsor 1991). It would be (in the Cuvierian tradition) a museum of comparative anatomy, where one could draw on many, many specimens and make comparisons and inferences based on the widest possible range of specimens. This project—part of that already mentioned worldwide movement to the building and developing of natural history museums (Richard Owen was just then campaigning to get the British Museum [Natural History] off the ground)—was finally completed later in the century under Agassiz's son, Alexander: the Museum of Comparative Zoology or Agassiz Museum. But it was Louis Agassiz's dream and that for which he would lecture incessantly in the public forum. Although he would have denied it vigorously, one senses that Agassiz was not entirely disappointed at the publication of the *Origin*—Darwin sent him a copy with a polite note—for it gave him full opportunity to mount the stage before large audiences and to talk on the topic of the day, making his points of rebuttal and underscoring the need for massive (and expensive) facilities to look at these issues in a full and professional manner. Agassiz's own opposition in itself showed that the matter was not yet resolved once and for all.

All of this was gall and wormwood to another Harvard professor, a native-born scientist, who had risen up from humble roots in upstate New York and through medical training (a route shared by Huxley and other evolutionists) had eased himself into a life of full-time science, by the time of the *Origin* being one of America's leading botanists (Dupree 1959). I refer to Asa Gray, a man so far within the Darwinian circle that he had been let into the great secret some years before the publication of the *Origin*. Gray was by nature your scien-

Asa Gray

tist's scientist, a man whose love was the private professional discussion or gathering, where experts in the field could assess data and evaluate hypotheses. He was a man for whom the real respect came from fellow scientists and who thought that ultimately the appeal to the public dimension was a little bit vulgar. The trouble of course with such an attitude is that if you seek out the professional respect and disdain the public forum, professional rather than public respect is precisely what you get. The big funds—monies provided by rich private donors or by enthusiastic state legislatures—just do not come your way. (Agassiz was getting $100,000 from the Massachusetts government alone, while botany had to grub around for $10,000 total.) Nor can you attract and support those flocks of students that were flowing toward Agassiz, or any of the other perks of the academic life.

Hence, when the *Origin* was published and Agassiz started declaiming against it, Gray was primed and motivated to start the counterattack, even if it meant going out into the glare of the public arena. And this is precisely what he did, taking on Agassiz and his antievolutionism before audiences at the American Academy of Arts and Sciences. (This was and is a New England–based organization. The National Academy of Sciences had not yet been founded. This was to come later in the decade and owed much to Agassiz's stimulus.) At the same time, Gray took to the pen, reviewing the *Origin* and making sure that it got a fair and full exposition in the American press. His essays were later to be published in a collected volume, *Darwiniana* (1876), which was to prove one of the more appealing and lasting publications from the evolutionary fray.

The Huxley-Wilberforce debate repeated itself across the Atlantic in Boston. But it was not truly the same debate, or at least there were significant differences. For Huxley, whatever he said later about agnosticism, a major motivation was his attack on the Church, not just the dogma but everything it symbolized. You really can think of this as a clash between science and religion. But in respects Gray was even more devout than Agassiz, who was (it will be remembered) moving from Christianity to Unitarianism. Gray was ever a devoted evangelical Christian—indeed, loyal and attached though he was to

Darwin, he never much cared for Huxley, whom he thought a rather vulgar man, with little appreciation of or sensitivity toward people's religious beliefs. Gray did believe desperately sincerely in evolution and bound this up with his religious belief: God has given us our powers of sense and reason, to understand His creation, and if He decided in His power and magnificence to create in a developmental evolutionary fashion, then it is for us to accept and glorify. (More on some of these points later in this chapter.)

But precisely because Gray was a Christian, whereas Huxley was indifferent toward natural selection because he was indifferent to design, Gray was so sensitive to and overwhelmed by design that he could not accept that natural selection working on nondirected variation could do the full evolutionary job adequately. To the despair of Darwin, who thought that the move gutted the very principle of which he was so proud, Gray ever supplemented natural selection with divinely guided variations: "We should advise Mr. Darwin to assume, in the philosophy of his hypothesis, that variation has been led along certain beneficial lines" (Gray 1876, 121–122). Interestingly, perhaps connected to the fact that as a Christian he was more committed to Providence than to Progress, Gray was not a great enthusiast for evolutionary progress. But probably the main factor here was that he was a botanist rather than a zoologist. "We have really, that I know of, no philosophical basis for high and low. Moreover, the vegetable kingdom does not culminate, as the animal kingdom does. It is not a kingdom, but a commonwealth; a democracy, and therefore puzzling and unaccountable from the former point of view" (letter to Charles Darwin, 27 January 1863; in Gray 1894, 496).

One hardly need remark that, whatever the motivation, the nonprogressiveness of evolution was a sharp stiletto in the war with Swiss transcendentalism, for whom the ultimate emergence of our species was the very point of God's creative efforts: "The history of the earth proclaims its Creator. It tells us that the object and the term of creation is man. He is announced in nature from the first appearance of organized beings; and each important modification in the whole series of these beings is a step towards the definitive term of the development of organic life" (Agassiz 1859, 103–104).

The Fossil Wars

The general lore is that as Huxley had vanquished Wilberforce, so Gray vanquished Agassiz. I am not sure that this is true of Huxley, and I am certainly not sure that this is true of Gray. All biologists are evolutionists now, so there is a temptation to think that what we believe true today must have been apparent to those working and debating back then. If you combine this with the fact that, for all that he thought humans the culmination of God's creative process, Agassiz held really rather repellent views on the independent creation of human races and hence the independent origin of whites and blacks, whereas as an evangelical Christian Gray was passionately opposed to slavery, the case for Gray over Agassiz seems definitive. But it is not always true that what we find plausible and convincing today was equally plausible and convincing back in the past, and certainly not so here.

For a start, Gray himself held views that would be anathema to today's evolutionists, as indeed they were then to someone back in England. Gray was a friend and he was fighting the good fight, but was the cost too high? "The view that each variation has been providentially arranged seems to me to make Natural Selection entirely superfluous, and indeed takes the whole case of the appearance of new species out of the range of science" (Darwin and Seward 1903, 1, 191). But even if you ignore this and allow that Gray won the battle, in the long haul it is more accurate to say that Agassiz won the war. It was he who had and trained the students. Every one of them may have become an evolutionist—they all did, including Agassiz's own son!—but the picture of change traces back to those lectures in Munich rather than to the teaching at Cambridge.

In fact, Agassiz had a somewhat uneasy and difficult relationship with his students. He would welcome them in, make them part of his family, overwhelm them with friendship and advice and instruction. But he could not let them go, nor even could he see that they might be capable of working on their own and thus deserving of public recognition for their efforts. In Hegelian terms, he could not realize that the master-slave relationships he had imbued in Europe do not translate readily into American terms. Eventually, his best and brightest students revolted. Forming the Society for the Protection of American Students from Foreign Professors, they upped and left, with bitter things felt and said on both sides. Yet they took with them Agassiz's teaching: progress, the search for underlying forms or patterns or "archetypes," and an indifference to natural selection.

Paradigmatic of these Agassiz students was Alpheus Hyatt: Maryland-born, military academy–trained, passionate naturalist, and directed by the master to a lifetime's study of marine invertebrate life. Also a man of considerable moral and physical courage, for—

breaking with his Confederate-sympathizing family and with Agassiz (who wanted his students to stay out of controversy and turn to science)—he joined the Union army, putting his boyhood training at its service. Yet although he broke with Agassiz personally, not only was his choice of material Agassiz-influenced but so also was his very research program. In particular, Hyatt was fascinated ("morbidly obsessed" might be a better description) with the possibility of degeneration. Could it not be that instead of uniform progress upward, sometimes evolution overtops itself as it were, and starts a slide downward? Perhaps when we reach a certain point indolence and degeneracy set in and instead of going forward, organisms relapse back into a kind of second childhood?

I should say that, although supported by appeal to the fossil record and given a firm Lamarckian (inheritance of acquired characteristics) causal backing—you get to the top and then get slack and so start to slide down, as your organs atrophy through non- or misuse— much of this thinking owed more to Hyatt's reading of social changes than to anything in real biology. As a schoolboy, before he went to Harvard, he was taken by his mother on a trip to Italy. What he saw there impressed and shocked and rather depressed the impressionable lad. All around was the glory that was—the monuments, the buildings, the statues, the pictures—and all around is the filth and decay that is. "The lazzaroni live, beg, starve, make love and shit upon the church steps and along the quai, which last being the most public is the place generally preferred for the last picturesque action" (Hyatt 1857, 6). Was this a universal law of nature? Apparently so. Think of a society that designs and builds huge and beautiful edifices—temples, meeting places, palaces, and more. Eventually, "the nation, having outgrown its strength, would begin to decline. The vast buildings would have to be abandoned, and smaller habitations would arise, in answer to the requirements of a poorer population. The architects, faithful to their inherited canons, but forced into simplicity, would gradually follow the decline, and record it in the structures of the decadence" (Hyatt 1889, 79).

This is an advance on Agassiz, but in basic respects it is right out of Agassiz, for the whole picture is one of internal movement up and then down, something that the Agassiz-like Hyatt found to be mapped in embryological development. Indeed, even the degeneration may have had an Agassiz source, for apparently it was the teacher who first suggested to the student that certain invertebrate groups might be worth of study precisely because they showed evidence of fall after their climb. Of course, Hyatt was an evolutionist, but here also there was something funny. Nobody is quite sure exactly when Hyatt became an evolutionist. You cannot really tell from the crucial papers! You can of course tell once the Lamarckism is added on, but this is old hat and not very Darwinian per se—it is not something that Hyatt was using as a research tool but rather added to make the picture complete. Nor was evolution something that Hyatt was deliberately hiding or anything like that. I have said that Hyatt was a man of courage, and he was not one to conceal his beliefs out of cowardice. The point is that evolution for someone like Hyatt really did not make that much difference to his science at all. It was rather a metaphysical assumption about the working of the world—according to natural law rather than miracle—than a tool of scientific inquiry.

This all ties in precisely with what I have been saying about the English, Huxley-driven situation. Contrast the happenings back in the 1950s after Watson and Crick had discovered the nature of the DNA molecule. Immediately a whole industry sprang up, trying to decipher the genetic code and working out how the information on the DNA molecules gets transferred into the building of the cell. Genetics changed dramatically with the double helix. This was not the way at all for Hyatt and evolution. Evolution was not functioning as a tool of professional scientific research, or at least only in a background sort of way. Evolution was rather a kind of basic way of looking at the world, a sort of secular religion, rather than something to be used as science in itself. And of course all of the stuff about progress and degeneration fit the bill precisely.

Nothing in this chapter so far makes American evolutionism that worthy of note. It is at best all a bit derivative, and even if you avoid simply making judgments based on today's knowledge, American evolutionism all seems a bit uninspiring. Things were not helped by Hyatt's being one of the world's foggiest and most confusing writers. Would that he had learned from Agassiz in this respect. Darwin, for whom Hyatt had terrific admiration and to whom he sent key papers, found him quite incomprehensible. It was not so much a question of not agreeing but simply of not following! But when this is said, let us not forget that no one was doing very much of real note in evolutionary circles in the years after the *Origin*. Other than for one or two students of the insects and similar fast-breeding organisms, natural selection languished unused, and at best people were into the business of spinning phylogenies from embryological analogies. The main function of evolutionary thought was to provide social and moral messages rather than insights about the living world.

Expectedly, the degeneration notion got picked up generally—whether from America or in parallel as it were—and we find that, as the century drew to a close, almost every evolutionist was worrying that social development had peaked and was now in a decline. This was a time when the failures of capitalism—poverty, slums, ill health, and overcrowding—were becoming apparent to all, and when the military arms race now on between countries like England and Germany bode exceedingly bad for the future. Just as religion bends itself to the time—original sin is a Christian notion that has had far higher prominence since such appalling events as the Holocaust—so also evolution proved capable of bending with the time.

But this was not all that there was to the American story, by any means. If you are willing to concentrate on evolution as path, on phylogenies, then fossils become more than just a side interest. They become absolutely central. And it was here, in the second half of the nineteenth century, that America proved its worth in ways that were beyond the wildest dreams of scientists before evolution came to town. It turned out that America, particularly the West (the Canadian West also), was a charnel house for the denizens of the past, and absolutely fabulous fossil finds were there for those who would look. Those who would look, although it helped also, in those days before government grants, to have a large private fortune to support the large number of assistants and helpers necessary to dig the remains from the soil and rock and to transport the spoils back to the civilized East.

Two men above all others were qualified to go fossil hunting in the West, and they did so with such vigor and enthusiasm and violent personal rivalry and success that their exploits are still today talked of with admiration or censorious disapproval—but always with respect for the abilities and achievements (Shor 1974). The first was Othniel Charles Marsh (1831–1899), who came into a fortune when he came of age, thanks to his maternal uncle George Peabody, a business partner of the great financier Junius Spencer Morgan. Entering college at a much older age than most students, Marsh ended by spending his whole life at Yale, where he was an unpaid professor of paleontology, with the money and leisure to devote all of his time to his science. Not an easy man with whom to work, more difficult and ever more suspicious of the motives of others as he grew older, Marsh was nevertheless a good manager: one who knew how to seize an opportunity

Othniel Marsh

when it arose and how to get the most from his employees and underlings. He became president of the National Academy of Sciences for twelve years and wielded very considerable power both within science and on the interface between science and government.

Marsh was not a man to be crossed, a fact that did not at all perturb the second of our fossil hunters, Edward Drinker Cope. Son of a wealthy Quaker, Cope resisted parental entreaties that he become a farmer and opted rather for the life of a vertebrate paleontologist (Osborn 1931). Unlike Marsh, for most of his life Cope had no university affiliation, although toward the end a connection was forged with the University of Pennsylvania. Unlike Marsh also, Cope did not have great managerial skills and he was positively naive when it came to money, eventually losing his great fortune (a quarter of a million dollars) in a mining fraud. He too was a difficult man when dealing with his own generation, and many is the spiteful or critical tale that one finds in the letters of the day. Part of this no doubt was predicated on Cope's voracious sexual appetite, the quenching of which led to an early death from syphilis. But, for all his faults, there was a manliness about Cope. He was raised a pacifist, but he was no coward, and in the course of his fossil searches would unflinchingly brave Indian territory when more prudent people stayed home. When he lost his fortune, he did not cry or whine, but busied himself with his work, as though nothing had happened. More importantly, he was a brilliant scientist, and although his speculations on evolutionary causes were little different or more imaginative than those of Hyatt—they were known as the American neo-Lamarckian school (Marsh stayed away from all causal speculations)—Cope could reconstruct a long dead animal from the most unpromising of fossil material. Personally, he was warm and charming toward the young, and the next generation of paleontologists felt always that they had learned much from him.

Marsh enters our story first, for it was he who made quite fabulous finds of fossil horses. In 1876, Huxley at long last fulfilled promises to visit the New World and came calling at New Haven. Huxley was committed to giving public lectures on evolution in New York, and in search of dramatic examples had settled on the horse, the evolution of which he and a brilliant Russian student (Vladimir Kovalevsky) had been tracing. Marsh's collections of equine materials quite staggered Huxley, and at once he revised his lecture, fully admitting that (although extinct in North America by the time of human occupancy) the horse had evolved in America and not in Europe.

> At each enquiry, whether he had a specimen to illustrate such and such a point or exemplify a transition from earlier and less specialised forms to later and more specialised ones, Professor Marsh would simply turn to his assistant and bid him fetch box number so and so, until Huxley turned upon him and said, "I believe you are a magician; whatever I want, you just conjure it up." (Huxley 1900, 1, 462)

Edward Drinker Cope

The result was one of the most famous and most reproduced pictures of the lineage of the horse from the four-toed ancestor to the single-toed living representative. Moreover, it came with a prediction that soon would be unearthed a five-toed ancestor, older than all known forms: "In still older forms, the series of the digits will be more and more complete, until we come to the five-toed animals, in which, if the doctrine of evolution is well founded, the whole series must have taken its origin." Within two months, Marsh discovered just such an animal, the famous five-toed Eohippus. Today, we know that the horse record is far, far more complicated than Huxley implied, with masses of branches and extinctions. But it was precisely the forceful simplicity of Huxley's demonstration that impressed. This was evolution that people understood. Evolution that convinced.

Then, when Huxley had returned home, the battle between Marsh and Cope began in earnest. Out from the West, in Colorado and Montana and Wyoming and other states, came reports of fantastical monsters, reptiles of truly gigantic size and shape, buried in the rocks but already poking out into the air. Both Marsh and Cope sent out teams to see and to excavate, and the specimens began arriving back in the East. Digging was frenetic, and on more than one occasion men from the two sides met and clashed and there were reports of

*The multitoed feet
of the earliest horse,*
Phenacodus
(ancestral even to
Eohippus)

fisticuffs. Certainly, neither leader was above subterfuge, concealing results from the other and trying to snatch prize specimens from beneath the nose of the other. Huge sums were spent. Cope put out at least $70,000 of his own riches. Marsh spent $200,000. And this at a time when Asa Gray was thinking himself lucky to get $1,500 a year.

Many specimens were lost through crude and amateurish methods of recovery and transportation. But the overall results were truly magnificent: Allosaurus, Ceratosaurus, Brontosaurus, Camarasaurus, Amphicoelous, Diplodocus, Camptosorus, Stegosaurus, and many many others. And these were just from one period (the Jurassic). Then they went after more recent specimens (from the Cretaceous). Amusing is the discovery of Triceratops, that fabulous monster with (as its name implies) three horns. At first Marsh described them as the horns of an extinct monstrous form of buffalo, *Bison alticornis.* Only later, when more complete specimens were unearthed, was he willing to assign it to the dinosaurs. By the time he had finished—just to give you some idea of the immensity of the labors that were involved—Marsh had fifty specimens of Triceratops. It is true that most of these were represented simply by skulls, but before you start downgrading the achievement, reflect that these on average weighed a ton each, and the biggest was three and a half tons. This was evolution with a vengeance!

It was also evolution—or rather evolutionary evidence—of a kind that people could appreciate. With the move to museums—and by the 1870s and 1880s more were being built, including the already-mentioned American Museum of Natural History in New York (founded by an Agassiz student)—the demand more and more was for visible striking evidence of evolution. People did not want fine-grained experimental fodder. No one was doing experiments anyway! What they wanted were striking, impressive demonstrations of evolution in action. Things to appeal to the emotions as well as—if not more than—the minds. And monstrous reptiles from the past did just that.

All of this was brought together and given its most polished presentation, at the end of the nineteenth century and the first decades of the twentieth, by Henry Fairfield Osborn, for many years director of the American Museum of Natural History (Rainger 1991). Another man of great wealth—his father was one of the great pioneers of the railroad system—Osborn devoted his life to paleontology and to the administration of the halls within which it could occur. Befriending Cope—for whom he had great admiration—Osborn snapped up his collections for the museum when the great fossil hunter fell on hard times. Then he went out and added more of his own, always conscious that there was an end to serve: the dignified amusement and the cultural and social education of the New York public.

I have hinted already at this important point, so let me emphasize it here. The great achievements in museum building at this time—Harvard, London, New York, and elsewhere—did not happen

*Henry Fairfield
Osborn*

just by chance or simply through civic or national pride (although the latter was certainly an important factor). On the one hand, all of those lower- and middle-class urbanites needed distractions and occupations for their free time. They needed distractions and occupations that would be moral and healthy and fulfilling in a spiritual and moral sense. Getting soaked in a gin palace was not the solution. Nor were such things as cock fighting or gambling or other traditional entertainments. Museums were a perfect answer—places of wholesome entertainment, suitable for the whole family, low cost (or free), and easily accessible in the city. On the other hand, all of those people needed—and this was especially true in cities like New York, with a large immigrant population—instruction in moral and cultural norms. At the most basic level they needed instruction in such things as hygiene and nutrition and at the more conceptual level they needed instruction in the proper ordering of the state and of the roles that we all play within it. Museums could offer this teaching—through displays about cleanliness and threats to health such as vermin and microbes. They could also teach about the great heroes of the past and present—the men (and very few women) who had made a contribution to the greatness of the country. They could tell of such things as the wildlife of the country and its other virtues and treasures, thus helping to forge a sense of pride in city, county, and country. And most of all, they could instruct about the nature of society and about the rightful place of those at the top and of those at the bottom. In short, they could tell something of progress and of how some peoples are rightly in control and others are not.

This was the philosophy of Osborn as director. This was the philosophy of his board of trustees. They were all rich men, powerfully established, concerned about immigration and degeneration, supportive of eugenics (the idea that you can improve humankind through selective breeding), wanting to maintain the status quo with the Anglo Saxon elite at the top and the Jews, Irish, Poles, Slavs, Italians, and—above all—the blacks down the ranks and forever staying there. Evolution was the perfect vehicle for their ends. It was interesting, it was fun, it was amazing—all of these things and more. And it preaches a message. Some have succeeded and risen higher than others. That is the way of nature and we must learn to accept it.

A master showman, Osborn put on wonderful displays of horse evolution—he even cadged the skeletons of famous race horses for his ends. And the dinosaurs. Generations of little East Siders were shipped up to Central Park (the American Museum of Natural History is halfway down the West Side), to stand in amazement before these monsters of the past. The dinosaurs blew people's minds away. They still do! A testament to the wonderful ways of nature and to the men who revealed them. Osborn, a student of Huxley in his youth—he had gone to Europe to study and, thrill of thrills, had once been introduced to Darwin—brought evolution as popular science, as secular religion, to its highest point.

Christian Reactions

"As secular religion"? Gray, we know, was an evangelical Christian. Hyatt nearly became a Catholic priest. Although Cope moved from the Quakerism of his youth, he never relinquished a deep conviction that God rules and cares for the world. And Osborn crossed from the Presbyterianism of his childhood to an Episcopalianism of middle age. There were certainly agnostics and atheists in America—agnostics and atheists for whom their evolutionism was an important part of their overall world picture. American Marxists were one such group. And more generally, there were voices who wanted to separate science and religion, arguing that the former looks ahead and the latter looks back. Some of the classic works proclaiming the warfare of science and religion came from American pens in the second half of the nineteenth century. But, America is a religious country—far more so than Europe, including Britain—and at this time we simply do not find the equivalent of Huxley, a major science-based evolutionist for whom evolution is a real religion substitute.

If religion was such a large element in people's lives—especially in American people's lives—what then were the responses by the religious toward evolution in general, and Darwinism in particular? Of course, for any category, there are always exceptions or people who do not fit exactly, but roughly speaking (in Britain and America especially) one sees three basic responses (Moore 1979). The first two came from groups who accepted, even welcomed, evolution in some sense. The first—following custom let us call them the Darwinians—were religious people (Christians) who more or less accepted evolution as is and extended this acceptance to natural selection, that is, to Darwin's own ideas and causal suggestions. Some, like Asa Gray, felt it necessary to modify Darwinism to allow for direction, but others did not even feel this. They happily accepted Darwinism raw, as it were.

And I stress "happily." There were people like John Henry Newman, the great convert to Catholicism (he ended as a cardinal), who were quite indifferent to science, who were prepared to give science what it claimed, and who then turned to other things. But there were also people who were interested in science and who positively welcomed Darwinism. Interestingly, these were often people of a more conservative or orthodox or high-church bent than otherwise. They were people who were interested in teleology and who saw in natural selection precisely the teleology-producing mechanism that they had been seeking. They were people (who, as I mentioned in the last chapter, were often Calvinists) who took very seriously the facts of cruelty and struggle and pain that are our fate on earth and who saw in natural selection God's way of deciding between sheep and goats. And they were people who saw in the unbroken law of evolution, not deism, but God's constant interest in and sustaining of—His immanence in—the creation.

The late-nineteenth-century Oxford theologian Aubrey Moore— a very high-church Anglican—was one of the more attractive and articulate of this number. He welcomed Darwinism with enthusiasm—with joy, even—seeing in the theory the proof definitive of God's constant care for and action in his creation. The Divine is no remote designer but one always and everywhere present and active.

> Science had pushed . . . God farther and farther away, and at the moment when it seemed as if He would be thrust out altogether, Darwinism appeared, and, under the guise of a foe, did the work of a friend. It has conferred upon philosophy and religion an inestimable benefit, by showing us that we must choose between two alternatives. Either God is everywhere present in nature, or He is nowhere. He cannot be here, and not there. He cannot delegate His power to demigods called "second causes". In nature everything must be His work or nothing. We must frankly return to the Christian view of direct Divine agency, the immanence of Divine power in nature from end to end, the belief in a God in Whom not only we, but all things have their being, or we must banish Him altogether. (Moore 1890, 73–74)

In America, George Frederick Wright, although later in life to become much more conservative theologically, urged Asa Gray to publication and argued himself that Darwinism threw up no new challenges for the man of god. "The student of natural history who falls into the modern habits of speculation upon his favorite subject may safely leave Calvinistic theologians to defend his religious faith. All the philosophical difficulties which he will ever encounter, and a great many more, have already been bravely met in the region of speculative theology" (Wright 1882, 219). Everyone has had a go at the true faith. Nevertheless, "The Calvinist has stood manfully in the breach, and defended the doctrine that method is an essential attribute of the divine mind, and that whatsoever proceeds from that mind conforms to principles of order; God 'hath foreordained whatsoever comes to pass'. The doctrine of the continuity of nature is not new to the theologian. The modern man of science, in extending his conception of the reign of law, is but illustrating the fundamental principle of Calvinism" (p. 220).

The second group, also in favor of evolution but a lot less directly Darwinian, consists of those generally known as Darwinistic. Generally, these were people who took a liberal Christian approach, often known as "modernism." They wanted to modify Christianity in directions that they thought more in tune with the modern world and thought. They tended to downplay the stern unforgiving aspects of Calvinism—such things as predestination would be ruled out completely, and original sin would fare little better—and providence tended to get fairly short shrift. Rather they wanted to get on the bandwagon of progress, and their Christianity was going to reflect this. They prided themselves on being more scientific than the scientists, which meant that they simply loved evolution. But of course the evolution they wanted was a user-friendly evolution—one where effort paid off and where the struggle could be played down—and an evolution that was firmly progressive.

In short, what they wanted was a Spencerian type of evolution rather than a Darwinian type of evolution, which was precisely what most Americans of all kinds wanted anyway. Remember how Spencer, in America particularly, was *the* philosopher of evolution. Henry Ward Beecher, brother of the novelist, charismatic preacher, adulterer, liar—religion is not the only great American tradition—put things well:

> If the whole theory of evolution is but a slow decree of God, and if He is behind it and under it, then the solution not only becomes natural and easy, but it becomes sublime, that in that waiting experiment which was to run through the ages of the

world, God had a plan by which the race should steadily ascend, and the weakest become the strongest and the invisible become more and more visible, and the finer and nobler at last transcend and absolutely control its controllers, and the good in men become mightier than the animal in them. (Beecher 1885, 429)

Then, in addition to these two evolution-friendly approaches, there was the third response: that which rejected evolution. Charles Hodge, professor of systematic theology at Princeton Theological Seminary, the leading Calvinist theological school in the United States, had no doubts on the subject. "What is Darwinism?" he asked in one of his books. "It is atheism," came the reply. Hodge exonerated Darwin himself from the charge of deliberate infidelity, but his theory simply could not be held by a believer. "God has revealed his existence and his government of the world so clearly and so authoritatively, that any philosophical or scientific speculations inconsistent with those truths are like cobwebs in the track of a tornado. They offer no sensible resistance" (Hodge 1872, 2, 15). Backed by his reading of Agassiz, "a giant in palaeontology," Hodge had little difficulty in rejecting evolution in any form, especially the Darwinian incarnation. Darwin's "theory is that hundreds or thousands of millions of years ago God called a living germ, or living germs, into existence, and that since that time God has no more to do with the universe than if He did not exist. This is atheism to all intents and purposes, because it leaves the soul as entirely without God, without a Father, Helper, or Ruler, as the doctrine of Epicurus or of Comte" (p. 16).

Hodge, of course, preferred the account of Genesis to the account of the evolutionists. But it is important to note that he did not reject evolution simply because it was science, nor did he accept Genesis simply because it was religion. The point is that (in his opinion) evolution failed as science and hence the way was open for someone to accept the account of Genesis as good history. Here Hodge was very much following the tradition of his church: Calvin, although concerned to stay with a literal reading of the Bible, realized that some interpretative work was needed. To this end, the great reformer had introduced his famous doctrine of "accommodation," one recognizing that the Bible is sometimes written in such a form as to make itself intelligible to scientifically untutored folk who would not have followed sophisticated discourse.

Moses wrote in a popular style things which, without instruction, all ordinary persons endued with common sense, are able to understand; but astronomers investigate with great labour whatever the sagacity of the human mind can comprehend. Nevertheless, this study is not to be reprobated, nor this science to be condemned, because some frantic persons are wont boldly to reject whatever is unknown to them. For astronomy is not only pleasant, but also very useful to be known: it cannot be denied that this art unfolds the admirable wisdom of God. . . . Nor did Moses truly wish to withdraw us from this pursuit in omitting such things as are peculiar to the art; but because he was ordained a teacher as well of the unlearned and rude as of the learned, he could not otherwise fulfil his office than by descending to this grosser method of instruction. . . . Moses, therefore, rather adapts his discourse to common usage. (Calvin 1847–1850, 1, 86–87)

Likewise Hodge. He accepted the geologists' claim that the earth must be very old and entertained seriously the rival hypotheses that either there was a very long period of time (unrecorded in the Bible) after the initial creation or that the six days of creation must be understood as six very long periods of time. He himself inclined to the second hypothesis, but the point is that Hodge—and his stand was definitive for many many people—recognized fully with Calvin that the Bible needs interpretation and cannot be used as a scientific text. If Darwinism is to be rejected, it must be on scientific grounds. Hence, even this third position was far from one of blind rejection of science, even though evolution did fail to find favor.

The Scopes Trial

It was not until the end of the nineteenth century and into the twentieth century that things started to get a lot tighter, with full-blown campaigning against evolution (Larson 1997). Now increasingly we find Christians inclined to take a much more literal stand on the meanings of the Bible. This came slowly, not all at once. The definitive conservative Christian position was given in a series of pamphlets published between 1905 and 1915, *The Fundamentals*—hence the term *fundamentalist* for those who take the Bible as the inerrant word of God. But even here opposition to science was not absolute. Some of the pamphlets even endorsed evolution! Not Darwinism, for natural selection alone could not do the job and was clearly atheistic, but some kind of theistic guided evolution. "A new name for 'creation'" (pp. 20–21). Most would not have gone this far, but the interpretation of "days" as long periods or an unmentioned lengthy gap between the creation of heaven and earth and the edenic creation was standard.

However, by now there were hard-line literalists: people who subscribed to a six-day creation some 6,000 years ago. Prominent among these absolutists were the Seventh Day Adventists, a sect starting in the middle of the nineteenth century, strongly committed to the Second Coming and the conflagration that would precede it (Numbers 1992). For them, such a literal reading of the Bible was needed as confirmation of visions of their founder, Ellen G. White, as well as support for their insistence on Sabbath observance (which for them falls on a Saturday). Unless one has a 24-hour day of creation, the biblical support for Sabbath observance becomes less secure. In addition, as people believing in the coming Armageddon, the universal Flood played an important part in their theology as something that, having happened, showed God's ability and willingness to act again in such a way if necessary. A kind of balance to and foretaste of the disruptions to come.

But even if this extremism then attracted no immediate great following, by the end of the 1910s opposition to evolution of all kinds was starting to rise. There were a number of factors here. One, undoubtedly, was World War I. Rightly or wrongly, many associated Germanic militarism (militarism of all kinds, in fact) with Social Darwinism. Hence, evolution was seen as directly implicated in the carnage in Europe—a carnage that many felt was no concern of America's, anyway. A second was the fact that evolution was becoming more of a personal threat to nearly everyone, thanks to an explosion in American secondary education. The numbers of children enrolled in such education shot up from 200,000 in the whole of America in 1890 to 2 million in 1920. Tennessee, of which more in a moment, jumped from 10,000 in 1910 to 50,000 in 1925. Evolution was no longer just a faraway phenomenon, concerning only professors. It was now coming into every home, through the textbooks. The evil was right there.

And third, and I suspect most important of all, the evangelicals saw who was the real threat to their way of life and thinking. In America, it was not the agnostics and atheists: these were and are no real threat; they are a minority not to be taken that seriously. The real threat was the liberal Christians. The conservative believers saw this kind of Christianity as representing everything they loathed. It did not help (although it was hardly any surprise) that people like Shailer Mathews, dean of divinity at the University of Chicago and a leader of the liberal Christian wing in America (known as Modernism), had endorsed America's participation in World War I as a Christian duty. And evolution as we know was a centerpiece of this kind of Christianity. The temples of science, places like the American Museum of Natural History, dedicated to evolution, were just the sorts of places that the liberal Christians (Osborn, for instance) were building and endorsing. With reason, evolution was seen as part and parcel of a whole philosophy, a way of life, that was resented and disliked and to be opposed.

With the war over and with the campaign against alcohol brought to a successful conclusion—Prohibition was enacted—attention could be turned to evolution, and so in the 1920s we see attempts to make illegal the teaching of evolution in state-supported schools. Tennessee was one such state, and this led directly to the famous Scopes monkey trial. The interest then was intense—the press coverage was enormous—and it has continued to be so, thanks particularly to an account in a best-selling history—*Only Yesterday: An Informal History of the Nineteen-Twenties*—and then later a wonderful play (1955) and film (1960): *Inherit the Wind*. What gave an edge to the whole affair was that Darrow was denied the opportunity to call his own witnesses in favor of evolution—the case after all was over whether or not Scopes had broken the law, not whether evolution was true. Darrow had therefore put Bryan on the stand as an expert wit-

The prosecutors in the Scopes trial (William Jennings Bryan is in the center)

ness on the Bible and had apparently made Bryan into a fool, as the former politician stumbled around trying to keep some semblance of consistency between his religious beliefs and then-standard science. Although evolution may have lost the immediate battle (in fact, the conviction and penalty of $100 were overturned on a technicality on appeal), it won the war. The nation—the world—laughed at Tennessee and at the fundamentalists. Never again was right-wing Christianity to challenge science in such a way.

In fact, as so often happens, real life is not exactly like the myth. The trial took place in Dayton, Tennessee, a town that set up the trial in the first place, thinking that the subsequent publicity would be good for business. Scopes had let himself be prosecuted deliberately, for the American Civil Liberties Union (working hard to define itself as a body needed in America) that organized and bankrolled the defense was looking for a case to test the constitutionality of the law. He was not the regular biology teacher but the physical education teacher, subbing—and there is some question as to whether he ever did actually teach evolution. Bryan was a big-name figure, but not necessarily everyone's choice for prosecuting attorney—he was not an experienced trial lawyer. Darrow, who was, was certainly not everyone's choice for defense attorney—his non-Christian views made many on the defense side very uncomfortable (some because they disagreed with him, and others because they thought that his reputation would hurt their cause). And the verdict's being overturned was precisely not what the defense wanted. They needed a conviction to fight all the way to the Supreme Court and to challenge the law constitutionally. As it happens, in the absence of such a challenge, the Tennessee law remained on the books until the 1960s.

John Thomas Scopes being sentenced

What about the overall effects from the viewpoint of religion and science? Memory, reinforced by book and film, is that fundamentalism went down to defeat, never to rise again. The reporting—especially that of Mencken—was so savage that no one again could take such Christianity seriously. In fact, there is certainly evidence that virulent fundamentalism, linked directly to anti-Darwinism, seems to have peaked with the Scopes trial and did go into decline; although the extent to which this was direct cause and effect is another matter. What is not the case is that Bryan was made quite the fool that he appears in the movie—apart from anything else, the key scene where he is made to look stupid through his subscription to a literal reading of "days" was simply not true. Bryan always believed that the days were periods of time (a fact that upset the Adventists).

> Bryan: I think it would be just as easy for the kind of God we believe in to make the earth in six days as in six years or in 6,000,000 years or in 600,000,000 years. I do not think it important whether we believe one or the other.
>
> Darrow: Do you think those were literal days?
>
> Bryan: My impression is that they were periods, but I would not attempt to argue as against anybody who wanted to believe in literal days. (Numbers 1992, 80)

General opinion among the fundamentalists after the trial was that Bryan had acquitted himself well and that a good job had been done.

Finally, it should be noted that in some respects the fundamentalists certainly won the war, for the textbooks were immediately gutted of controversial evolutionary material—and then things stayed that way for many years thereafter. The popular *Civic Biology* by George W. Hunter, used by Scopes, was dropped by the Tennessee Textbook Commission. More broadly, a six-page section on evolution was dropped from the edition for southern states, and the author set about revising the text for general use. The explicit discussion of evolution was trimmed and concealed and material modified, with explicit charts vanishing. The word *evolution* itself vanished, and Darwin was no longer described as the "grand old man of biology." Relatedly, Darwin's "wonderful discovery of the doctrine of evolution" became "his interpretation of the way in which all life changes" (Larson 1997, 231). This was no unambiguous victory for evolutionism.

Looking back over the years, what should we say? For myself, I cannot say these occurrences—gutting evolution from the textbooks—

were right and proper. Indeed, as one who is an ardent evolutionist—a Darwinian even—I deplore them. But equally, I cannot say that I am surprised or that I am entirely unsympathetic to the fundamentalists. Evolution after Darwin had set itself up to be something more than science. It was a popular science, the science of the marketplace and the museum, and it was a religion—whether this be purely secular or blended in with a form of liberal Christianity. I do not think that it had to be, but it was. When believers in other religions turned around and scratched, you may regret the action but you can understand it—and your sympathy for the victim is attenuated.

I do not say that evolution as religion was always a bad thing. Indeed, at the social and moral level, we have seen that it can be entirely admirable. But let us not pretend that it was not what it was and that right and decency was all on one side. As is usually the case in these things, both sides had their saints and their sinners, their people of reason and their people of emotion. Most may have pulled back from the extremes, but the polarization was more than simply one of black and white, chalk and cheese, science and religion. As always when you are dealing with history, when you dig beneath the surface, things become a lot more complex and interesting than appears at first sight.

Further Reading

At the conceptual level, James Moore's *The Post-Darwinian Controversies: A Study of the Protestant Struggle to Come to Terms with Darwin in Great Britain and America, 1870–1900* (Cambridge: Cambridge University Press, 1979) is a really detailed account of the science/religion relationship in Britain and (even more) in America in the years after the *Origin.* But the social questions are also important, and especially museums are significant now. Mary P. Winsor's *Reading the Shape of Nature: Comparative Zoology at the Agassiz Museum* (Chicago: University of Chicago Press, 1991) is excellent on the early years of museum building after the *Origin,* and Ronald Rainger's *An Agenda for Antiquity: Henry Fairfield Osborn and Vertebrate Paleontology at the American Museum of Natural History, 1890–1935* (Tuscaloosa: University of Alabama Press, 1991) is not only strong on the museum scene at the end of the century but gives you a real insight into the persona and activities of Henry Fairfield Osborn, the leader in American evolutionism from the late nineteenth century right up to the 1930s. Rainger is particularly good at showing how much evolution back then, palaeontology particularly, was a vehicle for social messages of one sort or another.

Ronald Numbers is the leading authority on science/religion relationships in American history. Having come himself from a fundamentalist background, he was particularly well prepared to write his magisterial work on the history of literalist readings of Genesis. *The Creationists* (New York: A. A. Knopf, 1992) digs into its subject with a vigor and penetrating understanding that one rarely finds in works of scholarship. Highly recommended! At a more specific level, Edward J. Larson (Numbers's student) has recently given us a detailed and brilliant account of the Scopes monkey trial: *Summer for the Gods: The Scopes Trial and America's Continuing Debate over Science and Religion* (New York: Basic Books, 1997). As I explain in my text, it is quite amazing how much myth has grown up around that event. A major reason for this quasi-fictional status must lie at the feet of *Inherit the Wind,* especially the movie starring Spencer Tracey and Frederick March and with Gene Kelly as a wonderfully cynical newspaper man.

For all that it is fictionalized, the movie—which is easy to find on video—is well worth watching. It does raise most interesting questions about the science/religion relationship. You should be aware that it dates from the height of the Cold War, when Americans were defending the virtues of democracy in the face of the external threat of communism. At the same time—thanks to the witch hunting activities of Senator Joseph McCarthy—they were thinking about themselves, trying to establish to what lengths people should be allowed to dissent in a free society. Why defend democracy if this is equated with total conformity? The writers are using the Scopes monkey trial to explore issues that they found important rather than simply to give a historically accurate account.

Finally let me recommend to you the science fiction novel, *The Time Machine: An Invention,* by the English novelist H. G. Wells. (It was first published in 1895 and is easily available in editions today. There is also a good movie version with Rod Taylor.) Wells was the student of Huxley and started life as a science teacher. Hyatt was not the only one worried about degeneration. As mentioned in the text, by the end of the nineteenth century, with the rise of militarism and the unsolved problems of industrialism, many were obsessed with the prospect of inevitable degeneration and downward slide. Wells picks up on this worry and explores it in the form of fiction. Thanks to his machine, the time traveler goes forward into the future and finds that humans have sunk into two races or species: the Eloi, warm, friendly, childlike, useless; and the Morlocks, industrious, intelligent, vile, underground-living cannibals. Can this really be the fate of us all? Is there even worse beyond that? These are Wells's themes in a story that is as fresh today as it was then.

Cold War Warriors:
Darwinism and Genetics

It was too bad you couldn't be at Princeton, where we had a kind of gladiatorial combat from which both sides finally emerged apparently uninjured, so far as each side thought of itself, but demolished, so far as each side thought of the other. At the end, Dobzhansky held out his hand for me to shake and I grasped it firmly, saying "I think you may in time come around after all," at which everybody laughed and the meeting broke up. (Beatty 1987, 289)

Thus a letter from the Nobel Prize–winning geneticist Herman J. Muller to a student, about an encounter with the leading American evolutionist, the Russian-born Theodosius Dobzhansky. Through the 1950s, they had battled in an increasingly bitter fashion, over the nature of evolution and particularly over the amount of heritable variation one might expect to find in any wild population of organisms. Dobzhansky had rallied his forces, his own students mainly, and had run experiment after experiment on populations of fruit flies (Drosophila), subjecting them to radiation and trying to assess the effects. Muller, who had won his prize precisely for his work on the effects of radiation, had responded through his students, critiquing the work of the Dobzhansky group and devising experiments of his own that proved precisely the opposite of what his opponents claimed!

There was a fair amount of name calling here, for while Dobzhansky supported (what we shall see was) the fairly straightforwardly described "balance" hypothesis, he succeeded in getting Muller's option labeled the "classical" hypothesis, with the connotations that it was something old-fashioned and outdated. But underneath were some deep convictions, far more than mere science. We were now in the frozen depths of the winter of the Cold War, and this affected the real positions, as did absolutely fundamental convictions about the nature of humankind and its future. But to unpack all of this, we must go back to the beginning of the century and to the birth of genetics.

Population Genetics

We know that a major scientific problem with the theory of the *Origin* was the lack of an adequate theory of heredity. Darwin opted for a blending view of characteristics and their causes. What we now see was needed was a particulate theory, where the units causing organic features, passed on from generation to generation, are held entire, no matter in what individual combinations they may appear, in any particular

Gregor Mendel

organism. Unknown to all, the right approach was even at that time being formulated by a Moravian monk, Gregor Mendel, but the work was not noticed and he died obscure. Then, at the beginning of the twentieth century, with renewed interest in heredity, his work was brought to light and developed. Together with discoveries in the nature of the cell (cytology), the basic ideas of heredity (now called "genetics") were fused together in a satisfying overall picture, thanks particularly to the work of the American biologist Thomas Hunt Morgan and his students (one of whom was Muller) at Columbia University in the second decade of the last century (Allen 1978).

The "classical theory of the gene" located the units

of heredity (the genes) on thread-like entities (the chromosomes) in the center (the nuclei) of the basic building blocks of organisms (the cells). The genes are the units of heredity, that is to say, they are the units that are passed on in each generation, in sexual organisms via the sperm and the ovum, carrying the blueprint as it were for the new organism. It is believed that the chromosomes come in pairs and that the genes are matched across chromosomes—the particular place on the chromosome (common to all members of the species) being known as the "locus," and a gene form that can occupy a particular locus being known as an "allele." The genetics is Mendelian in the sense that

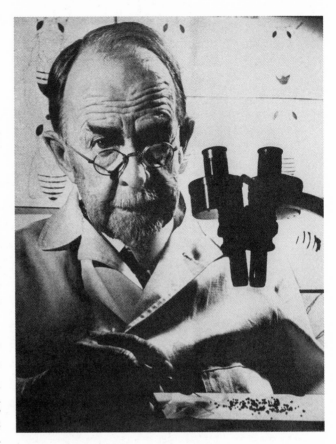

Thomas Hunt Morgan

each parent contributes equally to the new offspring, one and only one allele from each locus being transmitted. Which particular allele is transmitted is random, not in the sense of being uncaused but in the sense of being equiprobable and the choice not being a function of the efforts of the organism or the nature of the mate or the needs of the possessors or whatever.

How do you account for blending—skin color? How do you account for nonblending—sexuality? Genes are the units of function as well as of heredity—it is the genes, in combination with the environment, that cause the grown individual. The genes of the individual are known as the "genotype," the genes of the species are known as the "gene pool," and corresponding to the genotype we have the physical organism, the "phenotype." (There is no such corresponding term for the gene pool.) Paired alleles can be identical (this is called a "homozygote" with respect to that locus, or "homozygous") or different ("heterozygote," "heterozygous"). Sometimes the effect of one allele swamps the other allele, so that the heterozygote looks like homozygote of the swamping allele. In this case, the swamping allele is said to be "dominant" over the swamped allele, which is in turn "recessive."

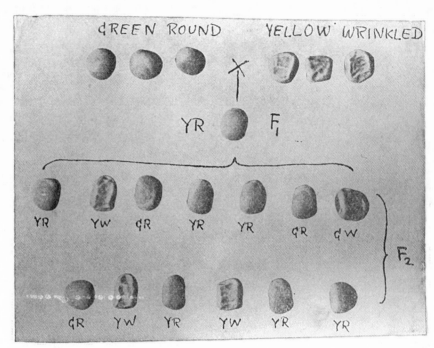

Mendelism as illustrated at the beginning of the twentieth century by William Bateson, the first British champion of the new science of heredity (the main point is that breeding may mask characters for a generation or two but does not destroy or blend them away)

What this all means is that a characteristic might be hidden for generations, only reappearing when identical recessive alleles get mated up again. What it also means is that a characteristic can be very rare indeed but will persist in the population so long as selection does not eliminate it. It will not get swamped out. Early geneticists did not realize this, but two mathematicians showed that original ratios will always stay the same as will the proportion of genotypes (the two homozygotes and the heterozygote), so long as no other factors are disrupting things. This simple ratio is known as the Hardy-Weinberg law, after those who found it.

The genes are very stable. From generation to generation, they change only rarely. Such changes as do occur are not uncaused. Muller won his Nobel Prize for showing how radiation can bring on changes. But they too are random in the sense that you can only say statistically how many changes there will be, and they do not occur according to the needs of the possessor. Most gene changes affecting the phenotype are harmful or deleterious. The gene is therefore the unit of change or "mutation." The early Mendelians somewhat naturally concentrated on large differences, so the idea grew up that significant changes are always largish, which led to the Mendelians seeing evolution as going in jumps or steps, from one variation to the next. They were therefore "saltationists" (as people like Huxley had been before them) and much inclined to play down the significance of natural selection. In this they were opposed by another turn-of-the-century group, the biometricians, who were working with statistical techniques trying to calculate the variations one finds in natural populations and who were the first group after Darwin actually to start taking natural selection seriously as a mechanism of change (Provine 1971).

In fact, although the debate between the Mendelians and biometricians was fierce and deadly—the leading biometrician dropped dead at 45 from stress—it did not last long. By about 1910, people were starting to realize that mutations can have very small effects as well as large ones, and it was understood that Mendelian genetics and Darwinian selection can be complements making the whole picture rather than rivals or contradictories. But as I have explained to you, these were not good days for evolutionary studies, at least not as a practicing professional science. People had ideas and intuitions but generally did not follow them up. It was to be another 20 years, around 1930, before mathematically inclined evolutionists put together the genetics and the Darwinism in one integrated theory: Mendelian genetics generalized to populations with the effects of such causes as mutation and selection factored in. With reason, this subject is usually known as "population genetics."

Three names are usually associated with this major advance in evolution's history. In England there were Ronald Fisher, one of the greatest statisticians of all time, and J. B. S. Haldane, a biochemist and mathematician. In America, one had Sewall Wright, an agricultural geneticist who had worked extensively on the blood lines of cattle, but who by 1930 was on the faculty at the University of Chicago. Today, we know that there was related work going on elsewhere, in Russia particularly. After the Soviet Revolution, people were looking for practical, low-cost science, and genetics fit the bill entirely. Unfortunately at the time this work was little known elsewhere, al-

Trofim Lysenko

though it had an effect on Dobzhansky, who left his homeland in the 1920s, never to return. Even more unfortunately, by the 1930s, Stalin had fallen under the spell of the charlatan Trofim Lysenko, who promised quick and easy—and totally fallacious—ways of obtaining favorable genetic results, and that was the end of that. Russian Mendelian approaches crashed never to rise again (Joravsky 1970).

Formally, the English and the American population geneticists produced identical work. They used different techniques—Fisher, for instance, used powerful classic mathematics, whereas Wright invented his own pragmatic techniques for problem solving—but they got the same answers from the same premises. However, as historians now realize, in intent there were major differences. For the moment, I will concentrate on the American picture, although in later chapters I will swing back to some of the English ideas. Fortunately, not only for my exposition but also for the men who followed him and who did not possess his mathematical skills, although Wright first presented his theory in formal style, at its heart was a pictorial metaphor. This was the famous "adaptive landscape."

Wright (1932) invited us to think of an area of land, with hills and valleys and plains. It is three dimensional: left and right, forward and back, up and down. Think of the surface as made up of points that could be occupied by different genotypes. Two organisms, very similar and just different in one or two alleles, would be next to each other, whereas two organisms with many differences would be far apart. Now the third dimension, up and down, is the kind of Darwinian dimension of fitness—if an organism was very much better at surviving and reproducing than another it would be higher, and if not, then lower. One would expect fairly smooth curves, because one or two allele changes would make only small differences to survival and reproduction ability.

This metaphor of the landscape—with organisms occupying spots on the surface—was the heart of Wright's theory. Initially, one would expect to find organisms clustered around the peaks of the landscape—not all of the peaks, but some of them. How then does evolution take place? Here, Wright's background in animal breeding became very important. He knew that the way that breeders have optimal success is not by trying to change the whole group at one time. Rather, you look out for features that you think particularly desirable, and you isolate them, trying to get them confined and spread through a small subgroup. Once you have done this, then you start to try to spread it through the whole group, by selective mating and choosing. But fragmentation and isolation are the initial key.

In real life, Wright thought that species tend to be divided into small subgroups. Obviously, however, such fragmentation and isolation on its own cannot do everything. Somehow one has got to get change taking the subgroups away from their shared, uniform past. Here Wright introduced what is now known as "genetic drift" or the "Sewall Wright" effect. If populations are fragmented into small subgroups, then one can show that within these subgroups selection might not be effective (even though it is at work), because the random factors of breeding might overwhelm it. In other words, change might come about through chance, and new features that have a higher fitness could appear. In terms of the metaphor, groups might wander down the sides of hills under the influence of drift, and then shoot up other hills thanks to selection—to peaks that were higher than the ones they left. Then these could thrive and perhaps swamp out everyone else.

Wright called his theory the "shifting balance theory" of evolutionary change, and this name often puzzles people because it is not obvious to what the term *balance* applies. In Wright's opinion, what you have always is a balance between forces that are leading to fragmentation and differentiation (drift and the like) and forces leading to recombination and uniformity (selection and so forth cleaning up afterward). In other words, everything is in a state of fluid balance or, to coin a phrase, "dynamic equilibrium," with

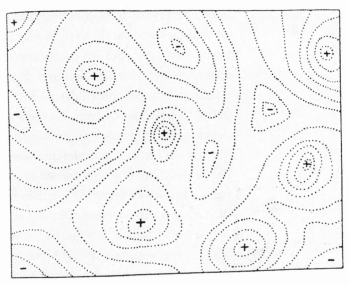

Sewall Wright's adaptive landscape

the forces toward heterogeneity squaring off against the forces toward homogeneity.

Now of course we have heard this kind of talk before—the terms are those of Herbert Spencer. But why should one be surprised? We know that Herbert Spencer was by far the most important evolutionist in North America, and in fact Wright was a student of a man who was an ardent Spencerian. (I refer to L. J. Henderson, one of Wright's professors at Harvard, where Wright was a graduate student around 1915.) And frankly, for all of the talk about selection and fitness, there is really nothing very Darwinian about the adaptive landscape metaphor. The chief force of change, certainly the most creative force of change, is genetic drift, which is about as non-Darwinian as you can get: something that Fisher, who was an ardent Darwinian, kept pointing out nonstop. Moreover, for Wright as for Spencer, what really counted was progress. You might think that the landscape is a bit like a Lyellian water bed—as one peak rises, so another falls. But this was not really how Wright saw things. He thought that the landscape was pretty rocklike and that over time real progress will occur. Certainly he did not think that the arrival of humans was pure chance, and in fact he had some pretty funny personal ideas about how everything is evolving upward so eventually we will all be part of one eternal mind.

But there was a big difference between Spencer and Wright. The Victorian had worn his values and his culture and his ideology in a very public fashion. He was preaching a doctrine, a secular religion, and it was there in full view for all who would read and listen. Wright was trained as a careful, professional scientist—one did not study science at Harvard to become a theologian, secular or otherwise. Furthermore, significantly, Wright was in a rather insecure branch of science. Today, genetics is a pretty top-dog sort of science—millions are spent on the Human Genome Project and some of the brightest minds go into the molecular biological business. Almost every year it gets a Nobel Prize or two. But back in the early years of the twentieth century, genetics was new with promises but no triumphs. The people who did it and supported it were agriculturalists—Wright spent the first 10 years of his career at the U.S. Department of Agriculture—and we all know where they tend to stand in the pecking order of academia. Above education and sociology, but not by much.

Hence, for all that he had a whole parade of private, Herbert Spencer–type values, Wright was absolutely not going to let these come to the surface of his professional work. In the tradition of Cuvier (of whom I suspect that Wright had never heard), for his own private subjective reasons, he was intent on pushing forward his science as objective. So whatever depths there may have been beneath the adaptive landscape, and I suspect that there were many and that they went down a long way, the population genetics of Sewall Wright represented a way of doing evolutionary biology that had not been seen hitherto. It was science of a much more professional standard.

The Synthetic Theory of Evolution

In the 1930s and 1940s, things now really started to move forward (Cain 1993). The key figure was Theodosius Gregorievitch Dobzhansky, to give him his full name. As a youth he read the *Origin of Species* (in Russian translation) and was at once converted to evolutionism, with a strong sympathy for Darwin's ideas. He trained as a biologist, making great trips across (prerevolutionary) Russia, specializing in that common little insect, the ladybug. But in the 1920s, Dobzhansky moved on a scholarship to America (never to return to his homeland) and, being located in the laboratory of Thomas Hunt Morgan, switched to the study of chromosomal variations in fruit flies. Clearly destined for big things, in 1936 Dobzhansky was invited to give a prestigious series of lectures in New York City, and the following year these were written up as *Genetics and the Origin of Species.*

Theodosius Dobzhansky

Unlike his American colleagues, who tended to be city types, Dobzhansky knew from his early training that there are simply masses of variation in wild populations—one finds differences between individuals and differences between groups. The standard uniform type is a fiction of the laboratory geneticist's imagination. Dobzhansky knew moreover that you get gradations from group to group: rarely if ever do you get abrupt changes. And he was keenly aware of adaptation, realizing that since Lamarckism is false, natural selection is really the only game in town. "A biologist has no right to close his eyes to the fact that the precarious balance between a living being and its environment must be preserved by some mechanism or mechanisms if life is to endure. No coherent attempts to account for the origin of adaptations other than the theory of natural selection and the theory of the inheritance of acquired characteristics have ever been proposed" (Dobzhansky 1937, 150).

But when the time came, it was not Darwin who really provided the inspiration and foundation for Dobzhansky. In 1932, at an international genetics congress, Dobzhansky saw a poster display of Wright's shifting balance theory. Although he was himself completely devoid of any mathematical ability whatsoever, Dobzhansky knew a good idea—a good picture—when he saw one. Wright's adaptive landscape, with its peaks and valleys, with groups of organisms either sitting on the tops of the peaks or subject to factors that were moving them from one peak to another, was the causal theory that Dobzhansky needed and within which he could place his knowledge of organisms in the wild as well as of experimental subjects in the laboratory.

> Each living species or race may be thought of as occupying one of the available peaks in the field of gene combinations. The evolutionary possibilities are twofold. First, a change in the environment may make the old genotypes less fit than they were before. Symbolically we may say that the field has changed, some of the old peaks have been levelled off, and some of the old valleys or pits have risen to become peaks. The species may either become extinct, or it may reconstruct its genotype to arrive at the gene combinations that represent the new "peaks." The second type of evolution is for a species to find its way from one of the adaptive peaks to the others in the available field, which may be conceived as remaining relatively constant in its general relief. (p.187)

We have, thought Dobzhansky, a group (like a species) "exploring" the slopes of a mountain peak, working in some way through "trial and error" until at last it escapes from its home base and moves across a valley and shoots up the side of a neighboring mountain.

The theory in Dobzhansky's book is therefore a funny synthesis. It is in essence Wright's Spencerian shifting balance theory, but to this is added both a deep knowledge of the real world of organisms and at the same time a keen appreciation of key Darwinian ideas, especially those of adaptation. What Dobzhansky does not offer is a synthetic unifying vision, as we find in the *Origin of Species*—he makes no effort to cover the wide range of topics that Darwin thought essential to his case. There is no mention of paleontology in *Genetics and the Origin of*

Species, nor embryology, nor many of the other subjects that interested Darwin and that he thought so important. The real emphasis is on speciation, not a subject on which Darwin dwelt at length. Expectedly, given Dobzhansky's time in Morgan's lab, there was much discussion of such factors as chromosomal variation and of how it can and cannot become important when groups split and new reproductively isolated groups (species) are formed. And this led straight to an absolutely vital underlying assumption of Dobzhansky's whole case, namely, that the way to understand major evolutionary changes is through the study of minor changes—changes so minor that you might not normally notice them or think them significant. "Experience seems to show . . . that there is no way toward an understanding of the mechanisms of macro-evolutionary changes, which require time on a geological scale, other than through a full comprehension of the micro-evolutionary processes observable within the span of a human lifetime and often controlled by man's will" (p. 12).

Strange hybrid though it may have been, *Genetics and the Origin of Species* proved to be an absolutely seminal publication. Here was an attractively written and reasoned work on evolution that all could understand—the mathematics was kept to a minimum!—and that could inspire a young researcher and offer a program leading to a career as a professional evolutionist. But there was more than just this, for at this point Dobzhansky showed himself to be a master at organization: if he did not himself want to cover the spectrum of evolutionary topics, then he was ready and very willing to bring others into the arena, urging them to work alongside him, filling out the picture of evolutionary change—a picture that went back ultimately to Wright's adaptive landscape metaphor. First there was Ernst Mayr. An immigrant like Dobzhansky, trained as an ornithologist and systematist (classifier), Mayr left his native Germany, traveling west to the American Museum of Natural History (still in the early 1930s under the directorship of Osborn), where he became curator of birds. Drawing on a vast knowledge of nature's denizens and combining this with a sensitivity to geographical conditions and variation, in 1942 Mayr produced his masterwork, *Systematics and the Origin of Species.* Most dramatic of all of the instances on which Mayr drew were the so-called rings of races where interbreeding subpopulations of organisms circle the globe, finally touching but unable to interbreed at such meeting points. Here was natural, gradual variation before one's very eyes—not just evolution (evolution as fact, that is) in the making, but the process of evolution (evolution as cause) showing itself. It simply was not possible to think that evolution proceeds by jumps, saltations: the touching sub-populations blended one into another without a break or a step, even though the end populations were genetically isolated from each other. More causal speculation than this was beyond Mayr's scope of inquiry, but it was the landscape metaphor that was assumed and that was in turn confirmed.

More theoretical was the paleontologist of the group, George Gaylord Simpson. He was unique among the "synthetic theorists" (as their theory came to be known) in having a facility with figures. He could go behind Wright's pictures, understanding the mathematics, and thus modifying what was a theory for populations in action over a few short generations into a theory that dealt with populations (at the physical or phenotypic level, for nothing was known of genes) over long periods of time—millions of years in fact. Particularly noteworthy in Simpson's *Tempo and Mode in Evolution* (published in 1944, having been delayed somewhat because of the war) was his treatment of horse evolution. Simpson showed how the mammals had been bush and tree browsers, then they started to move toward a grazing lifestyle (somewhat incidentally because of other changes, particularly toward a larger overall body size), then at some crucial point the horses had split, with some going right back to browsing and others moving across the valley and right up the path of Mount Grazing. As it happens, at some

The evolution of the horse as envisioned by George Gaylord Simpson

point after this the browsers went extinct, but this was an event only after all the exciting action had occurred.

Finally there was botany. Dobzhansky had deliberately canvassed the field, looking for someone to write on plants from the perspective that he and his friends were exploiting. His first choice let him down, and when he found a substitute, G. Ledyard Stebbins, Dobzhansky had Stebbins stay in his own home. The geneticist fed the botanist pertinent information until Dobzhansky was convinced that Stebbins would complete the task, and complete it properly. And so, in 1950, *Variation and Evolution in Plants* made its appearance. There were of course major differences in Stebbins's work from that of the animal evolutionists. For a start, in botany you do get jumps, as new species are formed by chromosomal events virtually unknown in the animal world. For a second, hybridization (where members of different species breed and produce fertile offspring) is a well-known and common and important method of change. But, for all of these differences and more, the underlying story is the same. Peaks and valleys, and organisms struggling up the sides and sitting triumphantly on the top, or being displaced and having to start the evolutionary process all over again. The landscape model triumphed.

Forging a Discipline

These were some of the major intellectual moves that were made by evolutionists after Darwinian selection (together with other bits and pieces such as Spencerian progress) had been fused with Mendelian genetics. But there was more to be done than this. Thomas Henry Huxley had done it for physiology and morphology. He had failed to do it for evolution. Make a professional discipline of it, that is. Now finally, there was a group who felt (with some good reason) that they had an adequate theory—one that was ready for development through experiment, natural observation, the-

Fore-foot. Hind-foot.

RECENT.

Equus.

PLIOCENE.

Pliohippus.

Protohippus (*Hipparion*).

MIOCENE.

Miohippus (*Anchitherium*).

Mesohippus.

EOCENE.

Orohippus.

T. H. Huxley's reconstruction of the evolution of the horse foot (based on specimens of O. Marsh) oretical amplification, and more—and who wanted to spend their time as professionals working on it. They did not want to spend their time as museum-based priests of a secular religion. So Dobzhansky and his coworkers deliberately set about making a professional science of evolutionary biology—not just a second-rate enterprise tracing hypothetical phylogenies with too little information and too much imagination, but a real discipline that was causally based and that took seriously experiment and theory.

The theologian and mathematician Blaise Pascal once asked about belief in the existence of God. He concluded that one has an asymmetrical situation: if God exists then you had better believe in Him, and if God does not exist then not believing in Him does not really matter. Hence, the sensible thing is to believe in Him. And if you complain that you cannot believe in Him, then go through the motions and you will be surprised how faith will come. This is known as "Pascal's Wager"—Pascal's branch of expertise was probability theory and he is offering you a bet or a wager you really ought not refuse. Founding a scientific discipline is a bit like Pascal's Wager. If you go through the motions, then you and others will start believing in it. And to this end, you need good university jobs, you need students, you need journals (preferably with lots of esoteric language), you need associations (that you and your pals are in and others are not), you need grants and other monies, you need supporters, and you need to shove it to your enemies and detractors.

The synthetic theorists achieved all of these ends and more. For a start, they moved into plum university posts and once there brought their friends in too. Dobzhansky got a job at Columbia. Stebbins got a job at Berkeley. Mayr got a job at Harvard, and before long he was campaigning (successfully) for Simpson. Dobzhansky had masses and masses of students and postdoctoral fellows that he treated like his children, supporting, guiding, encouraging, scolding. They worshipped him—there was no talk here of societies for the preservation of native Americans from foreign professors—and fanned out to carry the word. A journal, *Evolution,* was started, with Mayr as the first editor. Firm guidelines were put in place. The obvious esoteric language was mathematics, and even though Dobzhansky and Mayr would not have known a symbol if their sisters had married one, care was taken to see that their students were properly trained, and associates with mathematical skills were dragooned into coauthoring papers. Dobzhansky wrote a whole series of Drosophila articles with Sewall Wright: articles of which he understood the first lines and the last lines and absolutely nothing in between.

As importantly, causal, experimental work was encouraged. Path tracing was shown the door, or rather the pink rejection slip: "Your manuscripts have been scrutinized by two readers and both of them report that they consider them unsuitable for publication in *Evolution*. I have tried to get some detailed criticism for you (as you asked) but there seems to be nobody in this country now who is interested in phylogenetic speculations" (letter from Mayr to F. Raw, 2 August 1949; *Evolution* Papers, American Philosophical Society). I should add on a personal note that I dug this letter out of the meticulous files that Mayr, as editor, kept for those early years of the journal. Very atypically, I could find no record of the negative referees' reports. I challenged him on this, suggesting that the two referees might have been Ernst Mayr in the morning and Ernst Mayr in the evening, at which he just smiled.

An association was started: The Society for the Study of Evolution. President: G. G. Simpson; Secretary: E. Mayr; Council members: S. Wright and Th. Dobzhansky. And then there was the question of grants. Fortunately by the time things really got going, World War II was over and the U.S. government was realizing that it needed to subsidize basic science. The National Science Foundation was begun, and the evolutionists were right up at the head of the line with their begging bowls outstretched. They were not always successful, but they got some nice juicy grants, and at the same time had the satisfaction of seeing that others were denied. Simpson (like Mayr) squirreled away every piece of paper, including all of the referee's reports he wrote on grant applications. Of one man, at that time the editor of *Evolution*, for all that he and Simpson were at personal loggerheads (Simpson had troubles with personal relationships, to put matters euphemistically), the report reads: "This application is first-class in every respect." Of another, a brilliant systematist who had been rather rude about the synthetic theorists: "His approach is narrow-minded and shows consistent lack of thought into biological, as distinct from strictly mathematical, aspects of the problems considered." Merit rating: "Questionable"!

What about the Cuvier tactic, to which we have seen Sewall Wright sensitive? What about the need to come across as serious, objective scientists? The Scopes trial was not that long before and even if the evangelicals were now taking a low profile, there were still many happy to make of evolution a lot more than mere science—the spirit of Thomas Henry Huxley still roamed the land. Indeed, rather more than this—the genes of Thomas Henry Huxley were still active. His grandson Julian Huxley (brother of the author Aldous Huxley) was a prominent British evolutionist, and although he wrote more sober works—his *Evolution: The Modern Synthesis* appeared in 1942—his forte was inspirational material, blending science and value in an unabashed, neo-Spencerian fashion. He was an ardent humanist; you do not need to turn many pages of his 1927 *Religion without Revelation* to guess his intent and lifelong convictions: evolution as progress, evolution as popular science, evolution as religion. Although an ardent atheist, in the 1950s Julian Huxley became enthused with the speculations of the Jesuit priest-cum-paleontologist Pierre Teilhard de Chardin and, sensing a kindred spirit, wrote the foreword to the English translation of Teilhard's major opus, *The Phenomenon of Man* (1959).

As you can imagine, this kind of activity did not sit well with sober scientists—the English Nobel Laureate Peter Medawar wrote a scathing review of Teilhard and of Huxley's involvement (Medawar 1961). As you can imagine also, the synthetic theorists were tense about all of this and determined not to be tarred by the same brush. This does not mean that they did not have values—even that they did not have yearnings to treat evolution as a secular religion. There was not one of them who had not turned to evolution in the first place to find the meaning of life, hoping especially to discover the implications of an evolutionary approach for our own species. There was not one of them who was not a gung-ho progressionist, and Dobzhansky—a deeply committed Christian—was no less supportive of Teilhard than was Huxley. (Simpson, the paleontologist, knew and liked Teilhard, although as a sometime Presbyterian disapproved of Teilhard's womanizing—or rather of the hypocrisy of someone who took a vow of chastity and then broke it, flagrantly.)

But the synthetic theorists knew that, if they were to upgrade their science, they had to keep their evolution-as-professional-science separate from their evolution-as-secular-religion (not so very secular in Dobzhansky's case). So what they did was to write two series of books! The first series was the professional series: lots of talk about models and causes and quantification and so forth. Not a whiff of culture or social values. Then there was the second series: openly written for the general reader (if it did not say this on the title page, then it said it in the preface), with the same science as before supported with lots of nice illustrations and the mathematics removed (not much work here!), and with a couple of final chapters on life and its total meaning.

Simpson was the paradigm. His *Tempo and Mode in Evolution* (1944) is sufficiently pious and straight-faced in its serious intent that it could pass for a church sermon. Then in 1949 comes *The Meaning of Evolution: A Study of the History of Life and of Its Significance for Man,* a book that Simpson admitted meant much to him, with a subtitle that tells all. Four years later (1953), we are back with a revision of *Tempo,* retitled *The Major Features of Evolution,* as strict and straight as you could ever wish. But, oh, how those evolutionists did let it all hang out when they were allowed to! For the paleontologist Simpson when writing in popular mode there were two major social directives. First, there was the need to improve and promote knowledge— knowledge in itself, as a good.

> The most essential material factor in the new evolution seems to be just this: knowledge, together, necessarily, with its spread and inheritance. As a first proposition of evolutionary ethics derived from specifically human evolution, it is submitted that promotion of knowledge is essentially . . . both the acquisition of new truths or of closer approximations to truth (metaphorically the mutations of the new evolution) and also its spread by communication to others and by their acceptance and learning of it (metaphorically its heredity). (Simpson 1949, 311)

Next we have personal responsibility, leading to individualization and thus to integrity and dignity.

> The responsibility is basically personal and becomes social only as it is extended in society among the individuals composing the social unit. It is correlated with another human evolutionary characteristic, that of high individualization. From this relationship arises the ethical judgment that it is good, right, and moral to recognize the integrity and dignity of the individual and to promote the realization or fulfillment of individual capacities. It is bad, wrong, and immoral to fail in such recognition or to impede such fulfillment. This ethic applies first of all to the individual himself and to the integration and development of his own personality. It extends farther to his social group and to all mankind. (p. 315)

Fully to understand Simpson's thinking, especially his high valuing of responsibility and dignity, we must take note of the context and time within which he was writing. The Cold War was frozen in a seemingly endless winter. Worse, science in the Soviet Union was subject to dreadful pressures. I have mentioned that, even in the 1930s, biology was firmly under the thumb of charlatans led by the agriculturalist Trofim Lysenko, promoting neo-Lamarckian pseudo-techniques intended to raise wheat harvest yields to dramatic new levels. Using his friendship and connections with Stalin, Lysenko continued right through the 1950s to harass and persecute those who dared raise a voice against him. For all of his difficulties with personal relationships, Simpson was a man of the highest moral ideals: he had volunteered for dangerous action in World War II, despite the fact that he was over the enlistment age. It was natural, therefore, for him to take the crusade for democracy and freedom as a personal issue, and you will not be surprised to learn that, having established (as he thought) a biological basis for dignity and responsibility, he moved straight to a vehement condemnation of the totalitarian systems that spread from Eastern Europe right across Asia. At the same time, he made it clear that there was an alternative: the society within which he and his fellows were able to work so freely. "Democracy is wrong in many of its current aspects and under some current definitions, but democracy is the only political ideology which can be made to embrace an ethically good society by the standards of ethics here maintained" (1949, 321). And then added to this was a repeat affirmation of the significance of evolutionary biology for providing foundations and standards: "It bears repeating that the evolutionary functioning of ethics depends on man's capacity, unique at least in degree, of predicting the results of his actions. A system of naturalistic ethics then demands acceptance of individual responsibility for those results, and this in fact is the basis for the origin and function of the moral sense" (1969, 145–146).

Material similar to that of Simpson can be found in the popular writings of the other synthetic theorists. Stebbins (1969), during the late 1960s, while a faculty member at UC Davis, even went so far as to argue that it is good biologically to have a small group of radicals upsetting an otherwise complacent population!

The Classical-Balance Dispute

The mark of really good science, productive fertile science, is that it gives you lots to do. With every problem you solve you get a couple more questions thrown up. The last thing you need is just to sit polishing a theory like a prized antique. Fortunately, the synthetic theory showed its worth and more. All sorts of exciting problems and questions and anomalies kept coming to the surface. This really was dynamic stuff. For Dobzhansky (and hence for the others) the main move was, in the 1940s, to a much more selective stance than hitherto. This change came about because, for all that it might have pretended otherwise, Wright's shifting balancing theory was never very Darwinian. The key causal mechanism is genetic drift, supposedly occurring when a small population is fragmented from its fellows and subject to the vagaries of mutation and breeding. It is claimed that this is enough to fuel a slide down the side of an adaptive peak until, reaching the bottom, the group can then move up the side of a neighboring mountain. Although he was always much keener on selection than was Wright, in the first edition of *Genetics and the Origin of Species* (1937) Dobzhansky had accepted (without critical comment or question) much that Wright argued. But then, just a few years after the first edition was published, Dobzhansky became increasingly convinced that drift is far less significant than he had thought previously. Correspondingly, selection must be yet more important (Lewontin et al. 1981). Drift leads one to expect that there will be differences between populations, both at the genetic level and at the level of the chromosomes. In these respects, isolated populations will move all over the place. What drift does not lead one to expect, and in fact what drift leads one positively not to expect, is any kind of systematic variation in genes or chromosomes. One does not, for instance, expect to find that there are cyclical differences in chromosomal variation, from one season of the year to another and then repeating itself in subsequent years. Systematic change of this kind is a sign that natural selection is at work rather than purely random factors.

Studies of natural populations of fruit flies in the American West showed Dobzhansky that there are indeed such cyclical variations tied to the seasons, and he concluded that natural selection must be at work. Then his initial observations were backed by experimental studies that showed that variation in temperature and humidity and the like are just what is needed to bring about systematic variation in populations. Some chromosomal variations are clearly better adapted to some conditions than are others: there is no absolute perfect form, suitable for all climates. It is all a question of the relative conditions. The question for Dobzhansky now was how one is to explain this variation in populations, and how in particular such variation is held from one generation to another. Why does selection not simply wipe out all variations so a population is relatively uniform throughout? But then, how could selection operate to raise the levels of first one form and then the levels of another form? Mutations provide the raw material necessary for variation, but simply waiting on required variations when selective needs arise is hardly adequate. One needs some theory or mechanism showing how variation can be held within populations, so that there is always material to work on when new selective needs arise.

After considerable thought, Dobzhansky finally decided that selection itself keeps up the required variation. The key mediating mechanism is "balanced superior heterozygote fitness" (Dobzhansky and Wallace 1953; Dobzhansky and Levene 1955). This supposes that the heterozygote is different from both homozygotes and that, for various reasons, the heterozygote is favored by selection over the homozygotes. (More technically: that the heterozygote is fitter than either homozygote.) When one has a situation like this, then it can be shown easily that the heterozygote keeps reappearing in populations. Indeed, under favorable conditions, there will be a balance between the ratios of heterozygotes and of homozygotes. And what this means is that the different genes or alleles (or chromosome variations) persist in the populations and are therefore always ready to be used when selective needs arise. But it is selection itself that is keeping the variation present in the first place.

To use a metaphor, the conventional view would have one waiting on the occasional mutation, which will probably be no use in any case. This is akin to having to write an essay for an instructor where the only source of material you have is monthly offerings of the Book of the Month Club. You can be fairly certain that whatever comes your way through the mail will rarely, if ever, be the material that you need for your essay. Suppose, however, that you have the balanced heterozygote fitness situation. This is akin to having a whole library at your disposal. Pretend that your instructor demands that you write an essay on dictators. If, when you go to the library, you find there is nothing on Hitler, there may well be something on Napoleon. And if there is nothing on Napoleon, then one can look further down the alphabet for something on Stalin. You can be certain that there will be something on some dictator that you can use for your essay. Reverting back to real life, suppose a new predator appears, threatening a population of organisms. There is a whole new set of selective pressures. The balanced heterozygote fitness position suggests that if there is no variation within the population allowing some members to escape through camouflage, then perhaps there will be something allowing some to escape through a more effective defense like a thick shell or skin. Or, if neither of these defenses exists, then perhaps there will be the capacity for a different kind of behavior, perhaps behavior inclining one to get out of the area and to move somewhere else that is predator free (Ruse 1982).

Balanced superior heterozygote fitness is a neat way of solving the variation problem, but is it true? Well, there are certainly some cases where it seems to be in action. The most famous case occurs in humans, particularly among inhabitants of certain parts of West Africa where malaria is a major health threat. There is a certain allele, the sickle cell allele, that confers a natural immunity on its possessors when it is in the heterozygote state (in other words, the bearer is fitter than the homozygote without the gene) but is lethal when it is in the homozygote state. There is a balance between the heterozygote's good qualities and the inferior qualities of the homozygotes. But as Aristotle was wont to say, one swallow does not make a summer. Because a balanced heterozygote happens occasionally, it does not follow that it is a near universal phenomenon as Dobzhansky and fellow supporters of the balance hypothesis supposed. It was at this point that controversy started to grow. To prove his case, Dobzhansky and his students (particularly Bruce Wallace) ran experiments subjecting fruit flies to radiation. This would cause mutation, and because the mutations would generally occur only in a few alleles, the overall effect would be to increase the heterozygosity in populations—by and large, one would not expect members of new generations to be at once homozygotes for the newly mutated genes. And Dobzhansky and associates claimed that, overall, the radiated populations were indeed fitter than the nonradiated populations, supposedly showing that the increased number of heterozygotes pushed the populations farther up their adaptive peaks. In other words, the balance hypothesis was vindicated.

I hardly need say that all of this is a bit inferential, to put it mildly. One is not exactly measuring heterozygosity directly, nor is one doing anything very much about individual organisms and their fitnesses. Muller disliked the balance hypothesis intensely—more on the reasons in a moment—and found little reason to accept Dobzhansky's findings. Muller (1949) wanted to argue (the classical hypothesis) that organisms in a population are more or less uniform and at their adaptive peak. Mutations generally are deleterious, heterozygosity is therefore discouraged (certainly heterozygositiy that has any phenotypic effect), and the only variations one finds generally are those very few new mutant genes that prove fitter than the old types and that are therefore moving through the population to become the norm. Muller therefore argued that the Dobzhansky experiments were flawed, and when he and his students (particularly the Israeli Raphael Falk) ran their experiments, they failed to replicate the Dobzhansky results. As seems to be almost the norm in these cases, the experiments that they ran were slightly different—they used stronger radiation rates, for instance—and as is certainly the norm in these cases, Dobzhansky and company complained that the Muller results were irrelevant and meant nothing!

Now, why was everybody so tense about all of this? Why not just sit back and wait for some good results to come in? Well, one reason is that good results do not come from sitting back and waiting. The way to get things done is by pushing, and this is what everyone was doing. More seriously, there were serious scientific issues—intuitions if you will—at stake here. As we have seen, Dobzhansky desperately needed a supply of variation above single mutations, or else selection as he saw it could not function properly. Unlike Muller moreover, he was a field naturalist, and so he really knew that whatever the theory might say, variation (however caused) does truly exist in natural populations. And obviously, he was properly sensitive to the workings of external causes, including selection. Muller, on the other hand, from a lifetime's experience of experimentally inducing mutations through radiation, knew how horrendous were the effects of most mutations. He just could not see how mutations could remain positively in populations.

> I look upon mutation as a random process, so far as the nature of its effects are concerned, much as Brownian movement is random, and I therefore find it a necessity to conclude that the effects are on the whole disintegrating rather than integrating, just as the effects of Brownian movement are to decrease rather than increase the free energy of a system. In other words, I regard the principle as being merely a logical extension of the second law of thermodynamics and I would as soon expect to see the average or usual effect of irradiation on later generations to be an improvement as to believe a man who said that he had perfected a perpetual motion machine. (Beatty 1987, 299)

These factors were important in driving Muller and Dobzhansky apart, but one suspects that there was more than this, and there was. We have seen already how the evolutionists were affected by the Cold

War, and here was another point of impact—a point entwined with another factor already encountered—the need to find funds to support research. Both Dobzhansky and Muller had reason to feel tense about the Cold War, and they were indeed. The reason for Dobzhansky's state of mind is obvious: as a native-born Russian with a deep attachment to his homeland and an emotional identification with the United States and a hatred of the Soviet system, he was very torn about what was then happening. The reason for Muller's tenseness is hardly less obvious: as a former communist who in the 1930s had gone to live in Russia and barely escaped with his life, he felt very ambivalent about a military arms race that seemed bent on destruction of the whole world.

One of the main features of the Cold War was the buildup and stockpiling of nuclear weapons: weapons that required testing, in those days massive amounts of atmospheric testing with the consequent radiation fallout. People were very worried about the effects, short term and long term, on the human population, and here Dobzhansky and his school stepped in, prepared to use fruit flies as models for humans. The Atomic Energy Commission accepted the offer, and so through the 1950s we find that Dobzhansky and his students were getting monies for their work from this organization. Although, in principle, Dobzhansky declared himself against testing—or at least worried by its effects—one need hardly say that his results, showing that radiation far from being deleterious can be beneficial, went well with his paymasters. They were happy to support research like that! Conversely, Muller, who was regarded with suspicion by military and civil authorities, precisely because of his communist background, tended in turn to be wary of anything that spoke well for these powers and was certainly very uncomfortable with results that supposedly negated the bad effects of a military arms race. Interestingly, he was not against nuclear testing as such: he had no love of the Soviet system, nor did he have any illusions about it. But he did worry about bad science's glossing over the ill effects of such testing.

There was yet another factor dividing Dobzhansky and Muller, perhaps the most important of all. *Eugenics* refers to the belief and practice of altering the genetic composition of the human species for supposedly good ends. It is a topic fraught with tension, if only because of the Nazis' efforts in that direction: efforts based on their pseudoscientific understanding of human biology and motivated by their vile racist ends. For good and ill, eugenics is a topic that has fascinated geneticists since the birth of their subject—indeed, motivation and finance for early genetics often owed almost as much to eugenics as it did to agriculture.

Both Dobzhansky and Muller had strong opinions on the subject, and it is clear that these opinions influenced thinking on the balance/classical dispute. For Muller, there was an ideal, and it is our obligation to see that we do not fall beneath this. Bad mutations are bad mutations and we should get rid of them. Most mutations, especially most new mutations, are bad mutations, and so eugenical efforts should be directed toward their elimination. For Dobzhansky, there is no ideal. Populations are varied, and this is the way nature intended them. Just as we have fruitfly geneticists, so also we have hewers of wood and drawers of water—we are better off this way and should not tamper with things. There is no such thing as a bad mutation in itself. It is all a question of context.

Parenthetically, I need hardly say that Dobzhansky's Christian commitments were working flat out here—we may well be genetically nonidentical, but in the eyes of God and of Theodosius Dobzhansky we are of equal worth. No wonder he felt so strongly about the balance hypothesis. Even though it was deep in the heart of his professional science, for Dobzhansky it was an absolutely key element in his religiously infused vision of humankind.

Molecular Biology

The most important single event in the history of twentieth-century biology was the discovery in 1953 by James Watson and Francis Crick of the double helical structure of the deoxyribonucleic acid (DNA) molecule. At once it was seen that this molecule was the underlying foundation of the Mendelian gene and that the information of heredity was carried along its back, coded by the order of its many sequential parts. At first, the synthetic theorists tended to be suspicious of, if not outrightly hostile to, the work and results of the molecular biologists. As individuals, these scientists were not much loved either. The new science and its practitioners were bright, pushy, and contemptuous of old-fashioned biology, and willing and able to say just this on every available occasion. Striking back, the synthetic theorists—Ernst Mayr particularly—claimed that molecular biology is narrow and limited, that its technical achievements conceal its spiritual aridity. They claimed that whole organism biology—evolutionary biology centrally—necessarily looks at questions and issues that a molecular approach misses. Extreme "reductionism," trying to explain every-

James Watson (left) and Francis Crick (right) demonstrate their model of the DNA molecule

thing in terms of the very small, is a waste of time. Distracting from the real problems, in fact.

In a way, looking back over a half century, this dispute seems about on a par with the dispute half a century earlier, between the biometricians and the Mendelians. And about as worthless. We see now that molecular biology was no enemy of evolutionary studies but a really good friend. Whether it has made any difference to the actual theory of evolutionism is a matter we shall discuss in a later chapter. We shall see that it has given systematists powerful new tools in their attempts to discern phylogenies. And, what cannot be denied and is pertinent here is that it has certainly given conventional evolutionists wonderful new techniques to cut through problems that were hitherto intractable. Nowhere more so than over the balance/classical dispute. In the 1960s, Dobzhansky's prize student, Richard C. Lewontin, was one of a number who developed the method of "gel electrophoresis." Essentially this is a technique that uses the different electrostatic charges on molecules to discern differences in the molecules themselves, and through this it was possible at last to measure directly the genetic differences between organisms. And through this one can shortcut all of the laborious methods used by Dobzhansky and Muller in their efforts to find genetic differences and similarities, going straight to the genes themselves for information about their nature.

Richard Lewontin

The results were really quite incredible. The balance hypothesis was vindicated to a degree that not even Dobzhansky could have envisioned. Absolutely massive amounts of variation were being held in populations. Take the organism so favored by population geneticists: the fruit fly *Drosophila pseudoobscura*. Lewontin discovered that from one organism to the next, there were simply huge genetic differences. As many as one-third of the genes that were investigated using gel electrophoretic techniques proved to have different forms. (They were "polymorphic" for different "alleles.") Taking any locus on a chromosome, on average 12 percent of the allele pairs were heterozygotes (Lewontin and Hubby 1966, 608). All in all, therefore, Dobzhansky apparently was as right as one could possibly be and Muller apparently was as mistaken as one could possibly be.

But, as I am sure you are now starting to realize, as in love nothing in science is quite that simple. The variation could not be denied. But the balance hypothesis also gives us a cause of this variation, namely that it is held in populations because of heterozygote fitness. This was still denied—no longer by Muller, for he died in 1967, but by his students and associates. Perhaps, they argued, there are other causes, or perhaps no cause at all! Perhaps, because it has no effect on the phenotype, much of the variation at the genetic level quite escapes the effects of selection. Gel electrophoretic techniques pick out molecular differences in genes, or more precisely they pick out differences in the substances the genes code for immediately. But perhaps these differences make no difference to the finished organism, in which case they would be quite neutral with respect to natural selection. Molecular variation might therefore just "drift" up and down in populations, giving the results obtained through gel electrophoresis but without the immediate supportive consequences for the balance hypothesis. The classical hypothesis rides again!

This issue is still not resolved completely, and I shall return to it. For now we can pull away. We have seen how Darwinian selection theory was integrated with genetics to make a full and satisfying evolutionary theory and how in addition this was then used as a base to build a professional scientific discipline. The cost was that all of those secular religion features that so many found so attractive had to be dropped—or at least, corralled off to one side. But note how this was done, not so much from conviction that evolution as secular religion was a bad thing, but because unless it was done the desired status of professional science would not be achieved. Whether this is a strategy that will come back to haunt evolutionists, as we come toward the present, will be the subject of later chapters.

Further Reading

William Provine is *the* historian of population genetics. He is not always easy to read but well repays the effort. His first book, *The Origins of Theoretical Population Genetics* (Chicago: University of Chicago Press, 1971), is an excellent account of the coming of Mendelian genetics and its integration into evolutionary studies. Then his second book, *Sewall Wright and Evolutionary Biology* (Chicago: University of Chicago Press, 1986), is an absolutely massive intellectual biography of one of this century's giants in evolutionary studies. What is really valuable about this work is that it treats not only of its central subject but also of others who interacted with Wright, most especially Theodosius Dobzhansky. The story of the always-fruitful, sometimes-uncomfortable relationship between Wright and Dobzhansky is itself worth the price of the volume. Then third there is a book Provine coedited with the senior evolutionist Ernst Mayr: *The Evolutionary Synthesis: Perspectives on the Unification of Biology* (Cambridge: Harvard University Press, 1980). The book is based on two conferences nearly thirty years ago, when the leading figures in the making of neo-Darwinism were still alive and willing to give their recollections of the main events and moves. Much more work is still needed on the development of twentieth-century evolutionism—some can be found in the pertinent chapters of *Monad to Man*—but thanks to Provine we now have a good grasp of the really important ideas and their fates.

The Lysenko affair will no doubt be an ongoing focus of scholarly interest. With the opening of archives in Russia, we will surely be learning much more about what happened and why. But already we have several accounts about what went on in Russian genetics and biology generally under the Stalin and Khrushchev dictatorships. David Joravsky's *The Lysenko Affair* (Cambridge: Harvard University Press, 1970) is a very solid discussion based on a great deal of evidence. You should supplement it with the work of Loren Graham, who is the authority on science in Russia under the Soviets. His *Science, Philosophy, and Human Behavior in the Soviet Union*, 2d ed.(New York: Columbia University Press, 1987) is excellent for its insights into the trials and tribulations of Russian biology as well as its triumphs (such as Oparin on the history of life, a topic to be discussed shortly in chapter 6).

In the Beginning: The Origin of Life

B ack across the English Channel and back in time to 1859. This
was the year of the *Origin,* but in France it made no waves. At
least, this was the case for a few years until a translation was published, a translation with an inflammatory preface by radical thinker
Clémence Royer—positivist, materialist, atheist, and female to
boot—that may or may not have stirred her countrymen but as sure
as anything sent shudders of discomfort and disapproval through a
reclusive naturalist across the Channel in the little Kent village of
Downe. But that year did see a major controversy in France. This was
between Felix Pouchet, director of the natural history museum at
Rouen, and the great Louis Pasteur, discoverer of the means of the
prevention of rabies and many other diseases as well as seminal inves-
tigator into the preservation of foodstuffs (pasteurization) and into
much, much else (including important studies on viticulture, especially on the properties of yeast).

The topic of the controversy was "spontaneous generation," the
one-step appearance of living organisms from nonliving (inorganic or
organic) material: flies, worms, grubs, or smaller beings coming (by
natural causes) from mud or slime or whatever (so long as it was itself
inert). Pouchet affirmed his belief in this venerable doctrine. In a
major work, *Heterogenie ou traite de la generation spontanée,* he argued
that the eggs of microscopic-sized organisms are produced in single
steps from other organic materials. And so effective was he in his case
that the French Academie des Sciences put up prize money for further studies on the subject. Which brought in Pasteur, who was determined to refute what he saw as a thoroughly false doctrine. To this
end, Pasteur was led to perform a dazzling series of experiments—
experiments of such a caliber that today they are often highlighted as
the paradigm of scientific excellence—supposedly showing that spon-

Louis Pasteur

taneous generation of a kind promoted by Pouchet is simply impossible. There is just no way in which life can be derived manually from nonlife.

And this would seem to be the end of the matter. Although then you might well ask how this can possibly be so. If indeed evolution be true, that is, if the fact of evolution—the rise of today's organisms from primitive forms—be true, then surely one is forced back to ask about the origin of those first primitive forms. From where did they come? Was the earth seeded from outer space, or is the living part of the basic fabric of the universe and as old as time itself, or (most obviously) did the living come naturally from the nonliving? It is to these and other questions—about the beginnings of life and about the history of its early forms—that we turn in this chapter. As always, we shall find that the story is more complex and interesting than it appears on the surface. For a start, you might think that Pouchet was in favor of spontaneous generation precisely because he was an evolutionist. In fact, he was nothing of the sort. And, until the end of his life, Pasteur—he who goes into every elementary textbook as the final assassin of the spontaneous generation position—thought it was an open question! But we get ahead of ourselves, so let us start back at the beginning.

Spontaneous Generation

The belief that life can come naturally and spontaneously from nonlife is indeed a venerable doctrine. It goes back at least to the ancient Greeks and to Aristotle. Down to the time of the Scientific Revolution (the sixteenth and seventeenth centuries), people were torn on the subject. On the one hand, it seemed so clearly true and backed by the evidence of the senses. Meat left unattended rots and in a very short time is swarming with maggots—remember, these were the days before refrigeration! Where could these living beings have come from, except from the nonliving? On the other hand, what was the biblical evidence for such ongoing creation of life? Surely, God had finished His work on the sixth day, and that was that? Was not man himself the culmination of His creative outpouring? Yet, if man was the final act of creation, did this mean that there were maggots in Eden? And parasites and other ugly and unwanted organisms? Surely not!

It was all a puzzle, but with the coming of modern science—particularly with the coming of instruments of magnification—at least some of the issues were resolved. Or rather some of the boundaries were drawn in and made clearer. The Jesuit-trained physician Francesco Redi, for instance, was able to show how many putative instances of spontaneous generation were the results of insects laying eggs, which later hatch. If meat is covered with muslin or some such cloth, so that flies cannot get at it, then the maggots simply do not appear. "Although it be a matter of daily observation that infinite numbers of worms are produced in dead bodies and decayed plants, I feel, I say, inclined to believe that these worms are all generated by insemination" (Redi [1688] 1909, 27).

However, showing that for every step forward there seem to have been two steps back, Redi went on to say that he did not thereby intend to deny the possibility of spontaneous generation. Indeed, he was quite prepared to accept that as a caterpillar can be transformed into a butterfly, so also a piece of the flesh interior to an organism can be transformed into a parasite. He denied that the inorganic can be turned into the living (this is known as "abiogenesis"), but he asserted that the organic can be turned into the living (this is known as "heterogenesis"). More particularly, he asserted that something that is living can be turned into something new and quite different. (This was not evolution, in the sense of transformation from one form to another, but something stronger, where a piece of a living organism is turned into a new animal.) "If the thing is alive, it may produce a worm or so, as in the case of cherries, pears, and plums; in oak glands, in galls and welts of osiers and ilexes worms arise, which are transformed into butterflies, flies, and similar winged animals. In this manner, I am inclined to believe, tapeworms and other worms arise, which are found in the intestines and other parts of the human body" (p. 116).

Theological arguments, however, continued strong against spontaneous generation. To revealed theological objections (that is, to objections based on the Bible) were added natural theological objections (that is, objections based on reason). John Ray, the English clergyman-naturalist, was adamant that the designlike nature of the living world altogether precludes the appearance in one step by natural causes of living organisms. Laws lead to randomness and to things not functioning. (Murphy's Law, that if something can go wrong it will, is a metaphysical fact of nature and not just an engineer's joke.) God designed the world to work, and there is simply no way that things could spring into immediate being and work as well as anything else.

But now came a powerful counter, giving spontaneous generation a whole new lease on life as the eighteenth century moved toward its climactical events, first in North America and then in the final years in France. This was the Age of the Enlightenment, when philosophers and others challenged the old ways of thinking, going counter to the repressive religion that had dictated so much to so many for so long. New confidence was found in the power of unaided reason—philosophical, political, social, literary, and scientific—the latter most particularly Newtonian scientific, as the great English thinker's ideas spread and were developed and applied. With the new strains of thought, theological worries (particularly revealed theological worries) about spontaneous generation started to abate, and at the same time Newtonian undercurrents gave the doctrine a whole new lease on life.

Newton's theory of gravitation (which, incidentally, scholars now trace to some very strange "alchemic" views about subtle powers pervading all of creation and being the key to the transformation of base metals into gold) suggests that the universe is not what it seems on the surface (Westfall 1980). Bodies can affect each other at a distance, even though nothing is touching, and other odd effects likewise are to be considered part of everyday science. Could it not be therefore that there are special powers akin to gravity, affecting and animating living creatures? Even though it may be that nothing living can come from the inorganic (abiogenesis), it may well be that the living can come from the organic (heterogenesis). Such at least was the opinion of the most powerful and influential naturalist of the age, Georges Leclerc, le Comte de Buffon, in France. He believed in a kind of internal living force, the *moule interieur,* which animates and forms the shape and function of organisms, and for him it was quite plausible to think that a primitive life-form could thus be formed from the matter of other such forms. Nor were he and his followers convinced otherwise by experimentation, for instance, by the celebrated work of the Italian Lazzaro Spallanzani, who boiled various broths and the like in sealed flasks, showing that once life had been destroyed it never reappears. The Buffonians' counterargument was that these experiments proved precisely their case, because Spallanzani's boiling had destroyed not just life-forms but the very potential for life that existed hitherto! The *moule interieur* is itself something that must be protected, or else it cannot perform its intended functions.

This all leads us into the nineteenth century, where we start to see a polarizing of positions that lasted to the Pouchet-

Georges-Louis Leclerc, le Comte de Buffon

Pasteur debate, and indeed beyond right down to the present day in respects. On the one hand, we have the successors of Buffon, who thought that spontaneous generation is a constant and common fact of life. In fact, there were two groups of importance. First, there were the evolutionists, notably Erasmus Darwin and Lamarck (who had, incidentally, been a protégé of Buffon). They saw the natural origins of life as part and parcel of their overall developmental position. And for them, natural origins meant spontaneous generation. Thus Darwin, in verse:

> Then, whilst the sea at their coeval birth,
> Surge over surge, involv'd the shoreless earth;
> Nurs'd by warm sun-beams in primeval caves
> Organic life began beneath the waves . . . .
> Hence without parent by spontaneous birth
> Rise the first specks of animated earth
> (Darwin 1803, 1, 231–248)

Lamarck had similar views. He thought that worms and the like are produced from warm mud by the action of electricity and heat and so forth. (Remember, this was just after the time when Franklin had performed his spectacular experiments with electricity, so it was in itself a rather fashionable possible force of natural change.)

Second there were the *Naturphilosophen,* those German thinkers who saw and stressed patterns or isomorphisms throughout the living world—things that we today interpret in an evolutionary fashion and call "homologies." These people were (like the French and English evolutionists) also developmentalists, seeing life going from blobs to complex forms, although in the early part of the century they tended to understand things in an idealistic fashion rather than in terms of actual organic change. They were primed to see patterns not only in the organic world but also in the inorganic world—crystals were a particular focus of attention—and it was but a simple move to link patterns in the inorganic with the organic. Combined with the otherwise apparently inexplicable appearance of living forms where none had existed before—parasites were the big favorite in this respect—it was an easy and ready move to argue that here the links truly did exist and that life sprang spontaneously from nonlife. "Those who defend the doctrine of spontaneous generation do so through experience; when one sees an organized being born without being able to discover either a germ, or any way by which the body had been able to reach the place of formation, one admits that nature has the power to create an organized being with heterogeneous elements" (Burdach 1832, 1, 8).

Empirical experience was clearly an important factor in the support for spontaneous generation at the beginning of the nineteenth century. But we know that there was much more than experience driving people—whether full-blooded evolutionists or not—in their enthusiasm for developmentalism. It would not be fair to say that the main factor was materialism in any simple sense—Erasmus Darwin and Lamarck were both deists, and the *Naturphilosophen* saw life force or spirit in some sense pervading the whole of nature. What was important was the upward-driving force that they saw throughout nature—the force of progress, moving teleologically (purposefully) up from the primitive blob, the monad, to the most complex, the man. Spontaneous generation was part and parcel of this, whether one believed (as did Darwin) that life was created at the beginning and then not again, or whether one believed (as did Lamarck) that life was being created all of the time, in a continuous fashion. And this all went with a progressivist radical view on life—one identified with the forces and philosophies of change and of reform. One saw oneself set against the conservative or reactionary elements in society—aristocracy, inherited wealth, the church, and much much more. Spontaneous generation therefore was as much a symbol of a radical view of life as a mark of an empirically justified scientific stance.

All of this on the one hand. On the other hand, we have those who opted for no less of an amalgam of scientific claim and justification together with religious and philosophical urges, albeit toward a conservative, religion-supportive view of life. Most obviously and definitively one finds this in the life and thought of our old friend, Georges Cuvier. He loathed spontaneous generation and thought it quite unsupported by the empirical facts. "Life has always arisen from life. We see it being transmitted and never being produced" (Cuvier 1810, 193). If evolution was judged impossible, you can imagine how much less likely would be spontaneous generation. But clearly, Cuvier's opposition was more than just this. At one level, he hated both Lamarck's speculations and *Naturphilosophie* because they represented the kind of speculative and sloppy science that he thought so threatening to the neat, tight, objective work he was promoting, both for its own sake and for his own sake as an important (although minority-religion) scientist in a conservative society.

At another level he hated evolution and *Naturphilosophie* because they violated his teleologically inspired view of the living world—one that, thanks to the conditions of existence (and its corollary the correlation of parts), gave Cuvier (as he thought) a predictive science, as one tries to fit parts of organisms into an end-driven purposeful functioning whole. And at a third level, he hated evolution and *Naturphilosophie* precisely because they represented radical and revolutionary elements in society, and everything for which he himself stood was on the side of stability, and the status quo, and the establishment. All of these levels and more came into play against developmentalism and hence were focused even more on spontaneous generation. It was a false doctrine in every possible way.

The Opposition Continues

These two positions—part empirical, part metaphysical, part political—dominate thinking from here on. In the 1840s, the supporters of spontaneous generation were dealt a heavy blow when finally people worked out the basic facts of parasitism. It was shown how organisms such as tapeworms take on several different forms, according to the hosts in which they are embedded: far from being generated at one fell swoop in one set of hosts, they come from other hosts where they pass unrecognized because they do not yet have their final forms. Although a person may never eat food containing a tapeworm, as is found in humans, these foodstuffs do actually have parasites that are transmitted to humans: parasites that are indeed tapeworms in potentiality if not yet in actuality. This discovery was obviously a severe thrust against the doctrine of spontaneous generation, although apparently it was not the death knell. Supporters still continued to have faith in its existence.

> I am at the very first struck by the great a priori unlikelihood that there can have been two modes of Divine working in the history of Nature—namely, a system of fixed order or law in the formation of globes, and a system in any degree different in the peopling of these globes with plants and animals. Laws govern both: we are left no room to doubt that laws were the immediate means of making the first; is it to be readily admitted that laws did not preside at the creation of the second also? (Chambers 1846, 19–20)

Enter Pasteur. Pouchet was desperate not to be labeled a radical because of his belief in spontaneous generation. He thought it was proven by the facts. Hence, his adamant denial of evolutionism. He had no desire to be tarred by that particular brush. But to no avail. Pasteur was determined to roll right over such ongoing claims about the empirical plausibility of spontaneous generation. In a series of celebrated experiments, he boiled sugared yeast to kill off the live contents, and then showed that the treated material remained sterile unless and until it was recontaminated. He showed that these results hold in different conditions. Even when the material is open to the outside air, so long as the openings are such that contaminants cannot enter (through being long and thin and curved), there is no appearance of life. Potential breeding grounds remain sterile. But as soon as air is allowed to enter freely, or other nonsterile substances are permitted to infect the inner material, fermentation begins almost at once. Only when life is introduced does life multiply. If life is barred, it seems to remain forever absent.

Celebrated then and celebrated now. Pasteur was a brilliant experientialist, and he and others recognized this fact. He did truly strike a heavy empirical blow against spontaneous generation. But the joy of his countrymen at his successes far exceeded the mere empirical. France in the 1860s was a deeply conservative society, with the monarchy, the aristocracy, the church in full force. The last particularly was setting its face resolutely against change or modernity. In a Papal Encyclical of 1864, Pope Pius IX denied explicitly that "the Roman Pontiff can and must make his peace with progress, liberalism, and modern civilization and come to terms with them" (Error 80, quoted by Farley 1977, 95). All of these sorts of factors played a major role in the reception of Pasteur's work and the canonical status it achieved. The committee set up by the French Académie des Sciences was deeply conservative, quite determined to find in Pasteur's favor, and no less willing to preach to one and all that the fatal blow had been struck against the radical doctrine of spontaneous generation. And when Pouchet and friends complained against the bias of the judges, the Académie set up a committee even more conservative and predetermined to find in Pasteur's favor! By this time, France had received the *Origin* with its radical new introduction, and the forces of the French scientific establishment felt (with reason) that the counterattack must be mounted and supported with every possible weapon. Pasteur's work was just what was needed, and so his results were trumpeted far and wide.

Brilliant though Pasteur's work truly was, the existence of extraneous nonscientific factors in its reception—a reception the success of which echoes down to the present—is amply attested by the fact that (as I mentioned at the beginning of this chapter) Pasteur himself did not truly believe that he had forever disproved spontaneous generation. He did not believe that one could get life from living material or even from just plain organic matter (heterogenesis)—his experiments showed this and his conviction was part of his overall thinking on fermentation. But work he had done early in his career on crystallization rather disposed him toward the possibility of life from the inorganic (abiogenesis). "Life is the germ and the germ is life. Now who may say what might be the destiny of germs if one could replace the immediate principles of these germs (albumin, cellulose, etc.) by their inverse asymmetric principles. The solution would constitute in part the discovery of spontaneous generation, if such be in our power" (Pasteur [1883] 1922, 1, 375).

Pasteur, however, kept quiet about these speculations until late in his career, when—a new, more liberal government being in power after the disastrous Franco-Prussian war—he felt free to make them public. Although, paradoxically, by this time the rest of the world was moving on beyond spontaneous generation. Such had not been the case when, at the beginning of the 1860s, Pasteur began his assault on Pouchet. In Britain and Germany, the rapid rise in the respectability of evolutionism, thanks to Darwin's *Origin,* led to an immediate enthusiasm for spontaneous generation that had not engulfed those countries to such a degree ever before. At once, people saw that evolutionism demanded answers about the ultimate origins of life, and the old ideas were brought out and polished to shine more brightly than they had ever done in earlier times. Pasteur was ignored, if indeed he was noted at all.

Interestingly and paradoxically, Darwin himself made no contribution to this enthusiasm for spontaneous generation. In the *Origin,* he said virtually nothing about origins, merely talking of the rise of life from "one or a few forms." I am not quite sure about the reason for this silence. Part, I strongly suspect, is that Darwin realized that it was a topic surrounded by controversy, associated with radical thinking. Naturally cautious by nature and determined to push his ideas in a nonthreatening manner (because then they would be more likely to be accepted), Darwin stayed away from ultimate origins because speculation would and could only harm his case. If there is a nasty gap in your knowledge, then your best policy is to say nothing and to say it firmly! Also, I suspect his silence was in part because—a problem that plagues evolutionists to this day—although evolution seems to demand answers about ultimate origins, there is little reason to think that the evolutionist is in any way capable of answering them. I do not mean that no answers can be given, or that the evolutionist as such is a bad or inadequate scientist, but rather that the problems and answers lie outside of his or her professional domain. The evolutionist is a biologist. Origins require chemistry, biochemistry in particular, and lots of it. There is really no reason why the evolutionist as evolutionist should be able to answer these questions, even though the evolutionist's work points to these questions as demanding answers. So here was another reason for Darwin's silence.

Also at this point remember that Darwin was trying to promote evolution as a potential professional science, and so here again was reason to stay away from speculation. And remember that his supporters—Huxley in England particularly, and Ernst Haeckel in Germany also—had other ends in view. They wanted broad metaphysical speculations, grounds for the secular world philosophies or religions that they were spinning and endorsing and promoting. For them, evolution was the popular science par excellence, and they had no hesitation in pushing its limits to the ultimate and beyond. They had no need of caution on origins or on hypotheses about spontaneous generation. Indeed, in the 1860s Huxley and friends even thought that

Bathybius haeckelii, *supposedly an example of early life but in fact a chemical precipitate*

they had new empirical evidence of its truth. Mud dredged from the sea was taken to be full of life or life-potential forms or particles.

> I conceive that the granule-heaps and the transparent gelatinous matter in which they are imbedded represent masses of proto-plasm. Take away the cysts . . . [it would] very nearly resemble one of the masses of this deep sea "Urschleim," which must, I think, be regarded as a new form of these simple animated be-ings . . . described by Haeckel. . . . I propose to confer upon this new "Moner" the generic name *Bathybius,* and to call it after the eminent Professor of Zoology in the University of Jena, *B. Haeckelii.*" (Huxley 1868b, 212)

Unfortunately the euphoria did not last and neither did *Bathybius haeckelii.* In 1876, it was discovered to be inorganic—a precipitate of sulfate of lime—and that was the end of that. Huxley withdrew his claim. And despite the fact that there were those who still wanted to defend spontaneous generation, although by now evolution had con-quered almost all that lay before it, one senses that the days of sponta-neous generation were coming to an end. On the one hand, just too many things that had been hailed as evidence for the belief had by now been shown explicable by other means or simply not supportive of the doctrine. On the other hand, and probably more important over-all, people were now starting to dig further and further into the elements of the organisms—first cells, and then cell parts, and then parts of these parts. With each move to a yet-smaller level, the intricate complexity of the stuff of life was reinforced even more strongly than before.

And with this, whether or not one put all or any of this down to God's design and activity, the improbability of things having come together spontaneously became less and less. It was not so much that people like Pasteur had given definitive proof that spontaneous generation was impossible—we have seen that Pasteur himself had hardly done that because he thought that it was always an open question—but rather that the weight of evidence about the nature of the living world made spontaneous generation less and less plausible as an explanation. In an era of cell biology, it simply did not make sense.

The Oparin-Haldane Hypothesis

But still an answer must be sought. If not spontaneous generation, then what? If evolution be true, then the Darwinian strategy of silence can only last so far. One must return at some point to the issue of origins. As indeed Darwin himself did in a private letter.

> It has often been said that all the conditions for the first production of a living organism are now present which could ever have been present. But if (and oh! what a big if!) we could conceive in some warm little pond, with all sorts of ammonia and phosphoric salts, light, heat, electricity, etc., that a protein compound was chemically formed ready to undergo still more complex changes, at the present day such matter would be instantly devoured or absorbed, which would not have been the case before living organisms were formed. (Darwin 1887, 3, 18; letter written in 1871)

This gradual (but natural) appearance of life was also a position endorsed by Herbert Spencer. He set himself entirely against spontaneous generation. "That creatures having quite specific structures are evolved in the course of a few hours, without antecedents calculated to determine their specific forms, is to me incredible" (Spencer 1864, 1, 480). As Spencer pointed out, reasonably, in a way spontaneous generation threatens to undercut the whole evolutionary enterprise. If primitive-yet-complex organisms can appear in one fell swoop, what is to stop more sophisticated organisms appearing in like fashion? And if these, then where does such generation end? Why bother with evolution at all? Yet, Spencer did not want to deny the natural appearance of life. It had to appear gradually, that is all. "The evolution of specific shapes must, like all other organic evolution, have resulted from the actions and reactions between . . . incipient types and their environments, and the continued survival of these which happened to have specialities best fitted to the specialities of their environments" (p. 481).

But it was to be another fifty years before people started to make a research program out of such speculations. Two people particularly, in the 1920s, are credited with the ideas that started things moving in the direction of gradual development of life from nonlife: first the Russian biochemist Aleksandr Ivanovich Oparin ([1924] 1967), and then the English biochemist and theoretical population geneticist J. B. S. Haldane (1929). Like Darwin and Spencer before them, although in some-

J. B. S. Haldane

what more concrete terms given the advances in chemical understanding, they postulated the emergence of the living from more simple inert substances, through natural evolutionary-type laws. Not that the program that they started has yet been brought to full fruition. There are still major questions about how life might have come naturally, even if (especially if) it takes a very long time. But for all that, work in the twentieth century on the origin of life became almost a full-time industry, and the researchers themselves certainly think that they have made significant advances, if not in the traditional direction of spontaneous generation.

I should say that whatever the nature of the advances—the merits of which we shall consider in a moment—some of the controversy and questioning that still swirls around the question is to a certain extent self-imposed. Famously, or notoriously, both Oparin and Haldane were Marxists—Oparin especially so, for he was a key figure in Soviet science and a major backer of the agricultural genetics charlatan Trofim Lysenko. You might think therefore that even though you no longer have to subscribe to an outmoded philosophy like German *Naturphilosophie,* if you are today going to take a naturalistic stand on origins, you must endorse a philosophy that many (perhaps including yourself) find thoroughly objectionable. Fortunately, although Oparin particularly was given to tying in his theorizing with dialectical materialism, the links are at best tenuous (Graham 1987). In fact, in the 1920s, neither Oparin nor Haldane was yet a Marxist, so the strongest possible connections simply are not there. And when in the 1930s, Oparin did start to put things in Marxist terms (Haldane never did so, for his contribution to the question was confined to a suggestive essay), much that he said could be translated at once into nontheoretical language.

Marx—or rather his coworker Engels, who was more interested in natural science than Marx himself—postulated a number of laws (or "laws") that supposedly govern the workings of nature. One is the "law of quantity into quality," as when cooled water does not simply get colder but turns into something new, namely ice. Oparin took this law to incorporate the fundamental truth of his whole approach to the origin of life question, inasmuch as he was suggesting that nonlife turns into life. But

The title page of the presentation copy of Das Kapital *from Marx to Darwin*

this is hardly Marxist, as such, for it was also Darwin's view—and famously, although the Englishman received from the author a copy of *Das Kapital,* he never cut the pages and read it! The same holds true of the supposed connection of the origin of life question to Engels's "law of the negation of the negation" (as with Newton's law, that to every action there is an equal and opposite reaction). Oparin took this as proof that life once started makes impossible the further creation of new life. But again this was Darwin's view, and so not Marxist per se. (As I have said earlier, in a way, I see Engels's work going back to that very German idealism the Marxists thought they were refuting!)

Having disposed of this ideological red herring—"red" in more senses of the word than one, and a great favorite of the evangelical Christian opponents to any scientific approach to the question of origins—we can turn now to contemporary thought on the beginnings of life. It is customary and useful to break down the Oparin-Haldane hypothesis (as it is generally known) into a number of steps. First, there is need of the right conditions for the (natural) creation of organic molecules—those that make the ultimate building blocks of life—from nonorganic molecules (from a warm prebiotic soup, as is sometimes said). If conditions were like they are today, with a 20 percent oxygen atmosphere, then (as Haldane pointed out) the molecules simply could not have formed and persisted. One needed a very different sort of atmosphere. But as it happens, this fits precisely with what students of the subject think in fact might have been the case. The earth is believed to be about four and a half billion years old and initially in a molten state. As it cooled over the next billion years, the oceans formed, but it was surrounded by an oxygen-free atmosphere. It was one rather heavy in such gases as methane (CH_4), ammonia (NH_3), carbon dioxide (CO_2), and hydrogen sulphide (H_2S). Moreover, it would have been an atmosphere permitting the passage of much ultraviolet radiation, for there would have been no ozone (O_3) layer to block it.

Next, there is the making of these elementary organic molecules. Even today, nearly a half century after it was first achieved, this is the most celebrated part of the chain. Back in 1953 a number of chemists—notably Stanley Miller, then just a graduate student—set up a relatively simple apparatus showing how, under what were presumed original conditions, complex molecules ("amino acids," the components of proteins, chainlike molecules that make the structure of the cell) would form quite rapidly. A little electricity (simulating lightning) or radiation would turn inorganic substances (such as

Tungsten electrode

5-liter flask containing gases

Condenser

Stopcock for withdrawing samples during run

Condensate containing amino acids

500-cc flask containing boiling water

The experimental apparatus of Stanley Miller, through which he made amino acids naturally

methane and ammonia) into organic molecules of the required sort. Since those first experiments, even more successes (including the creation of the composites of the nucleic acids, the templates of life) have been formed.

Third (in the Oparin-Haldane sequence), one has to get the individual organic molecules to link together into the chains that are needed for the maintenance of life—proteins and nucleic acids. In fact, the joining is no great problem. It is just that such chains tend to break apart rapidly before the job is finished. Here it is thought that naturally occurring clays may be significant causal factors. Organic molecules adhere to clays and can build up chains while at the same time resisting the urge to break apart. Already, experimenters have shown how quite long chains can be formed in such (presumably analogous to natural) conditions.

The fourth step is a lot more tricky. Now you have to get the long chain molecules to replicate themselves. In cells today, deoxyribonucleic acid (DNA, the modern molecular equivalent of the gene) reads off itself to make copies for new cells. Also, ribonucleic acid (RNA) reads off the DNA, and then this RNA acts as an information template to make chains of amino acids, proteins. But at the beginning you have a bit of a chicken and egg situation. Without the superstructure of the cell, made of proteins, it is hard to see how the nucleic acids could function. But without the functioning nucleic acids, you get no proteins! Perhaps, suggest some workers, the mineral clays continue to play a significant role. Crystals repeat themselves, building copies on templates. And sometimes errors get incorporated into the crystal patterns and get repeated. Could it be that originally it was crystals that were reproducing (no one says that they were alive), with organic molecules as it were piggybacking on them? Then the organic molecules themselves started to take over reproduction, and eventually they dropped their mineral supports (Cairns-Smith and Hartman 1986).

Other workers are suspicious of this hypothesis—their trouble is that it contains the claim that today there is no confirmatory evidence of what happened originally! They rather prefer to think that the organic chains themselves may have gone directly to reproduction—most likely through the medium of RNA, which is needed for proteins and which is in some organisms the only nucleic acid and thus capable of acting as a template for itself without need of DNA. Of course, there are questions about how this is to be done. Suffice it to say here that biochemists are trying with some success to get RNA molecules on their own to replicate themselves. No one has succeeded in getting this to work properly yet, but already one can get an RNA molecule to add bits of chains like itself to other such molecules. Of such tiny steps are great edifices built!

Even if this all works out nicely, more steps remain. One has to tuck everything away in a nice globular cell, for instance. It was here that Oparin put much of his energies, as also did a number of American workers, notably Sidney Fox (1988) and his colleagues. They strove with some success to show how some organic molecules can be made to form self-contained spheres (like the outer shells of cells); how they can maintain themselves, even budding off to form other spheres; how such shells can keep and even promote differences between the inside and the outside, even selecting as it were certain compounds to cross over from the outside to the inside (while barring others); and how some of these compounds can be precisely the kinds of molecules (like ribonucleic acid molecules) that one would expect to have been preserved and cherished in new or protocells.

Enough! There is more, much more. Cells are very complex entities, certainly the cells of higher organisms. (A distinction is drawn between "prokaryotic" cells that roughly speaking are from simple organisms and "eukaryotic" cells that roughly speaking are the cells from more evolved organisms.) These latter, eukaryotes, contain not just the centers (the nuclei) where you find the DNA, but also other bodies (like mitochondria and ribosomes) that have various functions. It has been suggested by Lynn Margulis (1982) and others, with some considerable plausibility, that perhaps the eukaryotic cells were formed by incorporating prokaryotic cells: not so much cannibalism but in a form of symbiotic relationship. And then after that, you have the development of sexuality, something that is thought to be closely linked to the appearance of eukaryotic cells: virtually all major groups of organisms with eukaryotic cells have sexuality—and those organisms without sexuality in such groups are thought to have been sexual and then (for various selective reasons) to have lost it.

There are a lot of steps here and almost all of them are tentative—they require a measure of faith, not necessarily "faith" in the religious sense but in the sense that one might say one is making a gamble or prediction on what one thinks are reasonable grounds. But are they "reasonable grounds," or rather if you add everything up together are they reasonable grounds? One or two steps you might swallow. But six or seven or eight steps? Surely this is all a bit like gambling on all of the winners on a day's racing card. You might pick two or three winners, but would anyone want to put their money on picking all of the winners for that day—however good the odds? One prominent critic of all things evolutionary—Alvin Plantinga, North America's most distinguished philosopher of religion—is so contemptuous of the work thus far performed on the origin of life question that he cannot even bring himself to write on it. He speaks of hypotheses about the origin of life as "the most part mere arrogant bluster," adding that "given our present state of knowledge, I believe it is vastly less probable, on our present evidence, than is its denial." Indeed, so contemptuous is he of such claims that he finds that he cannot bring himself to "summarize the evidence and the difficulties here" (Plantinga 1991).

Can things really be this bad? Obviously the stand you take here is going to depend on a number of things, and most of them are not going to be purely scientific. If (with most scientists) you are firmly committed to the belief that natural explanations can be found for all physical (including organic) phenomena, then you are going to think that natural origins of life are reasonable, no matter what the gaps. If you are firmly committed to the significance of supernatural forces— divine interventions or miracles—then I suspect that you are probably going to think that God had a role here. Perhaps you will think this even if you are an evolutionist. If you are somewhere in the middle, then presumably you are going to end up somewhere in the middle!

But is it reasonable to be a naturalist? Are not scientists turning their heads away from the truth—out of ignorance or prejudice or whatever? (Do not discount the powers of indoctrination and prejudice. It would be a brave scientist indeed today to admit that he or she was going to invoke miracles. I can just imagine the comments on the next grant application.) Let me make two remarks. First, even though it is surely true that the scientist's assumption that everything can be given a natural explanation is an assumption that goes beyond the evidence—how could it be otherwise?—it does not follow that it is an unwise or irrational or even a risky move to make. The fact of the matter is that time and again things that have seemed incredibly puzzling, surely defying scientific explanation, have succumbed to constant pressure and investigation. Think of the wonders of physics—the planets, for instance, and how they shine bright when they loop the loop in the heavens ("retrogress")—for many years a monstrous puzzle, but then explicable in terms of Copernican theory. Think of the strange distributions of animals and plants around the globe. For years people wondered about their causes, and now we know that it is because the continents move around the globe on massive plates. Think of diseases that seemed beyond doubt to be acts of God, but that now are almost commonplace phenomena, explicable through microbes or viruses or whatever. When acquired immunodeficiency syndrome (commonly known as AIDS) was first reported, no one had any idea of its cause, but investigation soon brought the answer in the form of the human immunodeficiency virus (HIV).

My point is that being a scientific naturalist is a good strategy because again and again it brings results, even in the most unpromising situations. It may not be a logically sure bet—there could always be a miracle around the next corner—but pragmatically it is a very sensible way to go. It is not just a "leap of faith" in the sense of going against the evidence. It is the opposite. It is going with the evidence and precisely what we mean by being "reasonable." But what about the particular case of the origin of life? Is this not a special case that makes the naturalistic approach highly unpromising? I do not see that this is the case at all. Remember, until about a hundred years ago, people were trying to shortcut the whole process with spontaneous generation. Eventually, that fell to the ground, but it has only been in the past 60 or 70 years that people have really tried to crack the problem in a way that (even in principle) stands any chance of succeeding.

And it is even more recently that researchers of the subject have had at hand some of the really relevant tools of the trade—detailed knowledge about DNA and RNA and proteins and such things, for instance. It is a massive problem that faces the origin of life researchers—indeed, part of their advance is to realize precisely how massive a problem it is. I for one would be suspicious (especially given the history of the all-too-slick spontaneous generation idea) if the claim were that the problem was now licked or close to being so. Big problems require lots of effort—time too—to get at big solutions. I would not expect more than a progress report from the battlefield, which is precisely what we get.

But there are reports and then there are reports. Not all progress reports are progressive in the sense of reporting on genuine advance. I would say that here, however, we are given just such a report. The researchers are making advances—on the self-synthesizing of nucleic acids, for instance—and feel confident that more such advances lie ahead. One is not just given a whole heap of questions, with no one having the slightest idea about how to crack any one of them. One does not simply have awe and mystification and nothing else. One has real work and real results. I simply do not see that one could ask for or expect any more at this point. It is a fallacy to think that, because there are many links to be filled and most or all are thus far not connected, this means that collectively the case is hopeless. The point is that the links are open to study and investigation and that they are yielding to pressure.

For this reason, while not wanting to pretend that more has been done than actually has been done, I would suggest that the researchers' faith that answers will come—naturalistic answers will come—is not misplaced. It is simply silly (and a sign of almost wanton ignorance) to say that the work thus done is for "the most part mere arrogant bluster."

Early Life

About 600 million years ago, there was a huge increase in the number of life-forms on this planet. After the "Cambrian explosion," the earth teemed with life and nothing was ever again quite the same. But what about the time before the Cambrian? If the first life came over three and a half billion years ago, then we have a vast period for which to account—a period five or six times as long as after the start of the Cambrian and the much more familiar modern era. Is there any evidence of past life? What about the paths that it took? Can we say something about causes? Even if you agree that natural selection was the chief motivating factor, is there anything to show why certain things happened at certain times?

I have mentioned in an ealier chapter how the pre-Cambrian was something that worried Darwin a lot, for at the time of writing the *Origin,* there was simply nothing at all in the record to show that there had been life.

If my theory be true, it is indisputable that before the lowest Silurian [today, called the Cambrian] stratum was deposited, long
periods elapsed, as long as, or probably far longer than, the
whole interval from the Silurian age to the present day; and that
during these vast, yet quite unknown periods of time, the world
swarmed with living creatures.

   To the question why we do not find records of these vast pri-
mordial periods, I can give no satisfactory answer. . . . The case
at present must remain inexplicable; and may be truly urged as a
valid argument against the views here entertained. (Darwin
1859, 307–308)

As it happens, not long after the *Origin* was published, a number
of strange objects were unearthed in the pre-Cambrian rocks of
Canada. These were identified by Sir William Dawson, sometime
principal of McGill University and doyen of Canadian geologists, as
primitive life-forms, and they were given the name *Eozoon canadense,*
the "dawn animal of Canada." (O'Brien 1970). Darwin picked up on
this at once, and the discovery duly made its way into later editions of
the *Origin,* supposedly filling the acknowledged major gap in life's history. Paradoxically, although perhaps by now the kind of move you are
coming to expect, Dawson—a lifelong opponent of evolution—also
focused on *E. canadense,* making it the linchpin of his case against evolution! He argued that since it occurred isolated from all of its fellows, it must have been created miraculously and placed in position
by Divine intervention.

   As it happens, everybody's house was built on sand—metamor-
phic sand. It was soon discovered that *E. canadense* is no genuine organism but an artifact of great heat and pressure on limestone. The
pre-Cambrian therefore seemed as empty as ever before, and this was
the way that things lasted right down and well into the twentieth century. But then the record started to open up in a major way, taking us
back virtually to the (presumed) beginning of life, over three and a
half billion years ago. Moreover, what is really exciting is that the
most primitive organisms seem to be the oldest and that the most sophisticated, those edging close to Cambrian forms, come in the last of
the pre-Cambrian deposits. Nothing is out of order. What happened
where, and who evolved into whom—working out the path of pre-
Cambrian evolution, that is—is of course another matter, and the fossil record is hardly good enough for that. Here one needs to turn to
cellular and molecular traces, trying to infer past connections and
phylogenies. And at least some of this seems to be possible. One quest

has been toward finding the latest common ancestor of all living organisms (the "cenancestor"). It obviously came fairly early on in the story, and today it is believed that it might have been as long ago as three and a half billion years, although (showing how crude things still are, as yet) it might be as recent as two billion years.

One major question about early life history—perhaps the major question—is about the move from life that is exclusively prokaryotic to life that is also eukaryotic. We have seen already that part of what was going on here was probably the symbiotic coupling of prokaryotes to make eukaryotes. Is there more evidence of the evolutionary emergence of eukaryotes from prokaryotes, and is there evidence of when and why it happened? There are several suggestive lines of evidence bearing on these questions, with the key factor being oxygen. First of all, there are the ways in which the two kinds of cells obtain energy: their metabolisms. Both kinds of cells get energy from glucose, but whereas prokaryotes work by fermenting their foodstuffs, simply breaking down the glucose, and getting energy that way, eukaryotes work by respiration, burning the glucose in oxygen, and thus releasing energy. The second mechanism, respiration, is far more effective than fermentation, which in itself is suggestive—one has a presumed move from the less to the more efficient. But more significantly, fermentation and respiration are not two completely different mechanisms. The one metabolism, respiration, follows on the other, fermentation, by adding on more steps: an

Pre-Cambrian life

oxygen-using phase. This is just what one would expect were evolution at work: building on what you have rather than starting anew.

Second, what about the coming of oxygen? We have seen that free oxygen in the atmosphere cannot have been present when life first formed, for it would have had a devastating effect on the beginning organic molecules. One would expect to find therefore, that although eukaryotes need oxygen, the prokaryote story would be different. As indeed it is. Some prokaryotes can tolerate and even need oxygen (which is what one might expect if the oxygen-needing eukaryotes are to evolve from them), whereas for other prokaryotes oxygen is a poison (which is what one might also expect). But where would the oxygen have come from, if there was none free at first? Presumably, as today, it would have come through photosynthesis, where organisms free up oxygen from carbon dioxide, thanks to the energy provided by the sun. And expectedly, we find that some prokaryotes can perform photosynthesis. This is within the power of the blue-green algae (known as "cyanobacteria"), which interestingly and significantly seem to have a metabolism halfway between that of fermentation (like regular prokaryotes) and respiration (like regular eukaryotes). Surely pertinently, the cyanobacteria function most efficiently when oxygen levels are around 10 percent, that is, about half of today's levels. One presumes that they evolved at a time when the oxygen level was not as high as it is today, but that they paved the way for the evolution of higher oxygen level–using organisms.

Dramatic evidence that cyanobacteria function best when the atmospheric oxygen levels are half what they are now, suggesting that cyanobacteria evolved when oxygen was at half its present level

Finally, what about evidence of the time of the arrival of the eukaryotes and the rise in the level of oxygen? There are factors relevant to both of these questions, and they come together with coinciding answers, suggesting that we can claim to know the whole picture. Larger fossils of a kind one would associate with eukaryotes are to be found from about one and a half billion years ago, and this is a point somewhat after the evidence points to the rise in oxygen levels. For instance, uraninite (UO_2) oxi-

Era	Period	Epoch	Events
Cenozoic	Quaternary	Pleistocene	Evolution of Man
Cenozoic	Tertiary	Pliocene Miocene Oligocene Eocene Paleocene	Mammalian Radiation
Mesozoic	Cretaceous		Last Dinosaurs First Primates First Flowering Plants
Mesozoic	Jurassic		Dinosaurs First Birds
Mesozoic	Triassic		First Mammals Therapside Dominant
Paleozoic	Permian		Major Marine Extinction Pelycosaurs Dominant
Paleozoic	Carboniferous	Pennsylvanian	First Reptiles
Paleozoic	Carboniferous	Mississippian	Scale Trees, Seed Ferns
Paleozoic	Devonian		First Amphibians Jawed Fishes Diversify
Paleozoic	Silurian		First Vascular Land Plants
Paleozoic	Ordovician		Burst of Diversification in Metazoan Families
Paleozoic	Cambrian		First Fish First Chordates
Precambrian	Ediacaran		First Skeletal Elements
Precambrian	Ediacaran		First Soft-Bodied Metazoans First Animal Traces (Coelomates)

Millions of Years Ago (vertical axis): 0, 50, 100, 150, 200, 250, 300, 350, 400, 450, 500, 550, 600, 650

Life's history

dizes (to U_3O_8) in the presence of more than a 1 percent oxygen atmosphere. Predictably, in rocks older than two billion years, uraninite is to be found, but it is absent from younger deposits. Conversely, iron rusts in oxygen. In deposits less than two billion years old, we find iron oxides. These are missing from earlier deposits.

Lots of questions still remain, but the answers are starting to come in. And importantly, the answers fit together. They are consistent and coherent. A unified picture of life's early history is starting to shine through. Charles Darwin, who admitted to so much ignorance, would have been pleased. We should feel the same way also. Origins are no longer quite the mystery that they were once.

Further Reading

I cannot speak sufficiently highly of John Farley's *The Spontaneous Generation Controversy from Descartes to Oparin* (Baltimore: Johns Hopkins University Press, 1977). His knowledge of the science is deep and profound, and his ability to move from work in one language to the next is simply staggering. He is simply superb on the developments in our thinking about life's origins from the seventeenth century right down to the near present. I have used his translations in my discussion. A brand new book is a great complement to this older treatment. Iris Fry's *The Emergence of Life on Earth: A Historical and Scientific Overview.* (New Brunswick, N.J.: Rutgers University Press, 2000) is strong not only on the science but also on the underlying philosophical elements that enter into people's thinking about such a difficult and challenging subject as life's origins. She is balanced and fair without being in the least boring.

A topic for which I have no room in my main text but which certainly impinges on the subject of the origin of life is that of life elsewhere in the universe. Did life start here uniquely on earth, or has it occurred again and again throughout the depths of space? Did life perhaps start somewhere else, and was our planet seeded from outside? These have been topics of fascination to scientists, philosophers, and theologians ever since the Greeks. An excellent trilogy of works covers the field. First there is Steven Dick's *Plurality of Worlds: The Origins of the Extraterrestrial Life Debate from Democritus to Kant* (Cambridge: Cambridge University Press, 1982), taking us up to the end of the eighteenth century. Then Michael J. Crowe carries the story through to the beginning of the twentieth century: *The Extraterrestrial Life Debate, 1750–1900: The Idea of a Plurality of Worlds from Kant to Lowell* (Cambridge: Cambridge University Press, 1986). Finally Dick again, in a truly magnificent work, deals with the whole extraterrestrial issue in the twentieth century just gone: *The Biological Universe: The Twentieth Century Extraterrestrial Life Debate and the Limits of Science* (Cambridge: Cambridge University Press, 1996). A popular version is *Life on Other Worlds: The Twentieth Century Extraterrestrial Life Debate* (Cambridge: Cambridge University Press, 1998). This is a really wonderful coverage of science and fiction and speculation and much, much more. It is interesting how the astronomers are so eager to argue for extraterrestrials, including humanlike forms, whereas the evolutionists are much more skeptical. Of course, the astronomers have an interest in seeing all of those rockets shot off into space. This is not an interest much shared by evolutionists.

7

"Going the Whole Orang":
Human Origins

I am a professional philosopher, and I supposed I should feel pleased—flattered even—when scientists turn to my subject for help. In fact, my heart sinks to my boots, for bitter experience has shown me that scientists turning to philosophy are up to no good. All too often, when scientists are a bit unsure of their ground they bolster the case with deep-sounding references to philosophy, preferably hard-to-follow German metaphysics with long words and the verbs at the end. In a court case in which I was once involved—later, I will mention the circumstances—I found that both sides were quoting the distinguished Austrian-British philosopher, the late Sir Karl Popper, in support of their cases. The other uses of philosophy are little better. If it is not being suggested as an occupation for aged scientists well past their prime, then it is hijacked as a weapon in bitter dispute: invariably when scientists want to put the boot into an opponent, they accuse their victim of inadequate—preferably naive—philosophical assumptions.

I began to suspect that something like this was up when, recently, I was asked to review a book on "paleoanthropology"—the study of human evolution—for the *Globe and Mail,* Canada's national newspaper. In the book, all sorts of philosophical charges were being flung around in the hope that some would stick. For instance, I learned that although "inductive approaches" (whatever they may be) are very seductive, they can nevertheless be deeply flawed: "Technological advances have inadvertently promoted inductive science, as they opened the door to the analysis of vast amounts of new data. With exciting new techniques to arrange and analyse data, more and more paleoanthropologists have focused on this aspect of the scientific process. We worry that modern science will become increasingly characterized as 'technology looking for a problem'" (Wolpoff and Caspari 1997, 352.) And so on and so forth, until I began to sense that I was not so much facing someone who was devoted to the high profession of Plato and Aristotle, as a person whose nose was out of joint because his pet ideas had been capped by a breakthrough. I was sure of my ground when I read further: "Many of the younger scientists in our profession have grown up with technologies and algorithms beyond the wildest dreams of their professors" (p. 355). This is less the lament of an epistemologist (someone concerned with the theory of knowledge) or an ontologist (someone concerned with the theory of being), and more the cry of someone hurt as the threat is that their life's work will be pushed aside.

And so I discovered this to be the case. There is a bitter debate that has lasted over a hundred years. At some times it seems to be going toward one side, and the others are going under with their outdated theories. Then, new discoveries and interpretations apparently correct the balance. Before long, the former winners are grabbing undigested chunks of philosophy and hurling them around with gay abandon. I am afraid that it looks as though we are set to fight over the topic for many a year yet. I refer to the fate of the Neanderthals, in-

Modern Homo sapiens *(left) and* Neanderthal man *(right)*

habitants of a once-peaceful valley in Germany, first unearthed in August 1856. They slept for a long time but they have not slept since. Are they human? Some have portrayed them as respectable citizens, hardly distinguishable from the chap next to you on the bus or subway. Others—including most gloriously the cartoonist Gary Larson—paint them as hairy hunched monsters, stupid and criminal. Like something from a Boris Karloff movie in the 1930s. Some have seen them as obvious ancestors, because they are so similar. Some have seen them as too different and stupid to be other than extinct. And yet others have seen them as ancestors (of others!) precisely because they are degenerate: "Ferocious gorilla like living specimens of Neanderthal man are found not infrequently on the west coast of Ireland, and are easily recognized by the great upper lip, bridgeless nose, beetling brow with low growing hair, and wild and savage aspect. The proportions of the skull which give rise to this large upper lip, the low forehead, and the superorbital ridges are certainly Neanderthal characters" (Grant 1916, 95–96).

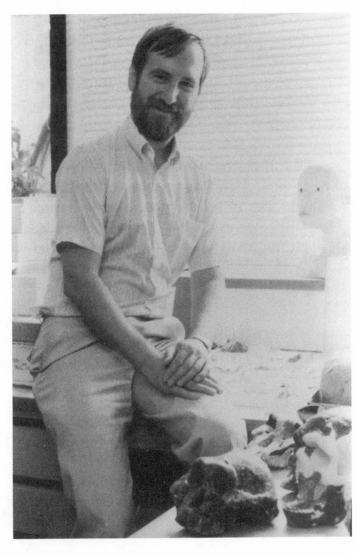

Christopher Stringer

This is the major still-ongoing clash about the Neanderthals. On one side is Christopher Stringer of the British Museum (Natural History) in London, champion of the "out of Africa" hypothesis. This claims that modern humans (our subspecies) evolved and moved out of Africa about 50, 000 years ago, moving in on the Neanderthals in Europe, making them go extinct. Of his opponents, Stringer speaks of them as being confined in "straightjacket," unable to accept or appreciate modern evolutionary biology, not to mention all of the ancillary sciences (Lewin 1989, 66). Then on the other side there is the University of Michigan paleoanthropologist Milford Wolpoff, author (with his wife) of the philosophical comments quoted above. He favors the "multiregional" evolutionary hypothesis, which sees humankind as having evolved in tandem throughout the world, with a certain amount of crossbreeding. Hence, there are Neanderthals firmly in our ancestral past—the ancestral past of modern Europeans, that is, for Neanderthals are found generally in Europe. Two people with two hypotheses, and if the tables seem one day to tilt a little one way, you can be sure that the next day they will seem to tilt a little the other way. Both camps are still very much alive.

But we are ahead of ourselves. Let us go back to the beginning of the nineteenth century.

The Antiquity of Man

Cuvier set the background position. Although in the eighteenth century there had been talk of "pongos" and "jockos" and other fabulous creatures, he found no fossil evidence of humans or humanlike crea-

Milford Wolpoff

tures. We are modern and appear in Europe after the last catastrophe, which you may remember that Cuvier identified with Noah's Flood. This was a conclusion that was welcomed by all, especially by those who had no intention of admitting anything to the vile evolutionary doctrines. People like Adam Sedgwick, Darwin's old teacher and friend, could allow that the earth is old with previous now-extinct inhabitants. Humankind came after all of this, and it is them that the Bible describes and discusses and explains. Christianity is a story about our relationship with God, and what happened before is irrelevant with respect to faith and those things that really matter. Pre-Adamite men are no more supported by science than they are welcomed by religion.

The first break in this picture came in 1847, when the French customs officer Jacques Boucher de Perthes described stone tools (axes) found in northern France in deposits also containing the remains of now-extinct animals (Oakley 1964). This all rather implied that humans go back some considerable time and that (as he saw it) we may not indeed be the first humanlike species. Boucher de Perthes's work, *Antiquites celtiques et antediluviennes,* attracted little attention for over a decade. Then, in 1858, the trained English geologists William Pengelly and Hugh Falconer explored Brixham Cave near Torquay in Devon, also finding tools and the bones of extinct animals, and rapidly opinion swung toward recognition of the Frenchman's achievements. Popular books, including Charles Lyell's *Antiquity of Man* (1863), together obviously with the acceptance of evolutionism that was just then occurring at a rapid pace, completed the demolition of Cuvier's conservative rejection. Mention has already been made of the fact that Thomas Henry Huxley at once went to the heart of the evolutionary issue, and, in his *Evidence as to Man's Place in Nature* (1863), on comparative grounds he argued strongly for our simian ancestry. It is true that there is more gap between humans and the nearest ape than between successive apes themselves, but there is more gap between the highest and lowest apes than between humans and apes. There is no question but that, as Lyell wrote worriedly in a private notebook, picking up on the penultimate organism in Lamarck's evolutionary scheme, we simply must "go the whole orang." We humans are part of the primate evolutionary picture.

But what about the "missing link"? Everybody knew what link this referred to, and everybody knew how important its discovery was going to be. Humans had evolved, there was no question about that: how and where and when were the key questions. Darwin and

The frontispiece of
Huxley's Evidence
as to Man's Place
in Nature

SKELETONS OF THE

GIBBON. ORANG. CHIMPANZEE. GORILLA. MAN.

*Photographically reduced from Diagrams of the natural size (except that of the Gibbon, which was twice as large as nature), drawn by
Mr. Waterhouse Hawkins from specimens in the Museum of the Royal College of Surgeons.*

Huxley rather favored an African origin for humankind. The great
apes live now in Africa, and the homologies between them and us
were precisely what these two men were stressing in their efforts to
convince people of the facts of human evolution. The more apelike
we could be made or the more humanlike they could be made, the
tighter the conceptual links and the greater the case for evolution as
fact. But when it came to paths, most other people had different
ideas. The racist progressionism of the late nineteenth century saw
white European humans as clearly superior to other races, especially
to blacks. It was argued—by Spencer and his followers particularly—
that the colder climates required more effort to survive than did the
warmer climates, and hence protohumans advanced more rapidly
(through Lamarckian inheritance following effort) in the colder climates than in the warmer climates. So, it was thought by many that
Africa could not have been the home of human evolution—certainly
it could not have led the way.

Asia became the favored origin of humankind—all those grassy
steppes seemed tailor-made for the evolution of humans out of the
trees and up onto two legs. Most influential were Ernst Haeckel's
writings and, inspired by them, toward the end of the century the
Dutch doctor Eugene Dubois found "Java man," pieces of a skull with
a smaller cranial capacity than today's humans and yet apparently, on
the basis of a thighbone also found in the deposit, an upright walker.
He named this being *Pithecanthropus erectus,* although today we put it
in the same genus as ourselves, *Homo erectus* as opposed to us, *Homo
sapiens.* Haeckel seized at once on the significance of Dubois's discovery, and as can be seen from the diagram given in a little book he

penned (revealingly entitled *The Last Link: Our Present Knowledge of the Descent of Man*), he had no doubt but that it represented the very piece of evidence long awaited (Haeckel 1898).

If one were just giving a rational reconstruction of the history of the discovery of human ancestral fossil remains—that is, if with hindsight one were just looking back at what happened or what one thinks ought to have happened if everyone were rational (that is, as rational as oneself!)—then one might expect that the next moves would have been directed to the finding of humanlike fossils even older than Java man. After all, the chimpanzee in Haeckel's picture is one of today's organisms, and no one claims that it was also our ancestor. Rather, the picture is intended to hint that our ancestor was chimpanzee-like in significant respects. And indeed, history seems to fit this reconstruction rather exactly, for the next major discovery in the 1920s in South Africa was of precisely an organism with upright stance but a far smaller brain. Raymond Dart, newly established professor of anatomy at Witwatersrand University (in Johannesburg), discovered and described this animal, informally known as Taung baby (because it was a juvenile) and officially classified in a different genus from humans, *Australopithecus africanus*.

But real history has a nasty way of taking on a life of its own and not following the path that rational people think that it should. Dart's discovery was opposed right from the beginning, being considered rather just an ape and not at all significant for the story of human evolution. It was not until the 1940s, when the climate started to change

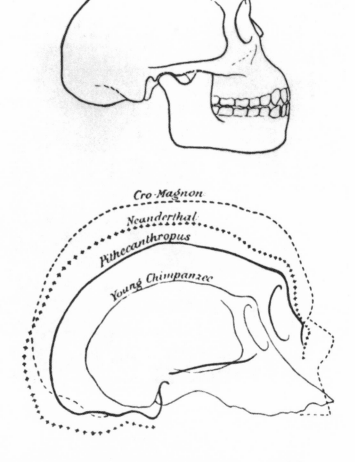

Haeckel's illustration comparing the brain of Cro-Magnon man (an early specimen of modern humans) with other brains

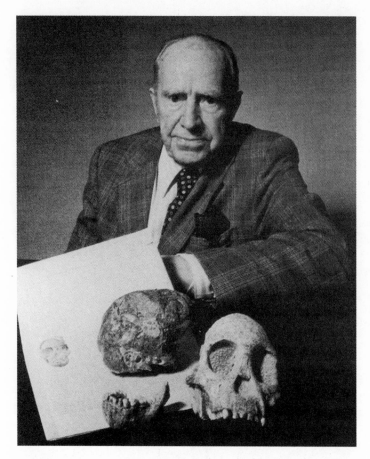

Raymond Dart with
Taung Baby

(and there was the discovery of more fossils, including parts pointing unambiguously to upright walking), that Dart's *Australopithecus* was recognized for the significant finding that it really was. Why the delay? One major factor clearly was that Dart's animal came from Africa, and most informed people were looking to Asia. The thought that we humans might have evolved in Africa was altogether too horrendous to contemplate. Taung baby just did not fit in, and anyone who knows anything at all about science will realize that prior convictions and expectations are a far more significant factor in observation than any thing out there in the real world. (That is an exaggeration, but not too much of one. No one denied that Taung baby existed or that it had the features that it had. The question was rather about what these features represented.)

Also there is no doubt but that a lot of people simply did not like the smooth upward rise that Haeckel showed and that Taung baby would seem to confirm. There was agreement, of course, that we had evolved and that ultimately we had evolved from beings with small brains. But there was a desire to push this back as far as possible. People did not want to be too closely associated with the apes—even the Neanderthals were now out of favor and portrayed as highly brutelike and not all respectably human. Indeed, white people (who were after all making the running in paleoanthropology) did not want to be too closely associated with their fellow darker humans and wanted many years of evolution independent from other groups. Fortunately, there was what was thought to be good evidence for this position of long-time separation—evidence that we know now to be one of the most notorious scientific frauds of all time.

I refer of course to the Piltdown man, or Piltdown Hoax, as it has been known since it was uncovered in the early 1950s. In southern England, around 1912 (the exact date of first discovery is clouded in mist), an amateur archaeologist, Charles Dawson, unearthed pieces of skull and jaw that seemed to confirm that precisely the required sorts of humans had lived and thrived, long before the present. These were humans with massive brains—virtually as big as ours in fact—and yet clearly primitive in other respects, particularly in the lower face and jaw. Conferring authenticity, Arthur Smith Woodward, a curator at the British Museum (Natural History), became involved in the discoveries, as well as the then-young French priest/paleontologist Pierre Teilhard de Chardin. Quelling doubters, a year or two later some really major pieces of evidence came to light. (Supposedly Dawson found these new fossils in 1915, although they were not announced by Woodward until after Dawson's death in 1916.)

We now know that it is hardly surprising that Piltdown man had a brain as big as ours, since the key skull was in fact a human skull!

Piltdown Man

Nor was it surprising that the lower face was primitive and ape-like—the jaw and teeth that were recovered came from an orang-utan. The pieces were suitably shaped and stained, and then the awkward bits (precisely those bits that would cast doubt on the brain and jaw being from the same animal) were broken off and thrown away. As I said earlier, "anyone who knows anything at all about science will realize that prior convictions and expectations are a far more significant factor in observation than anything out there in the real world. (That is an exaggeration, but not too much of one.)" Precisely.

The remarkable thing about Piltdown man was not that the fraud was eventually uncovered but rather that it lasted as long as it did. It really was quite a crude job. As soon as anyone looked, you could see all sorts of file marks and such things, including evidence of staining rather than weathering through time. And this was apart from physicochemical methods of dating materials. It ought to have been spotted early on, and indeed to their credit some people did always feel that it was highly and uncomfortably anomalous. But it fit precisely what most people were after—almost too patly one might say (especially when more relevant bits appeared almost to order)—and there are none so blind as those determined to see. And people were nothing if not this, especially English people, who were highly sensitive to the proud place that England now possessed in the search for human ancestors. The Germans might have those nasty Neanderthals, but fair Albion has been home to the greatest prize of all.

The possible identity of the perpetrator of Piltdown has filled more books than has the quest for the identity of Jack the Ripper—with about as much success. On the Internet, I found more information on the topic than I truly need for one lifetime. To be honest, the identity does not really matter, which I suppose is part of the attraction. Some of the suggested suspects rather boggle the imagination—although, unlike the Ripper, no one yet has suggested that the Piltdown hoaxer was the Prince of Wales. (The hoax may not have been a great work of art, but it required more energy and gumption than one generally associates with British royalty in the twentieth century.) One far-out suggestion is Sir Arthur Conan Doyle, the author of the Sherlock Holmes stories. He was a keen spiritualist and had a keen dislike of scientists who regarded his enthusiasms with contempt. Hoaxing them all like this would have been very satisfying. But motive alone does not make for criminal action, nor does opportunity. Teilhard de Chardin has been fingered by Stephen Jay Gould (1980b). However, as I shall explain later, the accusation probably tells us more about Gould than about Teilhard, who simply does not strike one as the kind of man to do something that required such systematic deception.

The most recent purported culprit is one Martin Hinton, a curator at the British Museum. He has been indicted on grounds of bits and pieces of supposedly incriminating evidence discovered in his effects after his death. But, it appears that he cannot have been the sole perpetrator—he was simply not around at some of the required times—and the evidence may not be what it seemed. (Particularly suggestive was a discovery in Hinton's effects of various chemicals that were needed for "aging" the orangutan jaw, but Hinton's chemicals do not match exactly the chemicals used on Piltdown.) General suspicion has always centred on Dawson, who had a bit of a reputation for being shifty, and probably this is not far off the mark. Woodward may well have been a dupe—it is interesting to note that his speciality was fish rather than humans.

The story continues and no doubt will continue to continue, so let us return to the main thread of our tale. Since the acceptance of *Australopithecus,* the last half century has seen massive efforts, richly rewarded, in tracing human origins—centered now almost exclusively in Africa. Thanks to the labors of fossil hunters at least the equal of the dinosaur hunters of the last century, we now have a reasonably good pattern of human evolution back for the last five million years.

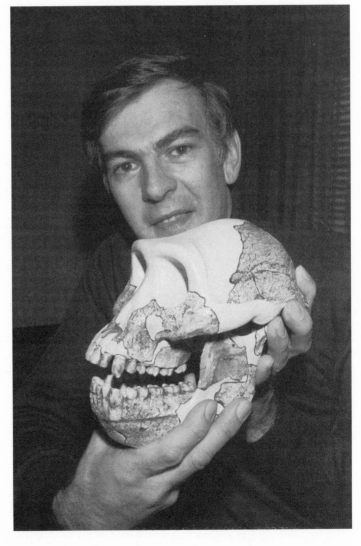

Donald Johanson with Australopithecus afarensis

Our earliest-known, direct ancestor seems to have been *Australopithecus afarensis,* represented dramatically both by more complete skeletons than we normally expect—notably "Lucy," the woman from Ethiopia—and by footsteps in drying volcanic ash in Tanzania. The animal was about half our height or a bit more, with a small ape-size brain of less than 500 cubic centimeters as compared to a human brain of around 1,400 cubic centimeters for a male and a bit less for a female. (The brain size of *Australopithecus afarensis* was ape size, but internal casts suggest that it was already not an ape brain.)

Most exciting of all, Lucy was undoubtedly and unambiguously bipedal. She walked up on her own two feet—she did not run around on all fours nor was she a knuckle walker like the great apes (who can run around very quickly, using their knuckles for support). Yet at the same time—terrific music in the ears of the evolutionist—it seems clear that *Australopithecus afarensis* was not as efficient a walker as are we humans. This does not mean that Lucy was an unstable hybrid, neither fish nor fowl. To assume so is to fall into the same kind of progressionist thinking as held sway at the beginning of the century. She was not an item on a directed line to humans. Had another meteor wiped out mammals two million years ago, she would still have been just fine. It was just that she was not fully human. And in fact, slight curvature of the bones of hands and feet suggest that she would have been much better at tree climbing than we tend to be.

After *Australopithecus afarensis*, the line split—some australopithecines went one way, evolving into more robust forms and eventually to extinction. Others, more graceful or delicate, went on to turn eventually into the human line, and down through several species of *Homo* to our own *Homo sapiens.* More on us in a moment, but first what should we believe about life before the Australopithecines? Here, as is well known, have come some of the most dramatic discoveries and changes of perspective. Until about 20 years ago, the firm conviction of paleoanthropologists was that we humans are a long way from the apes, comparatively speaking. It was thought that, probably, one needed to go back about 15 million years or so before one would find a common ancestor with the chimpanzees and gorillas and orangutans. Humans may or may not have evolved together, but we surely have evolved apart from the rest of creation.

The molecular biologists would have none of this (Pilbeam 1984). They had developed new techniques for assessing absolute dates, and by comparing the macromolecules of apes and men, they came to the conclusion that the ape-human break had to be much more recent—as recent, indeed, as five million years ago, which is really quite astounding when you think that Lucy is nearly four million years old. Expectedly these results—offered less as tentative suggestions and more as firm corrections—did not sit well with people who had spent their lives finding and interpreting fossils. How dare rank outsiders presume to tell them their business?! Wolpoff on Stringer is part of a grand tradition. "Unfortunately there is a growing tendency, which I would like to suppress if possible, to view the molecular approach to primate evolutionary studies as a kind of instant phylogeny. No hard work, no tough intellectual arguments. No fuss, no muss, no dishpan hands. Just throw some proteins into the laboratory apparatus, shake them up, and bingo!—we have the answers to questions that have puzzled us for at least three generations" (Wolpoff and Caspari 1997, 112). It just isn't fair! One can hear the plaintive cries of rejection and dismissal.

Fair or not, the molecular biologists won. Now it is accepted that although the break may be a little older than five million years, it is that order of magnitude. Moreover, although the evidence is still ambiguous, it could easily be that we humans are more closely related to the chimpanzees than they are to the other apes, the gorillas in particular. Although to our eyes chimps and gorillas may look more alike than chimps and humans, it could be that we have gone off on our own and the apes (in those similar-looking respects) have stood comparatively still. No matter what the details, we are a lot closer to the rest of the animal world than anyone dared think just a few years ago.

Causes

So far I have been talking more about the path of evolution, about phylogenies, than about causes—something one does rather dread broaching, for the discussion goes right off the subjectivity-emotion index, time and again. Indeed, one enterprising scholar has likened the causal tales told by students of human evolution to fairy stories, in a rather literal sense. Misia Landau (1991) draws on analyses of folk tales to show that common patterns keep reappearing. The hero starts in a happy initial situation that is disrupted by external forces—death or famine or the like. The hero then sets out on a journey to find salvation or the golden fleece or something similar and along the way has to struggle with forces and the elements, sometimes falling but eventually triumphing. So with the story of human evolution. We were happy apes up the trees in darkest Africa, minding our own business and happily surviving and reproducing. Then something happened. A drought is a favorite causal factor, and the home we loved was no more. We had to leave the trees and come down on the plains or savannah. But we were hardly suited to this, so we had to start evolving in a big way. We needed to be able to run around on the plain, so we developed bipedalism, jettisoning the now no-longer-needed adaptations for tree life. At the same time, things were tough out there on the plain—far more dangerous than up trees. We had to learn to cooperate, to get along with each other. What better way than through intelligence? So we humans (or protohumans) started the path up to full-time thinking ability. And now, finally, we have won. We have conquered the tasks set before us and achieved the goal, full humanhood. Our journey is ended.

Of course, you can run variations on all of this, depending on various factors. If, for instance, you incline to the view that encephalization (large braineded) preceded bipedalism (two leggedness), then you might well look for external factors other than drought as the stimulae for the initial evolution. Perhaps, for instance, it was all a question of new or different predators. And some writers are going to be more daring in their hypotheses than are others. They are going to be more inventive about the challenges supposedly faced and the solutions supposedly found. But the fairy tale—hero makes epic journey, conquering through trial, and arriving eventually at the promised land—persists. And a moral tale, too, especially as one can tie in some strands about the white race having had to travel farther and struggle more decisively than the members of other races—with the expected results.

> But if we know nothing of the wonderful story of Man's journeying toward his ultimate goal, beyond what we can infer from the flotsam and jetsam thrown upon the periphery of his ancient domain, it is essential, in attempting to interpret the meaning of these fragments, not to forget the great events that were happening in the more vitally important central area—say from India to Africa—and whenever a new specimen is thrown up, to appraise its significance from what we imagine to have been happening elsewhere, and from the evidence it affords of the wider history of Man's ceaseless struggle to achieve his destiny. (Elliot Smith 1924, 79)

Of course, today's paleoanthropologists deny vigorously that such approaches to causal factors are faults of which they are guilty. Although all of this may have happened in the past—undoubtedly did happen in the past—it is no longer true of today's work. It is far more objective and value free and so forth. After all, we are all Darwinian evolutionists now, so talk about "achieving destinies" is simply ruled right out of court. Darwinian organisms do not achieve destinies. If they are lucky, they survive and reproduce—for a time.

To which response—that paleoanthropology has changed and that with the coming of the synthetic theory it has become more scientific, and less simply a vehicle for telling one's favorite story—one can say that there have certainly been changes but that whether they are as absolute as some seem to think might be doubted. There is no question but that more attention is paid to fundamental biological principles and that new techniques have thrown up all sorts of new ways of finding pertinent information. But at the same time, values and culture still play a major role in the pictures painted and stories told by students of the human fossil past. Let me not exaggerate. We do know some things now that were not known before. Thanks to the fossil evidence, we know now that humans came down out of the trees and that only then did the brain start to explode up to three times its original size. And we know that there had to be some large selective pressures at work here, if only because brains take a huge amount of energy to run. Selective advantages that cows and horses, or chimpanzees and gorillas for that matter, have not found in their interests (or within their abilities) to follow or satisfy. So any pictures of human evolution that do not fit in with these constraints have to be false. Of this we can be certain.

But after this, there is huge scope for variation and inventiveness. Probably climate did have something to do with our leaving the trees and becoming denizens of the plains, but there are major questions as to why it all happened. Why, for instance, did the other apes not come down to the ground like us? And what was it on the plains that made it so attractive to be bipedal? Was it foodstuffs, and if so what kind? Was it seeds, as has been suggested, and did our hands evolve to pick and eat these seeds? Or was it the need to move around the plains to find food that was less evenly distributed than it was in forests? Walking is an efficient way of traveling—certainly, if the option is going on your knuckles all day, walking has its virtues (Lewin 1989, 68).

Move on to about two and a half million years ago. *Homo* was making its appearance now, and here we get the first human-made tools as well as the beginning of the really massive expansion of the brain. What is the cause? In the 1960s, the popular hypothesis was that of "man the hunter." Little groups of early members of our genus would set out hunting with their tools; catch, kill, and cut up their prey; and then eat it. Brains were needed for this exercise, for obviously we had to depend on skill for the hunt, not being fast and furious like other mammals, and with the coming of a meat or partially meat diet, brains could grow that much bigger because meat is a very rich food and can support organs that are high energy cost. By the end of the 1970s, this hypothesis, at least in its crude form, was starting to fall right out of favor. In its place was coming the hypothesis of man as scavenger—early humans just followed around behind big animals, and when they got into trouble, or when something else killed them, we would move in to grab our share and more. We were a kind of primate jackal.

In addition, there was now a lot more emphasis on food sharing, and females started to take a more prominent active role. The hunters (as in modern societies) were taken almost universally to be male—females therefore had a passive or noneffective role in early human life. Shades of Charles Darwin! With scavenging, and with associated food gathering, it was a whole family activity. Why the new perspective? At least some of this change of viewpoint was fact driven. Increasingly sophisticated studies of teeth and of bones, for instance, could tell that there had to be much more to diet than meat— vegetable matter was very significant. Then there was work being done on the possible modes of travel and life of the early hominids— archaeological studies of where they cut up their meat, for instance, trying to work out lifestyles. Did one kill and eat? Or did one kill and transport and eat? And if the latter, did one kill and cut and transport and eat, or did one kill and transport and cut and eat? Lots of questions like these, which were being tackled using molecular and microscopical and comparative and other studies.

At least some of the change was derived by changes elsewhere in evolutionary biology. The whole question of cooperation was becoming a big thing in Darwinian studies, and these undoubtedly slopped over into paleoanthropology. I shall be looking at cooperation in the next chapter so need say nothing here, except to remark that (as one might expect) if things get hot in one area of evolutionary biology one expects fully that workers in other areas will take note and see if there is anything in it for them. And some of the change was simply driven by ideology. The 1970s was the time when the feminist movement really got up a head of steam, and in an area like paleoanthropology—which has its full share of women workers—one could have predicted that "man the hunter" would get little sympathy. Which it did not! "Woman the gatherer" was an almost perfect counter—here, if anything, females were doing all the real work of collecting seeds and other small food stuffs, and men basically parasites, as always (Zihlman 1981).

Scavenging and gathering, a gender-reciprocal, food-sharing hypothesis, was a natural outcome from this polarization, appealing to those who wanted to acknowledge the significance of the female role in human life and evolution but yet did not want to relinquish entirely the important role of males in this picture. Here now we had a happy balance, with both men and women providing foodstuffs and sharing. What could be nicer? On the one hand, all of that aggressive stuff about hunting now takes a back seat—at best, we men have a rather low role in the meat-gathering business. Although one that requires intelligence. A perfect job for professors, as one might say, rather than for he-men in plaid shirts. On the other hand, the new male now takes his place along with his mate (there was also some stuff brought in about the virtues of sexual fidelity), sensitively sharing his bounty with hers. Those who do not think that such an approach is drenched with social values are as naive as the people writing it. I am not saying that it is bad. I am not saying that I could or would want to do better. I am saying that this approach was the way that it was and looks fair to being for the future.

The Neanderthal Controversy

Let us pick up now on recent human history and on the Neanderthal question. *Homo habilis* goes back about two and a half million years; *Homo erectus* appears about one and a half million years ago and lasts until about 500,000 years ago, at which time we start to get the appearance of *Homo sapiens,* or rather a group of *H. sapiens*–like organisms often known informally as "archaic sapiens." Around about 150,000 years ago we get the appearance of Neanderthal man, and he lasts until around 35,000 years ago—found mainly in Europe but with some in the Middle East. All told we have about 200 specimens, beginning with that first discovery a year or two before the *Origin* in the Neander valley in Germany (hence the name, since the German for *valley* is *Thal*). Modern humans, that is, *Homo sapiens* like us, were at one point thought all to come after Neanderthals, but now the thinking is that our remains date back almost as far, and there is evidence in some places that modern humans lived together with Neanderthals, without interbreeding—or at least without interbreeding enough to wipe out differences. (A new skeleton, apparently a modern human/Neanderthal hybrid, has just been discovered [Duarte et al. 1999]. This does not prove that hybridization was common or that the offspring were fertile. And indeed the meaning of the discovery itself has been hotly disputed.)

How and in what respects were Neanderthals different from us? This question reveals much of the difficulty of the whole Neanderthal problem: those who want to argue that we are descended from Neanderthals tend to minimize differences, whereas those who argue that we are not descended from Neanderthals tend to emphasize differences.

> His thick neck sloped forward from the broad shoulders to support the massive flattened head, which protruded forward, so as to form an unbroken curve of neck and back, in place of the alternation of curves which is one of the graces of the truly erect Homo sapiens. The heavy overhanging eyebrow-ridges and retreating forehead, the great coarse face with its large eye-sockets, broad nose and retreating chin, combined to complete the picture of unattractiveness, which it is more probable than not was still further emphasized by a shaggy covering of hair over most of the body. (Elliot Smith 1924)

At least some of this is pure fancy. Why on earth should Neanderthal man be covered by shaggy hair like a gorilla? Only in the author's imagination does this occur, but once done the Neanderthal comes out that much more apelike and different from us.

There are differences, and to be candid if anything these differences are such as to give rise to the ape-connection perspective. The Neanderthal is more robust and stronger than we and more significantly does have a face—the lower face particularly—which sticks out more. However, before you pack up and go home, thinking that the Neanderthals are definitely more apelike and could not possibly be our ancestors, I should also mention that if anything their brains tended to be larger than ours. Hence if brain size is a mark of progress, if anything we represent a step backward—although, as you can imagine, a good number of people have jumped in to warn against easy and facial identifications of brain size with intelligence. Often these have been precisely the same people who have been happy to accept and stress the significance of the difference in size between the human male and human female!

Two hypotheses: chief proponents and antagonists, Michigan's Milford Wolpoff and London's Chris Stringer. Remember, on the one hand, we have the "multiregional evolution" model or hypothesis. This sees *Homo erectus* as having evolved in Africa—the fossil findings on this seem to be definite—and then it was this species that traveled far from home, spreading at least through the old world, into Europe the one way and then toward Asia and up into China the other way. Once *Homo erectus* was in place, *Homo sapiens* emerged about 500,000 years ago or a little earlier—a significant point is that there is going to be no sharp dividing line and *Homo erectus* blends gradually into *Homo sapiens.* Then *Homo sapiens* kept on evolving, up through time to the present. By and large the separate populations kept separate, but there was a certain amount of gene flow—interbreeding between populations—thus ensuring that the populations did not go off and evolve into separate species and that there would be a substantial degree of continuity and uniformity in the form that this evolution took. You fit into this picture all of the fossil discoveries that have been made, and of course part of the picture is the Neanderthals being shown as the immediate ancestors of Europeans. Most Neanderthals are found in Europe, so most Neanderthals are now represented by modern-day Europeans. There may well be—there surely will be—some Neanderthal genes in today's Australian aborigines, but most Neanderthal contributions end up right in the places were we find their remains.

On the other hand, we have the "out of Africa" hypothesis. The beginning part is the same as the multiregional hypothesis. We start with the origins of humans in Africa—this was the home of *Homo erectus.* Moreover, it is agreed that *Homo erectus* went traveling around the Old World—Java man shows that that was the case. And these populations did go on surviving and evolving, but gene flow was insignificant or nonexistent and so there were different populations, perhaps well on the way to speciation. Meanwhile, back in Africa about the 500,000 or a-bit-more-years-ago mark, *Homo erectus* was evolving into *Homo sapiens.* Then at some point, around the 100,000 year mark, this population (or perhaps species now) starting moving out—at least some did, although others stayed at home. This group, *Homo sapiens,* spread around the world, and as it did it wiped out the populations of hominids already living there. How it did so is not in itself a matter of great moment—it was not necessarily through violence but could have been through disease or some such thing. Superior technology may have been involved. The point is that *Homo sapiens* did take over, and specifically in Europe this meant the end of the Neanderthals. They did not evolve into us, they are at closest related to us through *Homo erectus,* and they are now extinct.

These are very different hypotheses, starkly so, and would seem to lend themselves readily to test and comparison. One might think one is going to have a textbook case of science in action here; but, although one does in fact have a textbook case, it is rather one that shows just how difficult it can be to test and compare rival models, even when they seem unambiguously clear and different. Most obviously, one has the physical facts, that is, the remains of Neanderthals and the remains of modern *Homo sapiens,* and their relationships or nonrelationships. But as I have pointed out already several times in this chapter, people tend to interpret things in the ways that accord with their own hypotheses. Stringer has started with a number of modern techniques for classification—initially a statistical process known as "multivariate analysis" and more recently a newly refined form of systematics known as "cladistics"—and he finds clear differences between us and Neanderthals. He argues that we are not the same, that transitions are rare or nonexistent (although note the recent hybrid discovery mentioned above), and moreover (and expectedly) the real differences come between European Neanderthals and those Neanderthals found in the near-East and (very rarely) farther afield. Moreover, using increasingly sophisticated methods of dating, he argues that we do not find the Neanderthals giving way gracefully as it were to us, but rather that there is overlap and if anything the two groups evolve in different ways. Instead of converging as one might expect, the two groups stay apart or even move farther away from each other.

Wolpoff will have nothing of this. His philosophical remarks quoted at the beginning of this chapter are a warm-up to a knife through the heart of multivariate analysis, something that we learn he himself had used and discarded (or learned to regard with suspicion) long ago. We are told: "Multivariate techniques are attractive because they seem to give the data an opportunity to speak for themselves. However, there are many problems with the incautious use of these techniques that stem from a variety of sources" (Wolpoff and Caspari 1997, 353). And then: "The danger of using multivariate analyses to address the human origins issue is that the analysis presupposes the solution. When you plug your data into a statistical program, you will get an answer, whether you are using the appropriate statistics or not. It's like adding up the diameters of apples and oranges and taking the average. There *is* an average, but what is it an average of?" (p. 354). So much for that!

In this molecular age, can one use something from that kind of biology to throw light on the two hypotheses? Stringer thinks one can and in fact turns to one of the flashier (that does not seem an inappropriate term) scientific hypotheses of recent years. I refer to the so-called mitochondrial Eve hypothesis formulated by Allan Wilson and others at Berkeley (Cann, Stoneking, and Wilson 1987). Mitochondria are parts of the cell, outside the nucleus. They contain genes (DNA) and are passed on in reproduction. However, the peculiarity is that one gets all of one's mitochondria from one's mother and none from father. By comparing mitochondria in different people and by working out the rate of mutation (mitochondrial DNA mutates up to 10 times faster than nuclear DNA), one can work back to how long it has been since people shared the same great- great- and so on grandmother (the source of the original mitochondria). The amazing finding was that this female—immediately christened "Eve"—the uniting link for all humans on earth, seems to have lived less than 200,000 years ago. Now note what this hypothesis does not say and what it does say. It does not say that at one point there was just one human or hominid female on earth. It does not even say that the human species went through a major bottleneck with just a few members. It does say that, although we are all no doubt descended from many people, we are all descended from this female. (A good analogy is to think of surnames, in a case in which women took their husbands' names and so did their children. Think of four people: Jim White, Mary Brown, Fred Green, and Ann Black. Jim marries Mary and they have two sons. Fred marries Ann and they have two daughters. Sons marry daughters and there are four grandchildren. No bottleneck, all four original people equally related to the grandchildren, but all of the grandchildren with the name White. Hence, Jim is the Eve equivalent.)

Stringer seizes on this hypothesis and argues that it proves his point. Around 200,000 years ago we all had a shared ancestor, which means that we all come from one shared population—just what his hypothesis demands. It was after this that the migrations around the world began. Wolpoff is not convinced, contending that the quality of the work is a bit like the engineer's classical way of finding a solution: "Think of a number and double it. The answer you want is half the total." In any case, in his opinion, the Eve hypothesis is irrelevant to the debate. The multiregional hypothesis admits—insists on—gene exchange between populations. Eve could come at any time. *"Only if human groups were isolated after Eve's time would her age be of importance.* The finding that human populations were connected by low levels of genic exchanges means any age for Eve could be compatible with Multi regional evolution because her DNA type could potentially spread throughout the world at any time" (Wolpoff and Caspari 1997, 309).

What about the archaeological evidence? This is the really dramatic stuff, although by its very nature it is the most tantalizing—how much is lost, how do we interpret it, and so forth. The more you get away from human beings themselves, the more subjective things all become. But, the fact is that the evidence from archaeology—artifacts and so forth—really is very striking and does prima facie tell strongly for the out of Africa hypothesis. Tools start coming in with the arrival of *Homo.* For a long time, these are all pretty crude stone hand tools. What is remarkable is how little change there is for so long. Then with "archaic sapiens," we start to get a significant move in the direction of sophistication. But this is nothing to what we get 100 thousand years or so ago, and increasing as time goes on, intensifying 50 or so thousand years ago. Tools, materials, decorations, and so forth are just levels of magnitude above what they were before. It is very tempting to link this to the arrival of modern humans and to argue that even if there is a little bit of this among the Neanderthals it is because they copied us. It is even more tempting, if we can locate the earliest modern complex tools in Africa, because then there would seem to be some sort of causal connection between tool use and the subsequent migrations and successes of *Homo sapiens* (us).

Not that Wolpoff will accept any of this: "Africa may differ from other areas, but if it does so it is in the extent of its marked regionalization" (Wolpoff and Caspari 1997, 327). This means that you cannot expect to find, and do not in fact find, one culture swamping the human population and taking off from there. The most sophisticated "technologies are local" and moreover "on the whole they do not seem to reflect particularly more progressive behaviours. These and other similarities to much later industries and technologies are short-lived and disappear, hardly the pattern we would expect if they were heralding a new superior, pattern of behavior" (p. 327).

It is starting to be clear that nothing, simply nothing, is going to shift the protagonists at this point. The latest move has been to try to extract the DNA of Neanderthals and, after sequencing it, to compare it to our DNA (Krings et al. 1997). This would seem surely to give definitive answers and some think that it does, showing that we are significantly different from Neanderthals. Others, however, beg to differ. The impasse continues. But probably this is nothing very exceptional in science. The number of times that one side simply collapses and admits that it is wrong is rare indeed in science, as it is rare in real life. Perhaps the revolution in geology in the early 1960s, when people swung from thinking that the earth is stable and the continents unmoving over to thinking that the continents slide around the globe on big plates ("continental drift") is one such case—although there was really no question of two sides persisting. Rather, almost everyone switched over. In conflicts with two sides such as we have over the Neanderthals, we get more the persistence of the debate until people get tired or one side drops out (through retirement and death) or points on both sides are brought into an amalgamation in the middle.

My suspicion is that where you end on this debate depends a lot on where you start. Particularly, if you start from a biological side—a genetical evolutionary side—you are going to be uncomfortable with the multiregional hypothesis. It does not ring true that populations are going to evolve in parallel in time, separate but doing the same things. It is perhaps possible, given the right amount of gene flow, but does not seem likely. The needed assumptions are just a little too ad hoc and helpful to be entirely plausible. If you add to this the fact, stressed repeatedly by Richard Lewontin (1982), that the genetic variation between human populations is comparatively slight, then the case for a recent shared origin seems even stronger. (The mitochondrial DNA variation among humans is smaller than that among apes. This seems to imply that there is a lot of mixing and that for humans our shared ancestry is recent.)

Since I have made so much in this chapter of the ways in which people's extrascientific commitments cloud or at least predispose their judgments, it would be wrong not to mention that people like Lewontin would like very much for something like the out of Africa hypothesis to be proven true. They would then have a full historical reason to back their claims that humans are essentially all one under the skin and that biologically speaking racial differences are slight. On the multiregional hypothesis you can still argue that people are similar, but you start with the fact that they have been apart for a long time, so you are arguing to similarity rather than from it. You rather expect differences to be there. Whatever you might say, you can imagine Henry Fairfield Osborn in the background, nodding approval.

Language and Consciousness

We have gone this far in the chapter without yet mentioning what many people—every philosopher!—would think are the most distinctive and important aspects of our species: the facts that we can talk and that we are conscious. To a certain extent this is cowardice, or perhaps prudence, on my part. Language and thought tend not to get caught in the fossil record, so one had best be silent. But we cannot be completely silent, nor need we be. As you can imagine, language particularly has got caught up in the Neanderthal debate, with the out of Africa proponents arguing that the key difference between us and the Neanderthals is language—we have it and they did not, or at least not to the same extent—and the multiregional proponents arguing that this is unproven, untrue, and not needed anyway!

At least, let me modify things somewhat. No one today who takes evolution seriously wants to deny that human language is a deeply biological phenomenon, and no one who is not in some sense a Darwinian wants to deny that language has adaptive value in communication and so forth and that is why it evolved. Since the work of Noam Chomsky in the 1950s, it has been realized that languages are related with a shared "deep structure" and that they are not rational phenomena, but rather jerry-built, reflecting the constraints of biology and the vagaries of history. It is true that Chomsky himself opposes Darwinism for language, but his students and followers have shown precisely how language is the sort of thing put together by selection (Pinker 1994). But from here on we have difference and debate.

There are at least three ways in which you can approach the question of language. First there is the brain itself. This seems to imply that the growth of language has been a fairly gradual process, at least it does if you equate brain size with language ability. However, if you take organization into account, that is, the parts of the brain actually used in language, the traces left on the insides of skulls suggest that language may have come in a bound or leap, early on—certainly with the arrival of *Homo erectus,* and perhaps even with *Homo habilis* over two million years ago. Whatever else seems clear, by the time you get to *Homo sapiens,* and this includes Neanderthals, language was in play. It had evolved.

If second you go with archaeology, then the implications seem to be that language came in leaps and bounds. As we have seen, you get the development of some tools with the first hominids, *Homo habilis,* a jump with *Homo erectus,* another bigger jump with *Homo sapiens,* and then things go wild with the arrival of modern humans. The implication that has been drawn, especially since this does not reflect brute brain size, is that it reflects developments in language ability. What really marks us off from others, including the Neanderthals, is the fact that we have full and complex language abilities, which we use. Of course, you cannot use this claim as a piece of evidence independent from others in support of the out of Africa hypothesis. Already above we have seen appeal to the archaeological evidence in support of the hypothesis. But one can say that it gives an explanation of what was happening—why it was, in particular, that modern humans were able to succeed so well culturally and the Neanderthals were not. It was that we had language, or rather sophisticated language in a way that they did not.

But why, other than in this rather circular fashion, should we say that we had language and the Neanderthals did not? Here enters the third piece of evidence, supposed fossil remains of our actual physical

Comparison of ape and human vocal tracts

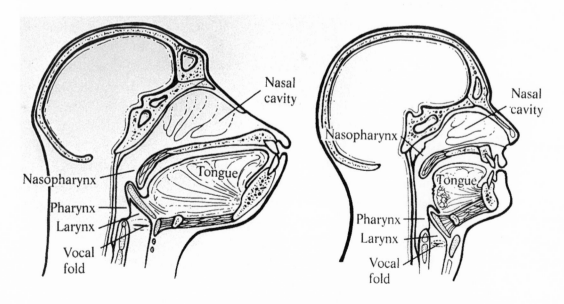

"Going the Whole Orang" • 195

language abilities—that is to say, evidence from the vocal tract: tongue, lips, larynx, and so forth. Of course, most of this does not fossilize, but there are some traces left in the bones of the skull. And in particular, although there is evidence that language does seem to increase with brain power, it is also possible that with the Neanderthals the language ability took a step back. This does not mean that they would be mute, or even back at the level of the apes, but that their language abilities would be far less than ours and possibly less indeed than that of some of their ancestors. (It could just be that Neanderthals came directly from *Homo habilis* so there was no question of their going back. It is just that we went forward and when the two populations met, the Neanderthals were behind us and also behind our ancestors who were older than the Neanderthals.)

And before you object that this cannot be so, it should be remembered that your denial may simply be based on an old-fashioned assumption about progress. You are saying, "Obviously language is a plus, and so the Neanderthals must have had it. They certainly could not have lost it, and given that they had large brains—larger brains than us, remember—they must have developed language abilities independently if not with us." But this is not necessarily so. Things change only if there are good reasons, and (for a Darwinian) good adaptive reasons. Speech of a sophisticated kind is no doubt a jolly good thing to have, but it does have its costs. In particular, grown humans are unable to speak and swallow at the same time, and (as every steak-house owner knows) there is the ever-present danger of choking given the kind of vocal apparatus that we have. Apes do not run into this problem because their vocal tracts are so designed that, although they cannot speak, they do not have openings that lead to choking.

The point is that there is a balance between sophisticated speech and being choke proof, and there was probably a crucial point in human prehistory when we moved across the divide: it was more in the favor of being able to communicate than to be absolutely protected against choking. Although as happens on these occasions, there would have been strong pressure, once the move had been made, to fine tune and perfect things—if you are going to be open to choking anyway, then at least communicate in the best possible way. It could just have been that the Neanderthal branch of *Homo erectus* never reached that threshold and so never made the move across. It evolved in its own ways under its own selection pressures, and then, when the competition came, it lost. (This whole question of Neanderthal speech is another highly controversial area. The claim that the Neanderthals did not have full linguistic powers has been championed by Philip Lieberman [1999]. The discovery of a key bone [the hyoid] possessed by Neanderthals as well as by humans supposedly tipped the balance away from human uniqueness [Arensburg et al. 1989]. The relevance of this finding has in turn been challenged by Lieberman [1994].)

However you interpret the role of language, every evolutionist agrees that the explosion of the brain in size had to be essentially adaptive. Whatever the cost, hominids with bigger brains are better adapted than hominids with smaller brains. But why? Exactly how the brain works and functions has always been a matter of significant debate and dispute. Many people today think that computers are a good analog for brains, and without necessarily making a simple identification—the brain is a computer made of meat, as one joker has said—these people feel that the functioning of the brain is much like the functioning of a computer, as the brain operates somewhat akin to a calculator in processing and using information. Extremely popular is the hypothesis that, as with computers, the brain is built on a somewhat modular pattern. This means that there is no one central mechanism doing everything all at once, in a generic sort of way, but rather there are different parts or units that are put together to perform different tasks. They are connected together to make the whole.

Of course, this does not address the ultimate question, namely, that of consciousness. As you might expect, there are divided opinions on this matter. There are those who, even today, want to deny that consciousness has any great biological significance. Others, relatedly, feel that consciousness is something very recently acquired, and so it cannot have been a major factor in human evolution. The average evolutionist, however, particularly the average Darwinian, feels extremely uncomfortable with such a dismissive attitude. Consciousness seems a very important aspect of human nature. Whatever it may be, consciousness is so much a part of what it is to be human that Darwinians are loath to say that natural selection had no or little role in its production and maintenance.

Whatever position is taken on evolution, no one is denying that consciousness is in some sense connected to or emergent from the brain. The question—at least the question that concerns Darwinians—is whether, over and above the brain, consciousness has some biological standing in its own right. General opinion (my opinion!) is that somehow, as brains got bigger and better during animal evolution, consciousness started to emerge in a primitive sort of way. Brains developed for calculating purposes and consciousness emerged and, as it were, got dragged along. Most Darwinians think that at some point, consciousness came into its own right. Perhaps, then, the causal connection was reversed, and brains were now dragged along, in order to make bigger and better conscious animals.

This raises the question of what consciousness actually does. Why should we not just have a nonthinking machine, which does everything? Is consciousness little more than froth on the top of the electronics of the brain? Is consciousness just an epiphenomenon, as philosophers would say? Slowly but positively, brain scientists do feel that they are groping toward some understanding of the virtues of consciousness, over and above the operation of blind automata. It is felt that consciousness may act as a kind of filter and a guide—coordinating all the information thrown up by the brain. Consciousness helps to prevent the brain from getting overloaded, as happens all too often with computers. Consciousness regulates experience, sifting through the input, using some and rejecting some and storing some. One important brain scientist, referring to this aspect of consciousness as access consciousness, writes as follows:

> Any intelligent agent incarnated in matter, working in real time, and subject to the laws of thermodynamics must be restricted in its access to information. Only information *relevant* to the problem at hand should be allowed in. That does not mean that the agent should wear blinkers or become an amnesiac. Information that is irrelevant at one time for one purpose might be relevant at another time for another purpose. So information must be *routed*. Information that is always irrelevant to a kind of computation should be permanently sealed off from it. Information that is sometimes relevant and sometimes irrelevant should be accessible to a computation when it is relevant, insofar as that can be predicted in advance. This design specification explains why access-consciousness exists in the human mind and also allows us to understand some of its details. (Pinker 1997, 138)

Still, you might complain that this does not explain consciousness in itself. Why do we have "sentience," as we might call it? Why do we have the capacity of self-awareness? To what was the seventeenth-century French philosopher René Descartes referring when he spoke of the *cogito,* as when he said, "I think, therefore I am"? Why is it that what is essentially no more than a bunch of atoms should have thinking ability? Why is it that I am able to write now and to think about what I am doing, and you are able to read what I have written: perhaps agreeing, perhaps disagreeing, perhaps liking what I say, perhaps disliking what I say, but certainly reacting in some fashion or another? I am afraid that at this point, we start to run out of answers. The Darwinian qua Darwinian is reduced to silence. This is not to deny the existence of consciousness. Anything but! "Saying that we have no scientific explanation of sentience is not the same as saying that sentience does not exist at all. I am as certain that I am sentient as I am certain of *anything,* and I bet you feel the same. Though I concede that my curiosity about sentience may never be satisfied, I refuse to believe that I am just confused when I think I am sentient at all!" (Pinker 1997, 148). The point is that as a Darwinian, that is to say as a scientist and an evolutionist, there seems to be no answer. At least, no answer at the moment.

At this time, perhaps it is best to turn to philosophy. Certainly, philosophers have thought much about the problem. Simplifying somewhat, we find two main approaches. On the one hand, there are the dualists. This group includes the great Greek philosopher Plato as well as Descartes, mentioned just above. They argue that consciousness is something altogether different from physical matter. They speak of it as being a substance in its own right: in Descartes's language it was *res cogitans* (thinking substances) as opposed to *res extensa* (material or physical substances). As the language implies, these people take thought or thinking as the mark of the substance of consciousness, as opposed to extension, which is the mark of the material or physical world. On the other hand, there are the monists. Most famously, there was the seventeenth-century Dutch philosopher Benedict Spinoza. He argued that when thinking of consciousness, there is no reason to think that one is considering a separate substance. Consciousness, in some way, is simply a manifestation of the physical world. Spinoza and his modern-day followers do not want to say that consciousness does not exist, or that it is simply material substance in a traditional way. Consciousness is obviously not round, or red, or hard, or anything like that. Rather, consciousness in some sense is emergent from or an aspect of material substance. In other words, the notion of material substance has to be extended, from red and round and hard, to include consciousness.

Most philosophers and scientists today are inclined to monism rather than to dualism. There have been relatively recent defenses of dualism by philosophers and scientists, notably by the philosopher Karl Popper and his friend the brain scientist John Eccles. Since both of these people would have thought of themselves not only as evolutionists but also as Darwinians, clearly one can hold both positions (dualism and Darwinism) at the same time. But there are serious problems with dualism, particularly about how one gets connections between material and thinking substance. Having distinguished them so firmly, it is hard to reconnect the two. For this reason, most Darwinians who think about these sorts of things are inclined to some kind of monism, or (as it is often known today) to some kind of identity theory. They think that body and mind are manifestations of the same thing, and that as selection works on one it affects the other, and as it works on the other it affects the former.

I hardly need say that all of these suggestions raise as many questions and problems as they solve. Philosophers and scientists are working hard toward answers and resolutions. But perhaps this is a point at which we might pull back from the discussion. The important thing from our perspective is that consciousness is a real thing. We are sentient beings. Moreover, consciousness is surely something subject to the forces of evolution, to natural selection in particular. More than this perhaps we need not say, or argue. As with the physical world, take it as a given. It is something wonderful, but commonplace, mysterious, yet familiar. All of these things and a great deal more. We must recognize that all inquiry must start at some point, and perhaps here is one such point. No one ever said that a scientific theory has to explain everything. Although some of my readers will now themselves be inspired to take up the quest. It will be an honorable task to set oneself.

Further Reading

There are lots of good books on human evolution. I really like the writings of Roger Lewin: *Human Evolution: An Illustrated Introduction,* 2d ed. (Oxford: Blackwell Scientific, 1989) and *The Origin of Modern Humans* (New York: Scientific American Library, 1993). A wonderfully opinionated account of the discovery of *Australopithecus afarensis* is *Lucy: The Beginnings of Humankind* by Donald Johanson and Martin Edey (New York: Simon & Schuster, 1981). It would seem that you need a massive ego to be a successful paleoanthropologist (student of human origins).

Don Johanson, the man who discovered Lucy, has that and more. The same is also true of Steven Pinker, who is not only a good psychologist but also a great writer. His *How the Mind Works* (New York: W. W. Norton, 1997) is detailed, informative, and at times very funny. Earlier he had taken on the question of human language in *The Language Instinct: How the Mind Creates Language* (New York: William Morrow, 1994).

In the text I make a somewhat exasperated comment about the World Wide Web, but truly for human evolution it really is invaluable. I do strongly recommend the sites on Piltdown (http://www.talkorigins.org/faqs/piltdown.html) and on Neanderthals (http://thunder.indstate.edu/~ramanank/index.html). The English novelist Angus Wilson wrote a terrific story inspired by Piltdown: *Anglo-Saxon Attitudes* (London: Secker & Warburg, 1956). He transforms the fraud into one about archaeology, but his novel is not only a great read but a penetrating insight into how a fraud might have started as a joke and then taken on a life of its own. I very much suspect that that is what must have happened back there at Piltdown. If I had done it, my first emotion would have been joy at having pulled it off; then horror at the damage I was doing to the subject I loved so much; and finally rank fear that someone might finger me.

8

Genetic Determinism?
Human Sociobiology Arrives

I n 1978, the eminent Harvard biologist Edward O. Wilson—the world's leading authority on the ants—was giving a talk at the annual meeting of the American Association for the Advancement of Science. Suddenly from the audience, a man carrying a glass of water dashed up to the podium, emptying it over Wilson's head. "There, Professor Wilson," he screeched to the noisy approval of a bunch of supporters, "now everyone can see that you really are all wet!" Even Cuvier, at his most combative, never thought of doing anything like that.

This was but an episode in a war that had now been going on for three years, pitting Wilson and his team against the opponents, several of whom were eminent evolutionists in Wilson's own department of organismic biology at Harvard. They were fighting over something that Wilson had labeled "sociobiology": more particularly, they were fighting over the implications of this sociobiology for our own human species. Wilson thought it was the most important move in evolutionary biology since the *Origin*. His critics, many of whom were Jewish and who loathed and feared any attempt to seek biological factors in human behavior and understanding, thought it bad science, morally reprehensible, and politically dangerous. If a little cold water could show the world the evil of Wilson's ways, then so be it.

Let us go back to Darwin and pick up the story there, bringing it down to the present and to the implications of sociobiology for understanding ourselves.

Social Behavior

Charles Darwin always recognized that behavior is as important a part of an animal's being as is its physical form. Biologically speaking, there is little point in having the physique of Tarzan if the only thing you are interested in is philosophy! Right from the beginning, in the *Origin,* Darwin acknowledged the significance of behavior and thought it as much an adaptation formed by natural selection as is any physical feature such as the eye or the hand. Indeed, the very first example that Darwin gives of selection at work in the *Origin* is of wolves hunting deer, and how the different strategies and behaviors might well lead to different physical features. Moreover, Darwin recognized that some of the most interesting and intriguing examples of behavior involve what one might call social behavior, where instead of working flat out to deprive or otherwise harm a competitor or fellow struggler for existence, one works to aid or help one's fellow, especially one's fellow species member. He was particularly interested in the hymenoptera (the ants, the bees, and the wasps), the paradigm of social animals, and in fact devoted a whole chapter to their study.

Now why should social behavior, adaptive social behavior, that is, be particularly interesting and challenging? A mother feeds her offspring. Surely there is no real problem here. If a mother does not feed her offspring, they will die. Although the mother may survive, her reproduction is as truncated as if she were sterile in the first place. Nor is there any real problem when you start to extend the range of social behavior. In a nest of ants, you find the workers (always female) helping the group by feeding the young, or going foraging for food, or acting as soldiers by defending the nest, or a number of other activities. The workers are after all helping their siblings by raising them: also aiding their mother, who is the queen of the nest. Why should there be any worry here? Or indeed, why should there be any worry when an organism helps any fellow species member? After all, surely selection has the good of the group at heart?

But this is precisely the problem. As we saw in an earlier chapter, in the eyes of Wallace, the codiscoverer of natural selection, the mechanism did work for the group. Characteristics, physical and behavioral, work for the group (meaning the species) as much as they work for the individual. In the eyes of Darwin, however, characteristics are adaptively directed toward the individual only, and the group not at all (Ruse 1980). The struggle for existence pits brother with brother and mother with daughter. One can circumvent this only if, in some sense, the social behavior—behavior, that is, that requires cooperation and working with others and perhaps even giving to others—benefits the individual. There is little point, for instance, in a mother harming her daughter—taking all of the food for herself—because then the mother harms herself. Her own reproduction is blocked. What then of the social insects, where one finds that cooperation has been driven to such a degree that the workers are sterile, giving their whole lives to the nest? How can they benefit, who have no offspring of their own?

Darwin was little worried about the sterility per se, for his knowledge of the agricultural world had shown him how selection can (as it were) work sideways, promoting features in nonreproductive animals (geldings and oxen and porkers) through their fertile relatives. But artificial selection is done for our ends. Who can benefit when there is no conscious intention involved? Eventually, Darwin decided that one could treat the whole hymenopteran nest as a kind of supraorganism, with the sterile members as parts of the whole: they exist rather as hands and eyes exist, not for their own sakes but for the sakes of the whole. Darwin was never really comfortable with this, however. But nothing more could be done on the problem, especially in ignorance of the proper principles of genetics (Richards 1987).

One thing that Darwin did always realize is that a significant—and to us humans by far the most interesting—social animal is *Homo sapiens*. We humans have made sociality our speciality. And Darwin was never loath to get right in there and speculate. It is true that there is little in the *Origin,* but the very first records that we have of Darwin's discussing selection (in a private notebook in the late fall of 1838) have him thinking about human evolution and about how some people are brighter than others thanks to natural selection! I do not think that Darwin became an evolutionist because he was obsessed with human beings—unlike quite a few other prominent evolutionists—but there is no doubt but that he thought that human evolution is an important part of the overall story. The *Descent of Man,* published in 1871, was written to deal with human evolution—with, it will be remembered, a particularly significant causal role being given to sexual selection.

The title page of The Descent of Man

THE

DESCENT OF MAN,

AND

SELECTION IN RELATION TO SEX.

By CHARLES DARWIN, M.A., F.R.S., &c.

IN TWO VOLUMES.—Vol. I.

WITH ILLUSTRATIONS.

LONDON:
JOHN MURRAY, ALBEMARLE STREET.

Darwin made it very clear that our social nature just as much as our physical nature (the two of course are very much combined) is the result of a selection-driven evolution. Some races (Europeans particularly) come out over others because they did better in the struggle: generally because the winners had a harder time in the struggle, thanks to the more difficult conditions in Europe than elsewhere, as in Africa. Males differ from females because of the different selective forces: not only do males have different physical characteristics but also they have different emotional and behavioral characteristics. The classes are stratified because of selective pressures. Remember how Darwin gives a long discussion of the virtues of capitalism—just what you would expect from the grandson of Josiah Wedgwood! And there is much more along the same lines. The Darwinian man is a social man is a biological man, and that means evolution through natural and sexual selection. We may have come out on top—Darwin thought that we did—but we are still part of the whole. In fact, for Darwin, coming out on top is precisely a matter of being, like everything else, part of the organic world: there was a race and we won. In this sense, the Darwinian picture is very much part of that progressivist world vision, set off against the Christian providentialist world vision, which latter judges us to have won because we were never part of the race in the first place. For the believer, we humans are the top because God made us that way, in His image.

Move the clock forward rapidly, through a hundred years. By the time of the *Origin*'s centenary in 1959, evolutionary theory in general had made major strides forward. Except in the area of behavior, social behavior in particular. It is true that a number of European workers, the "ethologists," were working on such issues as mate recognition and honey bee activity, but compared (say) to the activity in population genetics or systematics, the area was one of neglect. There were a number of reasons for this. Most obviously, behavior is much more difficult to study than something like morphology. If you are interested in anatomy, you can kill your subject, pop it into formaldehyde, and then pull it out and chop it up when you are ready. Behavior has to be studied on the job, as it were. You can try experimenting, but it is well known that experimental conditions can affect even the most basic of activities—consider how difficult it can be to get animals to breed in zoos (or to stop the breeding of other animals, quite reproductively isolated in the wild). And if you try to study in the wild, then costs and difficulties arise. It is one thing to study the eye color of Drosophila in the lab and quite another to measure breeding activity in a jungle or a desert.

Then again, going against the study of social behavior, there was the rise of the social scientists. They were young, insecure, and jealous of their territory. They were terrified that evolutionary biologists might come down, take over, and hang out a new shingle: "Evolutionary biology (sociology division)." So they resisted any attempt at a takeover or even collaboration. Studies were done on white mice or rats, generalized to other animals, and then it was declared that the uniformity showed that there was no need for a comparative approach! Learning behavior, for instance, was considered quite outside the evolutionary context. An animal could learn to avoid or welcome anything, in any way, at any time. The thought that perhaps one might be more receptive to learning in certain periods and not others was considered slightly silly. (I write now with some bitterness as one who was first introduced to foreign languages at the age of eleven, just the point at which we are now assured the biological door closes firmly shut.)

Then finally there was the human question. Here all sorts of factors worked against an evolutionary approach. Freud, for instance, was himself quite receptive to evolutionary ideas (Sulloway 1979). In his seminal works on human sexuality, he started by stating simply that some people are as they are because of their biology—no need of protective mothers and hostile fathers to do the work. Biology is self-sufficient. But his followers, from personal ignorance (not trained as was he in biology) or from arrogance (who needs biologists?) or from avarice (how can one justify high fees listening to moaning about mother when the genes did it in the first place?), cut out the biological component almost completely. Then the social scientists were full of all sorts of progressivist ideas about changing society, so long as we do the right things. The peak of self-deception was achieved by Margaret Mead, who, so eager was she to show that our Western sexuality has no reflection in innate human nature, allowed herself to be the butt of schoolgirl jokes about Samoan sexuality. Thanks to the influence of such studies (if one might so dignify them) as these, it became accepted wisdom that human beings are infinitely plastic—it is all a matter of the environment.

And as the century went on, hanging over everything was the terrible example coming out of Nazi Germany. In that land, there was the claim that humans are different because of their biology, and from this belief stemmed the most terrible actions and injustices. Jews, gypsies, homosexuals, Slavs, the insane, and more and more groups were judged biologically inferior and subjected to oppression and the lack of liberty and ultimately the final punishment, death. Who could think that a biological approach toward humankind could have any merit whatsoever? Even if it be true that biology might play some role, it has to be minor, and the risks raised by studying it far outweigh any potential benefits. There are some things that are simply best left alone.

But things did start to change, and what began as a trickle soon swelled right out into a torrent. First perhaps came the theory, and this had both a critical side and a positive side. On the critical side, the 1960s started to see a significant shift toward a Darwinian approach to (what became known as) the level of selection, as older assumptions were subjected to withering analysis. At the beginning of that decade, with very few exceptions, the automatic assumption of evolutionists was that natural selection could and did work at all levels—for the benefit of the individual, the group, or the species. An adaptation therefore might help you personally, or it might be of no value whatsoever to you as possessor but of great worth to other members of your species. The ethologists never doubted that this might be the case. Konrad Lorenz wrote a whole book on aggression, arguing that in fights between species members, constraint is always shown because otherwise the species would suffer. A dog will never knowingly kill another dog, because this would be bad for doggyhood in general. And others thought the same. A major work on animal population numbers ar-

Konrad Lorenz

gued that they are regulated by individuals because otherwise one might have overpopulation—bad for the group (Wynne-Edwards 1962). You might benefit from one or two more children, but what if everyone did the same? (There was often an interesting subcurrent, to the effect that humans uniquely seem not to obey group rules, to the detriment of all. We have no means of restraining aggression toward fellow humans, and clearly we cannot contain our sexual passions. We are the naked ape with blood-stained jaws.)

This was now seen as totally fallacious reasoning (Williams 1966). Natural selection has no forethought. It acts only in the present. If an organism benefits in this generation, then so be it, however disastrous the long-term consequences. Consider two species members, the one of which acts purely selfishly by having lots of offspring and the other of which acts purely altruistically by having but few offspring. In the next generation, there will be far more of the selfish member's offspring than the altruistic member's offspring, and so on down the line. Even though some 10 or more generations hence it might be better for all were the altruist to prevail, by then it would be too late. The selfish member's offspring would be the populational norm. The point is that, as Darwin realized, group selection simply cannot work. (In fact, one can show that under certain special circumstances, a group effect can overwhelm an individual effect, but such cases are few and far between.)

On the positive side, the early 1960s was just the time when theoreticians were starting to produce models, showing how individual selection can work and how in fact one can throw light on interesting problems, hitherto insoluble. For the point is that social behavior does occur, and animals do show altruistic inclinations and actions toward one another—usually (although not necessarily always) toward fellow species mates. If one cannot explain this directly through group selection and must therefore rely on individual selection, the question arises as to how this is to be done. And the answer, obviously, is that one must just show that in helping others one is helping oneself. Indeed, one must show that one helps oneself more by helping others than if one did nothing.

Now, in a way, one can follow through fairly directly on this insight. Mothers care for their offspring. Why? Obviously, because the offspring carry on the mother's line. (None of this necessarily happens at the conscious level. Rather, our genes make us do it.) Or let us put matters another way. Natural selection is a matter of making sure that one's units of heredity, one's genes, are represented in future generations. I am fitter than you if a higher percentage of my genes get through rather than yours. But it is hardly a question of my genes as such. Rather, it is a question of copies of my genes. And thus understood, we can say that a mother cares for her offspring because, by so doing, she is ensuring that copies of her genes are transmitted. If the offspring all die without issue, then the genes are stopped dead.

You can generalize this idea, which is precisely what was done by the English, then-graduate student William Hamilton (1964a, b). He reasoned that altruism—helping other organisms—will always pay if those organisms are bearers of the same genes as oneself. One is helping one's own genes in the struggle for existence, vicariously as it were. Or rather, he reasoned that altruism would pay if one could do more for one's genes through such altruism rather than otherwise. A distant cousin will have only a very small proportion of genes in common with you. Hence, there is little point in forgoing one's own reproduction for that cousin, unless you can have very few offspring yourself and that cousin can have many more offspring than otherwise. And this indeed suggests a simple little formula that governs the altruism relationship: essentially, altruism kicks in only when the benefit through help or altruism exceeds the reciprocal of one's blood connection to the beneficiary. As the blood relationship falls away, so it is necessary that the benefits rise accordingly.

Genius is not always recognized at once. Hamilton's thesis supervisor thought so little of his student's insight that he urged Hamilton not to use it in his thesis, for fear of failing! But slowly it was seen for the brilliant move that it is. And what did start the realization of its importance was that Hamilton applied his idea (known now as "kin selection") to that very problem that had stymied Darwin. How is that hymenopteran workers devote their whole lives to the good of others, without breeding themselves? Hamilton pointed out that (as was well known) the hymenoptera have a funny mating system. Whereas females have both mothers and fathers (they are diploid, meaning that they have the usual paired set of chromosomes), males have only mothers (they are haploid, having only one set of unpaired chromosomes). A queen is inseminated but keeps

W. D. Hamilton

the sperm, sometimes for many years. If an egg is fertilized then a daughter is born, but if an egg is not fertilized a son is born.

What this means (as you can see from the diagram) is that although mothers and daughters have the usual genetic relationship of 50 percent (just like humans), sisters are more closely related than normal (75 percent as opposed to the usual 50 percent). This implies, from a selective viewpoint, that female hymenoptera are better employed raising fertile sisters than fertile daughters. The altruism that workers show in the nest is preserved and cherished by natural selection, even though the workers are sterile! In the case of males, they are 50 percent related to mothers and to daughters (they have no sons), and so there is not the same urge to help. Notoriously, male hymenoptera are "drones," good only for breeding purposes. Interestingly (and expectedly), you do sometimes find that "sterile" workers will lay unfertilized eggs that hatch into drones—this is a move that one would expect given natural selection.

To give you some idea of how this thinking first met opposition and then conquered all before it, let me quote to you the full and generous account that Wilson gives of his first encounter with Hamilton's ideas. I do not know of quite anything that gives such a sense of the excitement of scientific ideas or of the way in which science is no re-

The relationships within a hymenopteran family

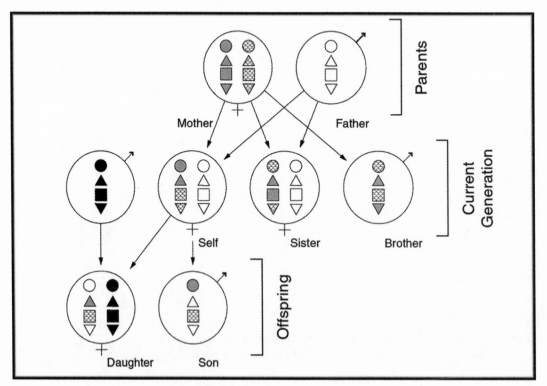

specter of status, only brilliance. Wilson explains that the year was back in 1965, and he was on a train carrying him south from his home in Boston to his field station work in Florida. Keep in mind that Wilson was a Harvard professor, in the same department as Jim Watson of double helix fame (Wilson got tenure before Watson!), and with good reason thinking of himself as the great man in insect biology, before whom all must defer.

> I picked Hamilton's paper out of my briefcase somewhere north of New Haven and riffled through it impatiently. I was anxious to get the gist of the argument and move on to something else, something more familiar and congenial. The prose was convoluted and the full-dress mathematical treatment difficult, but I understood his main point about haplodiploidy and colonial life quickly enough. My first response was negative. Impossible, I thought; this can't be right. Too simple. He must not know much about social insects. But the idea kept gnawing away at me early that afternoon, as I changed over to the Silver Meteor in New York's Pennsylvania Station. As we departed southward across the New Jersey marshes, I went through the article again, more carefully this time, looking for the fatal flaw I believed must be there. At intervals I closed my eyes and tried to conceive of alternative, more convincing explanations of the prevalence of hymenopteran social life and the all-female worker force. Surely I knew enough to come up with something. I had done this kind of critique before and succeeded. But nothing presented itself now. By dinnertime, as the train rumbled on into Virginia, I was growing frustrated and angry. Hamilton, whoever he was, could not have cut the Gordian knot. Anyway, there was no Gordian knot in the first place, was there? I had thought there was probably just a lot of accidental evolution and wonderful natural history. And because I modestly thought of myself as the world authority on social insects, I also thought it unlikely that anyone else could explain their origin, certainly not in one clean stroke. The next morning, as we rolled on past Waycross and Jacksonville, I thrashed about some more. By the time we reached Miami in the early afternoon, I gave up. I was a convert, and put myself in Hamilton's hands. I had undergone what historians of science call a paradigm shift. (Wilson 1994, 319–320)

In the spirit of Hamilton, other models were devised showing how sociality could be preserved given individual selection. The American Robert Trivers (1971) came up with "reciprocal altruism": this is a case of "you scratch my back and I will scratch yours." Here animals cooperate because they both benefit. The interesting thing about this kind of situation is that it can cross species boundaries, and Trivers gave interesting examples drawn from fish, where predatory species will nevertheless refuse to attack other fish that specialize in cleaning them of parasites. The predators get cleaned and the cleaners get a good meal. Both sides benefit, which would not be the case if the predators immediately ate the cleaners (or ate the cleaners after a cleaning).

And the English evolutionist John Maynard Smith (1982) systematized much of our thinking about social situations by making heavy use of game theory. He showed how selection can promote certain equilibrium situations, where everyone gets the most that is possible, given that everyone else is trying to do the same. These Evolutionarily Stable Strategies are what one finds when one has mixed populations, with different members trying to achieve their ends by different means (or where every member has alternate means to achieve the same end). Most famously, we have a species consisting of hawks and doves (that is to say, some members show hawklike behavior and other members show dovelike behavior, where these translate as fighting as opposed to fleeing). A population of hawks would just tear each other apart, and so a dove would be selectively favored. A population of doves would never threaten, so a hawk would be selectively favored. But given costs and gains (if the cost of fighting is slight, then being a hawk is better than if the cost of fighting is heavy), one can show that the population will achieve a stable equilibrium at certain ratios—different behavior will be held in the population by (individual) selection.

John Maynard Smith

This kind of theoretical thinking stimulated the empiricists: experimentalists and naturalists. Realizing that one would have to spend much more time in the wild or more care over experimentation than previously, the new models nevertheless inspired people to try to see if one could measure behavior in action and draw solid conclusions. One of the most successful workers was the English evolutionist Geoffrey Parker (1978), who made his mark through a series of papers stemming from his thesis project: the behavior of one of nature's less prepossessing members, the dung fly, *Scatophaga stercoraria* (Ruse 1996, 1999). Parker spent many long hours in fields, surrounded by herds of cattle, following the brutes around and waiting for them to defecate. He knew that the flies' reproductive behavior is focused on the waste that the mammals expel and leave behind, and he soon found that there are standard behavioral patterns followed by male and female flies. First, it is the males who fly in, looking about for fresh cow pats. Then the females arrive and are seized by the males, who mate with them vigorously. After this, the now fertilized females fly onto the pats and lay their eggs. Sometime later, the larvae hatch and bury down into the cow feces, thus able to feed abundantly from the rich nutrients within which they find themselves embedded. It was in the variations and elaborations on these standard patterns that Parker found much scope for scientific investigation: an opportunity that he exploited to the full, with diligence and intelligence.

Geoffrey Parker

To do science successfully, you need hypotheses to build models. With the interest in reproductive behavior, Darwin's mechanism of sexual selection seems the obvious tool of inquiry. Today, were one to suggest this, there would be no great surprise. But even thirty years ago, this was not so. For the century after the *Origin* and the *Descent,* for all that Darwin himself had championed sexual selection, it had never been a great success. We saw Wallace's unsympathetic reaction, and while few biologists shared Wallace's enthusiasm for spiritualism (which lay ultimately behind his rejection of the mechanism), ever fewer wanted to credit Darwin with having found in sexual selection a significant factor in evolutionary change. Indeed, it seems fair to say that for the first two-thirds of the twentieth century, sexual selection (if considered at all) was thought but a minor and not significant form of the general mechanism of natural selection. But with the move to a more individual-based perspective on the working of selection, sexual selection—which is an individual-versus-individual form of selection par excellence—started finally to come into its own. So perhaps after all it was no great surprise that, for all that he was working in the late 1960s, Parker's focus was very much on sexual selection, particularly on the competition between males, who in the dung flies outnumber the females by four or five to one.

Particularly interesting and significant was the distribution of the males, who had to choose a site carefully—fresh pats of dung tend to be far too liquid for safety—where they could be reasonably sure of finding a female and yet able to defend themselves against the needs and desires of other males. "Males should be distributed between zones in such a way that all individuals experience equal expectations of gain. Hence the proportion of females captured in a given zone should equal the proportion of males searching there, assuming that all females arriving are equally valuable irrespective of where they are caught" (Parker 1978, 219–220). What made Parker's work so exciting was the fact that his predictions about spacing held so exactly: observation and theory differed not at all in any significant way. Although the work could not be ended with just one set of findings, for Parker soon discovered that he was dealing with a fluid situation. As the first round of mating comes to an end, successful males must now balance their labors between guarding their females from other males and going off in search of new females. Hitherto unsuccessful males, meanwhile, must move from trying to find mates in their own right to trying to pry females away from successful (copulating or postcopulating) males.

As time goes by, the cow pats form a skin and thus are less hazardous for the flies—in particular, females can start moving toward the pats in order to lay their eggs. One expects therefore that the males will move from a general wide distribution around a field toward the cow pats. Parker found here that theory and findings were close but not quite as close as before. Perhaps the smells of new droppings crowd out the smells of older droppings and the males have to adopt strategies to allow for this: "However, this information about new droppings may be obtained by spending time in the grass upwind" (p. 225). One important assumption in all of this is that the females are able to sustain and use multiple matings—a one-time mating with a male does not exhaust a female's supply of unfertilized eggs. In fact, turning now to experiment in the laboratory, Parker found that the last male in any mating succession was by far the most successful from an evolutionary perspective. By sterilizing selected males, by encouraging multiple matings, and by counting the fertile eggs that females laid, Parker (1970) discovered that an amazing 80 percent of the eggs laid by any particular female were fertilized by the sperm of the last male to mate with her.

It is indeed truly the case that it pays a male to take over a female or—if he already has a female—to protect her from intruders and competitors. Apparently, there is a balance between protecting the female one has already and finding another female where one will be the final male: "In conditions of high male density during reproduction and with mating followed immediately by oviposition, in *S. stercoraria* evolution seems to have favored the optimum active *copula* duration with inhibition of separation so that pairing is extended for guarding the female during oviposition" (Parker 1970, 785).

Work like this—theoretically ambitious and predictively fertile and successful—convinced evolutionists that their theory was moving forward rapidly. It is not surprising that people began to think in terms of synthesis, and in 1975 Edward O. Wilson attempted just this. But Wilson's book, *Sociobiology: The New Synthesis,* was more than just a compilation. It was a manifesto. A call to arms. Speaking of Hamilton's work as revolutionary, *Sociobiology* is a flamboyant, oversized tome with lots of pictures. The title of the first chapter, "The Morality of the Gene," sets the tone, and the opening words continue in the same vein:

> Camus said that the only serious philosophical question is suicide. That is wrong even in the strict sense intended. The biologist, who is concerned with questions of physiology and evolutionary history, realizes that self-knowledge is constrained and shaped by the emotional control centers in the hypothalamus and limbic systems of the brain. These centers flood our consciousness with all the emotions—hate, love, guilt, fear, and others—that are consulted by ethical philosophers who wish to intuit the standards of good and evil. What, we are then compelled to ask, made the hypothalamus and limbic system? They evolved by natural selection. That simple biological statement must be pursued to explain ethics and ethical philosophers, if not epistemology and epistemologists, at all depths. (p. 3)

Although the pace of the book never slackens, as Wilson warms to his task the melodramatic language and imagery do recede somewhat. Having first shown how he sees sociobiology as a natural outgrowth of evolutionary ecology, Wilson turns to a detailed and comprehensive discussion of the causal factors behind animal sociality. We get a basic discussion of the principles of evolution and of genetics, coverage of the sorts of models introduced earlier in this chapter (kin selection, reciprocal altruism, and so forth), and—an area where Wilson himself is a world expert—much attention paid to methods of animal communication, especially chemical communication between insects using so-called pheromones (p. 231).

Then, after brief overviews of such topics as aggression, dominance, caste systems, sexuality, parental care, and the like, Wilson turns to what he obviously considers the real meat of the book: a survey moving upward through the animal social world from colonial microorganisms through insects and lower mammals right up to our own species: "Man: From Sociobiology to Sociology." And here we find (as we have surely been led to suspect all along) that the inclusion of *Homo sapiens* is no last-minute decision, something done for completeness, as it were. We humans in a way are the raison d'être of the whole book.

> To visualize the main features of social behavior in all organisms at once, from colonial jellyfish to man, is to encounter a paradox. We should first note that social systems have originated repeatedly in one major group of organisms after another, achieving widely different degrees of specialization and complexity. Four groups occupy pinnacles high above the others: the colonial invertebrates, the social insects, the nonhuman mammals, and man. Each has basic qualities of social life unique to itself. Here, then, is the paradox. Although the sequence just given proceeds from unquestionably more primitive and older forms of life to more advanced and recent ones, the key properties of social existence, including cohesiveness, altruism, and cooperativeness, decline. It seems as though social evolution has slowed as the body plan of the individual organism became more elaborate. (p. 379)

A paradox, but one that is a challenge rather than a barrier (p. 382). Tearing into the "culminating mystery of all biology," namely just how it is that humans have been able to stem the flow away from social integration, we learn that as humans evolved away from the apes, they reached a threshold. Arguing consciously with metaphors drawn from cybernetic thinking, Wilson reasons that at such a point a kind of feedback situation kicks in. There is suddenly an incredibly rapid and significant form of evolution, where it is appropriate to apply a kind of autocatalytic (self-driving) model of change. In a two-stage process, first humans got up on their hind legs and walked, thus freeing hands for tool use, and then sequentially there was an explosion of brain size with corresponding increase in mental power. This opened the way to a kind of cultural evolution, which in some sense takes us humans up and beyond our biology—although only in a sense, for Wilson makes it very clear that in other senses our biology remains (and always will remain) very important. If biology does not control the course of culture directly, then culture feeds back into the biology so that the genes in some fashion track the social. Either way, today and forever, much that we think and do is under genetic control—training and the environment are important but never all-important.

Had sociobiology—as, from now on, we can call the study of the evolution of social behavior—simply confined itself to the nonhuman part of the animal world, then although it would have been celebrated in biological circles, one doubts that it would have been heard of elsewhere. After all, dung flies do not have the sex appeal of dinosaurs. But with the move to humans, even though (or perhaps especially though) this was following in the grand tradition of Charles Darwin himself, it was bound to be controversial. And matters were not helped by works that followed up on Wilson's *Sociobiology*. First, there was a popular account of the whole new rising discipline, an account coming from the pen of a young English student of the evolution of social behavior. *The Selfish Gene* by Richard Dawkins (1976) was as provocative as it was flamboyant as it was compulsively readable. Through a brilliant use of metaphor—who can take group selection seriously after genes have been thus labeled "selfish"?—Dawkins brought home the moves and developments of this new branch of science in ways more vivid and compelling than would have been achieved by thick volumes of normal academic prose. In fact, Dawkins himself said little about the application of sociobiology to the human realm—his own discussion of the subject rather suggested that cultural evolution is something apart from biological evolution—but his examples and language spoke for the cause. You know perfectly well what he thinks of male/female differences after you learn that females have two choices in the battle of the sexes: either they can take the "he-man" strategy, trying to get themselves the strongest and sexiest male, or they can take the "domestic bliss" strategy, trying to get themselves a mate by providing the best home life.

Richard Dawkins

If all of this was not enough, Wilson himself then reentered the scene with a more popular book of his own. *On Human Nature* (1978), a work for which Wilson won the Pulitzer Prize, is an extension of the discussion of the last chapter of *Sociobiology*, given to exploring precisely how it is that biology yet impinges on human consciousness and action.

In the case of sexuality, for instance, we learn that male animals tend toward aggression whereas females toward being "coy" and to looking for males who will remain and help with child-rearing. "Human beings obey this biological principle faithfully" (p. 125). Nor is alternative sexuality overlooked. Perhaps, for instance, homosexuals are like worker ants: they themselves might not do so very well in the reproductive stakes, so their efforts are diverted into helping close relatives raise more offspring. Although, of course, all humans are into some forms of help or altruism: "Individual behavior, including seemingly altruistic acts bestowed on tribe and nation, are directed, sometimes very circuitously, toward the Darwinian advantage of the solitary human being and his closest relatives" (pp. 158–159). And so we come to religion. This is no afterthought but is central to Wilson's conception of the functioning human: "The highest forms of religious practice, when examined more closely, can be seen to confer biological advantage. Above all they congeal identity" (p. 188). In belonging to a group, we find meaning in our lives. At the same time, we further individual self-interest.

Critical Reaction

Enough! Although Wilson was genuinely surprised at the reactions his work invoked, one might say that whatever his other faults (real and imaginary), he was being dreadfully naive if he thought there would be no response at all. Social scientists surely were going to be made tense, and those for whom any kind of biological approach to humankind was highly suspect (especially Jews) were going to react negatively. And this is precisely what did happen, especially in America where these things were felt somewhat more deeply. Sociobiology, especially the human variety, was accused of just about every sin under the sun. What gave the debate—if one can thus dignify an all-out war of words and personalities—a particularly keen edge is the fact that among the most prominent critics of Wilson's vision of sociobiology were several of his colleagues at Harvard, including at least two in his own department: the molecular geneticist Richard Lewontin and the paleontologist and soon-to-be-famous popularizer of things evolutionary, Stephen Jay Gould. They were candid about what drove them. If Wilson's program works, then we are right back in the 1930s or earlier.

> Just as theories of innate differences arise from political issues,
> so my own interest in those theories arises not merely from

their biological content but from political considerations as well. As I was growing up, Fascism was spreading in Europe, and with it theories of racial superiority. The impact of the Nazi use of biological arguments to justify mass murders and sterilization was enormous on my generation of high school students. The political misuses of science, and particularly of biology, were uppermost in our consciousness as we studied genetics, evolution, and race. That consciousness has never left me, and it has daily sources of refreshment as I see, over and over again, claims of the biological superiority of one race, one sex, one class, one nation. I have a strong sense of the historical continuity of biological deterministic arguments at the same time that my professional mature research experience has shown me how poorly they are grounded in the nature of the physical world. I have had no choice, then, but to examine with the greatest possible care questions of what role, if any, biology plays in the structure of social inequality. (Lewontin writing in Schiff and Lewontin 1986, xiii)

Human sociobiology was accused of being false. How can one argue for the significance of the genes when culture clearly changes at rates that far exceed the speed at which genes can take effect? For instance, the rise, triumph, and fall of Islam took less than a thousand years, a mere blink in the evolutionary life of the genes. There is simply no way in which biology can have been significant in this event. In any case, there is nothing but ignorance on the part of the human sociobiologists in their speculations. Who is to say that there are "gay genes," making people into homosexuals? And is there any evidence that homosexuals do in fact help their relatives to have and raise more offspring?

Human sociobiology was accused of being unfalsifiable—a charge not entirely consistent with the one that it is false, but no matter. There is simply no way in which its flabby claims can be put to check. All exits are covered:

When we examine carefully the manner in which sociobiology pretends to explain all behaviors as adaptive, it becomes obvious that the theory is so constructed that *no tests are possible*. There exists no imaginable situation which cannot be explained; it is *necessarily confirmed by every observation*. The mode of explanation involves three possible levels of the operation of natural selection: 1. classical individual selection to account for obviously self-serving behaviors; 2. kin selection to account for altruistic or submissive acts toward relatives; 3. reciprocal altruism to account for altruistic behaviors directed toward unrelated persons. All that remains is to make up a "just-so" story of adaptation with the appropriate form of selection acting. (Allen et al. 1977, 24)

The *Just So Stories* were the fantastical stories made up by the English author Rudyard Kipling to account for the elephant's nose and other strange features of the living world. The critics claim that, just as it is silly to take seriously Kipling's claim that the nose resulted from a crocodile's pulling on a normal nose, so it is equally silly to take seriously the sociobiologist's claim about such things as sexuality and religion. No matter what counterevidence you produce, the sociobiologist will have an answer.

Sociobiology was (and is) accused of being sexist, racist, classist. It is argued that it is just not true that men are naturally aggressive and women naturally coy and retiring. This is all in the imagination of the evolutionists, and then read into nature—at which point it is read right back out and triumphantly held up as objectively validated! Even if it be true of the nonhuman animal world, the point about humans is that we are flexible and can escape our biology. No one can deny that many societies, including our own, treat women as inferior and that even women internalize this treatment and behave as if they are second in major, desirable characteristics to men. But this is culture and biology has no part in it. From the viewpoint of our genes, it is a level playing field.

Racism is another point of contention. Here, Lewontin—drawing on his expertise as a population geneticist—has had much to say. And bluntly, the conclusion must be that there is simply no evidence for the broadscale differences supposed by the sociobiologists. "Of all human genetic variation, 85% is between individual people within a nation or tribe" (Lewontin 1982, 123). Indeed we can put matters more strongly than this. Suppose there were a world holocaust and only Africans survived. We would have lost only 7 percent of human variation. In fact, "if the cataclysm were even more extreme and only the Xhosa people of the southern tip of Africa survived, the human species would still retain 80% of its genetic variation" (p. 123). Lewontin can hardly deny that some differences between peoples may have an adaptive basis. Take body shape, a plausible candidate if anything is. There are (as we know from other animals, including birds) good adaptive reasons for minimizing surface area in cold regions.

"Typically, the Eskimo has a large, chunky torso and short limbs, whereas the Dinka of Africa is tall and thin with very long arms and legs" (p. 128). Yet even this gets guarded treatment by Lewontin: "Although these trends seem to make good sense, there is no actual demonstration that they subserve greater survival and reproduction" (p. 128). And generally, Lewontin has nothing but contempt for those who would tie a strong link between human traits and personalities and abilities and our biology, our Darwinian adaptively shaped biology in particular. Not only can you not separate out genetic factors and environmental pressures but the very attempt is founded on a mistaken view of the way biology and nature interact. Genetic causes and environmental causes are truly "inseparable" (p. 68).

An African and an Eskimo compared

Classism (thinking social classes are found, not made) also is a major problem with human sociobiology. "Since the seventeenth century we seem to have been caught up in this vicious cycle, alternately applying the model of capitalist society to the animal kingdom, then reapplying this bourgeoisified animal kingdom to the interpretation of human society" (Sahlins 1976, 101). Darwin himself led the way, arguing that capitalism is a good thing, because then there will be people freed from toil and strife and able to devote their time to other things. Of course, this was grossly self-serving, and for every Darwin who did work there are a hundred parasites who do nothing for their livings. All that sociobiology does is give false justification to evil social and civil iniquities. "What is inscribed in the theory of sociobiology is the entrenched ideology of Western society: the assurance of its naturalness, and the claim of its inevitability" (p. 101).

There was more, but this will give you a good flavor of what things were about. Looking back, some twenty-five years later, it all has a bit of a quaint look about it. There is nothing academics like more than a good fight. After all, this is what we are paid to do! And I cannot say—I ought not say, since I myself was right in the thick of it—that there were no good points made or matters of real issue. There is no doubt about it that some of the work produced by the sociobiologists was sexist, at least as judged by the exacting conditions of political correctness that prevail in universities today. Wilson and his friends did rather assume that males are naturally superior and that females like it that way, and leave matters at that. They did jump way ahead of their evidence, and then congratulate themselves on a hard empirical slog well done. And they were determined not to let a little counterevidence stand in their way. To be candid, they were determined not to let a massive amount of counterevidence stand in their way.

On the other hand, they were by no means as guilty as the critics would claim. Negatively, no one then was always that sensitive about male/female differences. Positively, there has long been a tradition in evolutionary studies of taking a feminist stand, from Alfred Russel Wallace on. Sure enough, sociobiology produced its own feminist counter. Sarah Blaffer Hrdy argued (in *The Woman That Never Evolved* [Hrdy 1981]) that females conceal ovulation, so males do not know exactly when the females are fertile. Hence, males cannot be assured of the paternity of their social offspring unless they stay around and help. In other words, it is females who make the running in the battle of the sexes and the males who are led along on a string. Far from men being on top, it was the women.

Sarah Hrdy

In other respects also sociobiologists could clear themselves. It is true that a lot of their speculations about homosexuals were based on very little evidence, but how else do hypotheses start? As it happens, a number of people went out and worked hard on this very issue, and now there is some considerable evidence pointing to the fact that there may indeed be genes coding for homosexual orientation (LeVay 1996). And as far as racism is concerned, it is simply not true that sociobiologists went in for the "blacks got rhythm" sort of thinking. In any case, as they pointed out, in the twentieth century more harm has been done by those who think that you can change human nature through social engineering than through any beliefs in genetic engineering. Certainly, no sociobiologist thought we were invariably "genetically determined" to do what we do, as the critics often claimed.

Things have rather subsided now. Indeed, those who keep arguing are looked upon more with embarrassment than with respect. Quarrels grow old—they may not be solved, but they get boring. So let us now, in concluding this chapter, turn to the most important question of them all. Criticism and countercriticism, where stands sociobiology today? Where stands human sociobiology today?

The Contemporary Scene

Animal sociobiology has never really been in question, except it has been attacked as a support for human sociobiology. Let me simply make reference to one celebrated piece of research, a study of the dunnocks or hedge sparrows, small birds that live in the hedgerows and bushes of the English towns and countryside. The British ornithologist Nicholas Davies (1992) has discovered that they have the most remarkable set of sexual customs, something that would not be out of place within the covers of *Playboy*. They have breeding arrangements that go all the way from polygyny (where one male will have two or three mates) through monogamy to polyandry (where one fe-

Nicholas Davies

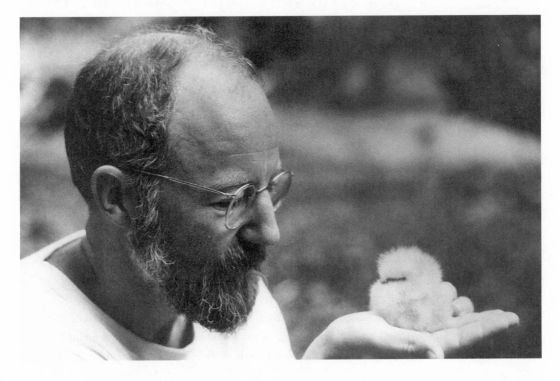

male will have two or three mates) and even to a form of polygynandrous relationship (the polite name for group sex, where several males mate up with several females).

Why? Because then there are selective advantages, given the particular circumstances. If the situation is such that a male can service two or three mates, and the food stuffs are there such that the females can benefit from having an alpha male (or the females and other males cannot prevent the male from acting as he does), then we get polygyny. And corresponding reasons for the other sexual arrangements. How can Davies be so sure that he is right? Because he has used the most modern of molecular techniques, so-called DNA fingerprinting (the very same that is now used in murder trials), to trace genetic relationships. He finds that he can track, just about exactly, the time that individual birds spend helping with offspring with the genetic relatedness of the males to these offspring. A male who has fathered the whole brood puts in the time to help at the nest—all of his time if it is his only brood, and proportionately if there are others. Conversely, other males give no help—except if there was a chance that they contributed to the brood. Just what one expects given individual selection.

Davies also goes on to discuss the question of parasitism.

The dunnock is a favourite host of the cuckoo in Britain, with about 2% of nests being parasites. Individual female cuckoos specialize on one host species. Experiments with variously coloured model cuckoo eggs show that the degree of host-egg mimicry exhibited by the different cuckoo gentes [*Gens*, plural *gentes*, means a particular group or race related by descent.] reflects the degree of egg discrimination shown by their respective hosts. Unlike other gentes, dunnock-cuckoos do not lay a mimetic egg, as expected from the fact that, in contrast to other hosts, dunnocks show no egg discrimination.

Nevertheless, dunnock-cuckoos still lay a distinctive egg, different in shade from the other cuckoo gentes. Experiments provide no support for predation as an important selective pressure. Either selection by secondary hosts, or by cuckoos themselves (for an egg which is cryptic in the nest) may be involved.

It is unlikely that dunnocks accept nonmimetic eggs because rejection is peculiarly costly for them or of less benefit than for other hosts. Experimental parasitism of species which have no history of interaction with cuckoos shows that before parasitism occurs hosts exhibit no rejection of eggs unlike their own. Dunnocks may, therefore, be recent victims of the cuckoo, lagging behind in their counteradaptations to a new selective pressure. (Davies 1992, 234)

You can see how questions are asked and solved, using natural selection, in a way that would have altogether delighted Darwin. Here is an extension of evolutionary thought—selection-based evolutionary thought—of the most exciting and fertile kind.

But what about humans? Do we really have any significant scientific advances, or is it all a question of hypothesis and supposition and wishing? Do we get anything more than "just so" stories? Let me tell you about one case where the sociobiological approach really does seem to have paid major dividends. It concerns murder or, as the authors call it, "homicide." Two Canadian psychologists, Martin Daly and Margo Wilson, have made an extensive study of homicide: because they are Canadian, they are particularly interested in the differences between homicide in Canada and homicide in the United States. What fascinates them—what fascinates Canadians particularly—is that here we have two countries, with very similar lifestyles, running right next to each other, and yet they have dramatically different homicide rates. The American rates are four to five times higher, or even more.

There are some fairly obvious reasons for this, the most prominent being the availability of guns. By and large, Canadians do not have access to guns, certainly not to handguns, the means by which so many Americans kill each other off (and, to be fair, themselves also). But when it comes to certain kinds of killing, even if the proportions are different, the patterns between Canada and the United States are similar, chillingly similar. In particular, Daly and Wilson concerned themselves with cases of parents killing children. This should not happen in the best-ordered Darwinian worlds—you are stopping your genes in their path. The psychologists hypothesized that perhaps what was happening was that stepparents were doing the killing—especially stepfathers (who are the ones more likely to be living with someone else's children). And the data proved their hypothesis in an incredibly strong fashion. "Daly and Wilson found that step parenthood is the strongest risk factor for child abuse ever identified. In the case of the worst abuse, homicide, a stepparent is forty to a hundred times more likely than a biological parent to kill a young child, even when confounding factors—poverty, the mother's age, the traits of people who tend to remarry—are taken into account" (Pinker 1997,

434). Why is this? "Stepparents are surely no more cruel than anyone else. Parenthood is unique among human relationships in its one-sidedness. Parents give, children take. For obvious evolutionary reasons, people are wired to want to make these sacrifices for their own children, but not for anyone else." The answer is obvious. "The indifference, even antagonism, of stepparents to stepchildren is simply the standard reaction of a human to another human. It is the endless patience and generosity of a biological parent that is special" (p. 434).

These are incredible findings. Moreover no one can accuse Daly and Wilson of twisting the facts to their own end, and even less can you claim that the findings are "obvious" and that you hardly needed a sociobiological perspective to find what they found. So remarkably strong were the biases of social science—supporting the belief that biology has nothing to do with family relationships—that neither in the United States nor in Canada did the authorities keep track of biological versus social parental connections. They simply did not know whether stepparents were more likely to commit violence. Hence, Daly and Wilson had to go out and gather their own data: data that did indeed prove precisely what they predicted. Moreover, the findings about stepparental abuse are backed by other findings about the nature of homicide and the people who do commit it far more often than others. For instance, it turns out that the real killers are young males, who have little to lose and much to gain by violence—precisely what sociobiology predicts. The new enthusiasm for locking people up for long periods of time may indeed have a significant effect on violent crime statistics. It is not that the perpetrators are cured by imprisonment or deterred by the threat of punishment. It is rather that when they get out, they are no longer all that young, with all of that testosterone pumping through their systems.

One example cannot be definitive. But Daly and Wilson's work—part now of a whole spectrum of studies—does show that human sociobiology can work and can throw incredible light on human nature and behavior. Nor is it easy to see that it is infected with all of the faults that critics found endemic of early exercises in human sociobiology. It is certainly not sexist, for instance, and neither is it racist—the figures seem to hold whatever the ethnic group—or classist—the figures are not affected by poverty, for instance. The work is falsifiable and as far as one can see, true not false. In short, a paradigm of good work on problems of social science. Only time can tell whether it will prove to be one of a very few such studies that really work or whether it will prove to be the norm. But for the time being, the future looks promising—one might even say "bright"—for human sociobiology.

Further Reading

A good place to start on the general theory of sociobiology is Richard Dawkins's sparkling book, *The Selfish Gene* (Oxford: Oxford University Press, 1976). I would call it a popularization but really it is more than that. The metaphors he uses, especially that of the title, have entered into the scientific discourse and stimulate researchers into looking at problems in altogether new ways. The human side of things is the subject of a provocative essay by Edward O. Wilson: *On Human Nature* (Cambridge: Harvard University Press, 1978), a work for which Wilson deservedly won a Pulitzer Prize. I myself wrote a quick survey of the field dealing not only with the science but with many of the philosophical undercurrents: *Sociobiology: Sense or Nonsense?* 2d ed. (Dordrecht, Holland: Reidel, 1986).

Recent work on human sociobiology often conceals its intent, preferring to go under such names as "evolutionary psychology" or "human behavioral ecology." A rose by any other name . . . One work, however you name it, that should be on your reading list is by historian Frank Sulloway. In *Born to Rebel: Birth Order, Family Dynamics, and Creative Lives* (New York: Pantheon, 1996), Sulloway argues that much of what we do and think relates to our order in the family and that this in turn relates back to our biology. He argues that first borns tend to be conformists and not open to new ideas, whereas those down the family order tend to be much more adventuresome and open to new challenges intellectual and physical. From here he branches out to explain not only the genius of Charles Darwin (fourth child and second son) but also the nature of the Protestant Reformation and the course of the French Revolution. I have to say that I buy very little of what Sulloway claims, but it is fun trying to show him wrong and if you are down the order you might find merit in what he claims. (The fact that I am a first born has nothing to do with my judgment. I simply do what my parents told me.)

I believe that some of the most interesting and significant implications of sociobiology will be for my own discipline of philosophy, especially trying to answer questions in what is known as epistemology ("What can I know?") and ethics ("What should I do?"). In my *Taking Darwin Seriously: A Naturalistic Approach to Philosophy,* 2d ed. (Buffalo: Prometheus, 1998), I explore some of these avenues in a preliminary sort of way. This has been a somewhat controversial book—most philosophers are not keen on the idea that evolutionary biology might be the key to unlocking the secrets of their inquiry. So for somewhat different perspectives turn first to Daniel Dennett's racy (albeit overly long) *Darwin's Dangerous Idea: Evolution and the Meanings of Life* (New York: Simon & Schuster, 1995), a book that managed to offend just about everyone (except me and Richard Dawkins), so it must be saying something right. Then look at *Unto Others: The Evolution and Psychology of Unselfish Behavior* (Cambridge: Harvard University Press, 1998), a work coauthored by philosopher Elliott Sober and biologist David S. Wilson. This is a book that tries to resuscitate the notion of group selection over individual selection, a project in my opinion on a par with King Canute's trying to stop the tide from entering. (Unlike Sober and Wilson, Canute knew that what he was doing was futile and was simply trying to show his sycophantic courtiers that he was not capable of miracles.) You may end by thinking that Sober and Wilson are right and Ruse and Dennett are wrong, but what I want you to see is how today philosophers of very different convictions are nevertheless turning to evolutionary biology for insight into their philosophical problems.

And in a related way, novelists are doing the same. *Enduring Love* (Toronto: A. A. Knopf Canada, 1997), by the English Booker Prize–winning novelist Ian McEwan, is a fascinating exploration of sociobiological ideas. The hero is a science writer who is obsessively tracked by a young man who suffers from a form of homoerotic obsession known as de Clérambault's syndrome. The story is the account of how the hero reacts to this pressure: not very well in fact, for he ends up losing his girlfriend and shooting (not fatally) his stalker. But in the course of the account, McEwan explores the ways in which we are all in a sense prisoners of our biology, leading half lives midway between reality and illusion—and how escaping from this state can be dangerous for ourselves and destructive on our relationships. At the same time, however, McEwan shows how this escape can move us to acts of true nobility, beyond our animal natures, and how real love can be achieved. (The title comes from Saint Paul's First Epistle to the Corinthians. "Love bears all things, believes all things, hopes all things, endures all things.")

9

Challenges to Orthodoxy:
Alternatives to Darwinism

The best-known evolutionist active today is Stephen Jay Gould, of Harvard University. His books are read and enjoyed by millions. But if he is looking for glory and praise from his fellow evolutionists, he is out of luck. You expect that philosophers will sometimes turn a little nasty. That comes with the job. The less we connect with the real world, the more choleric we become. But you do not expect such bile of the leading evolutionary game theorist, John Maynard Smith, a man whose boyhood years at England's leading private school (Eton College) reflect in the courtesy and charm he shows in conversation and in writing. Yet, writing in the *New York Review of Books*—a place, admittedly, where unbalanced emotion is the norm rather than the exception—he suddenly swung from his allotted task (a mild review of something on another topic) and started declaiming against Gould and his false and sloppy thinking.

> Gould occupies a rather curious position, particularly on his side of the Atlantic. Because of the excellence of his essays, he has come to be seen by non-biologists as the preeminent evolutionary theorist. In contrast, the evolutionary biologists with whom I have discussed his work tend to see him as a man whose ideas are so confused as to be hardly worth bothering with, but as one who should not be publically criticised because he is at least on our side against the creationists. All this would not matter, were it not that he is giving non-biologists a largely false picture of the state of evolutionary biology. (Maynard Smith 1995, 46)

Sometimes a dignified silence, however difficult, is a strategy preferable to all-out counterattack. Such is not Gould's way. Labeling people like Maynard Smith and Richard Dawkins as "Darwinian fun-

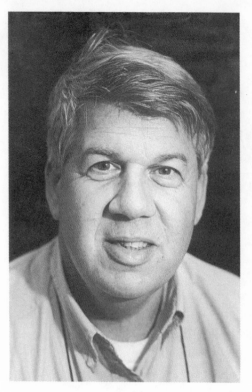

Stephen Jay Gould

damentalists," Gould lamented that although Maynard Smith has "written numerous articles, amounting to tens of thousands of words" about Gould's work, whereas those were "always richly informed," now alas he has been seduced into adaptationist fanaticism.

> He really ought to be asking himself why he has been bothering about my work so intensely, and for so many years. Why this dramatic change? Has he been caught up in apocalyptic ultra-Darwinian fervor? I am, in any case, saddened that his once genuinely impressive critical abilities seem to have become submerged within the simplistic dogmatism epitomized by Darwin's Dangerous Idea [i.e., all-powerful natural selection], a dogmatism that threatens to compromise the true complexity, subtlety (and beauty) of evolutionary theory and the explanation of life's history. (Gould 1997, 37)

As we shall learn, there are different levels to this quarrel, but let us start with the most obvious level—that of the science.

Punctuated Equilibria

The year of the centenary of the *Origin,* 1959, was the heyday of Darwinian natural selection. After years of neglect and denial, finally the significance of selection as a mechanism was being recognized, in America as well as Britain. Great and long were the celebrations, with honorary degrees being handed out like candy to all of the major figures in the field. It is therefore no surprise that, when Stephen Jay Gould began his career in the mid-1960s as a paleontologist, specializing in the evolution of snails (Gould 1969), he was an orthodox Darwinian. Confirming this, an earlier review paper on problems of relative growth showed how things considered nonadaptive can be fitted readily into a selectionist framework that can be extended to explain nonadaptive characteristics (Gould 1966). But in a sense, American Darwinism was always skin deep—remember how Spencer had been a far greater influence—and, for all that George Gaylord Simpson labored in Darwinian fields, paleontology was always on the edge of the pasture. The fact of the matter is that paleontology cannot use selection directly, as can the student of today's organisms, such as the sociobiologist. Selection is not a tool of research where you can go out and discover and test and come up with results. You are working at a distance—a very long distance—with evidence (fossils) that is spotty and incomplete and very very dead. You are always having to take somebody else's exciting ideas and see if they do anything for you.

Those who know of the self-confident personality of Stephen Jay Gould—he is not about to take a back seat to anyone—could have predicted that he would not tolerate this. Before long, Gould would be moving forward to make his own mark on evolutionary studies. This mark would make paleontology a central focus of attention, arguing that the evolutionist needs paleontology not just for establishing the fact of evolution and for ferreting out the path of evolution but also for discovering the true nature and full extent of the causes of evolution. Expectedly, in the early 1970s, this prediction came true. Together with a former fellow graduate student, Niles Eldredge, Gould began pushing forward a supposedly all-new perspective on the paleontological record—a perspective that Gould and Eldredge somewhat inelegantly labeled "punctuated equilibria."

The two young palaeontologists started with the fact that traditionally, the course of evolution is seen to be one of smooth gradual change. This is something that comes about simply because natural selection makes sudden change highly improbable. The only way in which organisms can stay in adaptive harmony with their surroundings is by changing only minutely in each generation. Therefore, any apparently sharp breaks in the fossil record should not be explained in terms of major jumps from one form to another but should be put

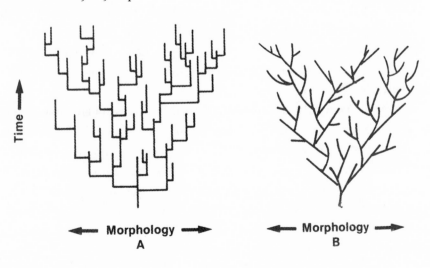

Time

◄— Morphology —►
A

◄— Morphology —►
B

Punctuated equilibria (left) compared to phyletic gradualism (right)

down to the incompleteness of the record and so forth. What El-dredge and Gould argued, to the contrary, was that the paleontologi-cal record is in fact much better and stronger than most people allow, and that hence a causal explanation must be found to explain this. One must accept that there are long periods of relatively little evolu-tionary change—periods of equilibrium, or stasis—broken, or punc-tuated, by rapid moves from one form to another. "The history of life is more adequately represented by a picture of 'punctuated equilibria' than by the notion of phyletic gradualism. The history of evolution is not one of stately unfolding, but a story of equilibria, disturbed only 'rarely' (i.e., rather often in the fullness of time) by rapid and episodic events of speciation" (Eldredge and Gould 1972, 84).

The controversial and exciting part of the Gould-Eldredge the-sis was that an explanation can indeed be found. And interestingly, at this point, far from wanting to break from conventional (American) neo-Darwinism or the synthetic theory, Gould and Eldredge argued that it is precisely this theory itself that has the resources to explain the paradox! To make their case, the paleontologists turned to the ideas of Dobzhansky's associate, the major ornithologist and system-atist Ernst Mayr. Some years previously, in order to explain specia-tion (the fact and causes behind new species), Mayr (1954) had pro-posed what he termed the "founder principle." According to Mayr, speciation results from a small group of organisms getting broken off or isolated from the main species population. Simply because of the new circumstances in which they find themselves, the members of this subpopulation start to evolve rapidly away from the parental form. In addition, argued Mayr, given the masses of genetic variation that occur naturally in any population, any small subpopulation will necessarily be atypical with respect to the whole group. There will therefore be a kind of shaking down as the members get used to each other and learn to do with much reduced genetic resources. Within the "founder population," there will be what one might call a "genetic revolution."

Mayr certainly thought of himself as being fairly orthodoxly Darwinian in his claims about speciation, although with hindsight one can see that what he was proposing was something much more in the spirit of Sewall Wright's shifting balance theory than Darwin's theory of the *Origin*. (Sewall Wright thought it was the shifting balance the-ory!) Mayr was arguing that a certain randomness, which occurs be-cause of the breaking off of the subpopulation, is the crucial factor in the forming of new species. One has, as it were, a kind of genetic drift writ large. But whatever the true lineage of Mayr's ideas, this hypothesis was highly congenial to Eldredge and Gould. It suggested that new species will form very rapidly, not in the neighborhood of their immediate ancestors, but in new areas. You have species A in one place and then, almost overnight as it were, you have species B somewhere else. This could just be the kind of jerky fossil record that Gould and Eldredge thought was the true story to be read from the rocks. "If new species arise very rapidly in small, peripherally isolated local populations, then the great expectation of insensibly graded fossil sequences is a chimera. A new species does not evolve in the area of its ancestors; it does not arise from the slow transformation of all its forebears" (Eldredge and Gould 1972). In addition, the two paleontologists liked the way that Mayr was making the dynamics of populations (rather than the dynamics of isolated individuals) absolutely central to the evolutionary process. In the eyes of these paleontologists, factors operating over large periods of time, involving groups of organisms, yield the crucial causal keys needed for a full understanding of evolutionary processes. Here, for all that they drew on Mayr, Gould and Eldredge were starting to stand against population geneticists in the Dobzhansky tradition: scientists who looked at microevents often involving just a few individuals.

Yet at this point, although Gould was starting to embrace some ideas with but a loose connection to real Darwinism, he was not presenting himself as a dramatic revolutionary. This was to change in the next decade as Gould began to take a stronger and stronger position, setting himself more and more in opposition to prevailing orthodoxy. Why did he do this? There were a number of reasons. Undoubtedly, one was the fact that in the 1970s Gould immersed himself in a huge reading program in the history of biology. This was in preparation for *Ontogeny and Phylogeny,* his major work that appeared in 1977. Part history and part science, *Ontogeny and Phylogeny* argued that traditional links between embryology and phylogeny are better taken than people in the twentieth century had been prepared to recognize. At the same time—and perhaps in major part because of his reading program—Gould was growing increasingly sympathetic to elements of German evolutionism. He responded particularly warmly to that tradition going back, through Haeckel, to the morphology of the early nineteenth century that had so upset Cuvier: *Naturphilosophie.* Gould embraced with enthusiasm the *Naturphilosophen*'s emphasis on form rather than function, their insistence that what really counts when studying organisms is the architectural nature of the underlying ground plan, or

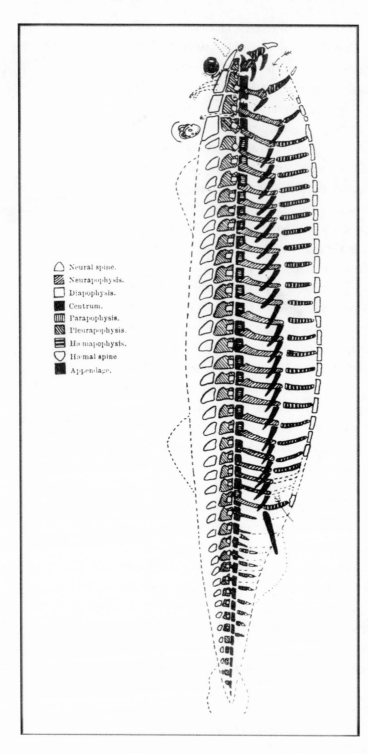

Neural spine.
Neurapophysis.
Diapophysis.
Centrum.
Parapophysis.
Pleurapophysis.
Hæmapophysis.
Hæmal spine
Appendage.

Richard Owen's
picture of the
vertebrate Bauplan
(which he called an
"archetype")

Bauplan (Russell 1916). He liked the turn to homology and the retreat from what the German thinkers regarded as a rather superficial cherishing of selection-caused functionality.

From this, it was but an easy step for Gould to move right into an attack on all-embracing adaptationism. Notoriously, in 1979, writing with a colleague in the department of organismic biology at Harvard, the population geneticist Richard C. Lewontin, Gould produced an article arguing that much to be found in the organic world bears little or no direct connection to adaptive advantage (Gould and Lewontin 1979). Gould, with Lewontin, argued that there are significant constraints on development: these constraints forming and molding organisms in nonadaptive ways. And, simply as part of developmental processes, even when selection is at work there are bound to be a great many nonadaptive byproducts. Much that seems to have purpose probably exists for no end-related reason whatsoever. With Lewontin, Gould drew attention to the triangular areas at the tops of pillars in medieval churches, things that they labeled *spandrels* (although it turns out that the true technical name is *pendentive*). These triangles—one finds them in St. Marks

Church in Venice, as well as on the roof of King's College, Cambridge—are often used as vehicles for wonderful mosaics or carvings. They seem therefore to have a direct adaptive function. But, indeed, they really are simply part and parcel of the architectural constraints that were involved in medieval church building.

Gould and Lewontin argued that, analogously, many organic characteristics have a no true adaptive significance. The human chin, for instance, seems to be something with a purpose. Surely, if naught else, it is part of the design of the face for sexual attractiveness. But, in fact, detailed study shows that the chin is really something that comes about simply as a result of trying to put together other adaptive facial features: the jaw and the teeth and so forth. Seeming purpose should never be equated simplistically with genuine purpose.

The spandrels of San Marco

In King's College Chapel in Cambridge, for example, the spaces contain bosses alternately embellished with the Tudor rose and portcullis. In a sense, this design represents an "adaptation," but the architectural constraint is clearly primary. The spaces arise as a necessary by-product of fan vaulting; their appropriate use is a secondary effect. Anyone who tried to argue that the structure exists because the alternation of rose and portcullis makes so much sense in a Tudor chapel would be inviting the same ridicule that Voltaire heaped on Dr. Pangloss: "things cannot be other than they are . . . Everything is made for the best purpose. Our noses were made to carry spectacles, so we have spectacles. Legs were clearly intended for breeches, and we wear them." Yet evolutionary biologists, in their tendency to focus exclusively on immediate adaptation to local conditions, do tend to ignore architectural constraints and perform just such an inversion of explanation. (Gould and Lewontin 1979, 583)

We are now at the end of the decade (1980). Gould was on a roll. He was mounting an all-out assault on the synthetic theory of Theodosius Dobzhansky and his colleagues. Gould (1980a) went so far as to argue that the synthetic theory is "effectively dead." At the same time, punctuated equilibria—which was now becoming more and more identified with Gould alone—was breaking entirely from any connections with conventional evolutionary thought. In particular, it was being presented now as an outright saltationary theory, that is to say as a theory where large jumps (presumably brought about by macromutations) are the key factors in evolutionary change. There was an expressed likeness for "hopeful monsters": organisms that take phylogenies directly from one form to another form. Drawing on his deep knowledge of evolution's history, Gould was bringing forward evolutionists from the past who were supportive of saltationism: evolutionists who, so Gould maintained, had been unfairly belittled or denied credit simply because they were out of tune with the ideology of the then prevalent Darwinism. The synthetic theory, so he claimed, was little more than an extension of nineteenth-century liberalism, with its fondness for gradual change rather than revolution.

As you might have expected, conventional evolutionists—those working on fast-breeding organisms and concerned more with microevolution than with macro changes—started to get very tense. Here was a very public evolutionist—Gould's *Ever since Darwin,* published the same year (1977) as *Ontogeny and Phylogeny,* was a runaway best-seller—telling the world that their theory was not true science but merely washed-up Victorian ideology. G. L. Stebbins, the botanist member of the cohort who put together the synthetic theory, together with Dobzhansky's student Francisco Ayala, wrote an influential paper pointing out that natural selection is sufficiently powerful to bring about all of the so-called saltationary changes that Gould was demanding (Stebbins and Ayala 1981). In addition, these critics argued that although selection may seem fairly leisurely in the eyes of an individual human, from the perspective of geological time it is more than sufficiently rapid to bring about any conceivable macro changes: both those recorded and those not recorded directly in the fossil record. In other words, as Darwin and his followers had always argued, the gaps in the record are as much artifactual as genuinely representative of things that truly happened.

Continuing their counterresponse, these doughty defenders of tradition pointed out that no Darwinian has ever claimed that the course of evolution is always as smooth and gradual as is implied by Gould's caricature of their theory. It has always been recognized that the pace of evolution is something that speeds up and slows down, according to many different factors. There are impinging conditions imposed both from without the organic world, geological factors, for instance, and impinging conditions imposed from within the organic world, competitors and the availability of desirable ecological niches, for instance. It is true that Darwinism demands that immediate change be gradual—there is indeed no place for hopeful monsters—but over the time scales recorded in the fossil record, there is no reason at all to expect uniformity. "Living fossils" such as horseshoe crabs have persisted over hundreds of millions of years. Other organisms have evolved very rapidly. And in any case, the saltationists of the past, worthy scientists though they may have been in their time, are now simply outdated and wrong.

Gould has never been one to acknowledge directly that he is mistaken or even that he is walking on dangerous ground. There was certainly to be no dramatic retraction of any of the claims that he had made when he was writing at his most vehement level. However, over the next decade—that is to say, through the 1980s—in many respects, Gould did start to pull back from the more extreme positions that he had taken or floated. Not entirely accurately, he now denied that he had ever made the extreme claims ascribed to him. In particular, he denied strongly that he had ever been an outright saltationist. Gould (1982) now started to argue that he was not so much against Darwinism as such, but that what he had been advocating and would continue to push for was a kind of expanded Darwinism. This would be a vision where natural selection and adaptation are indeed very important aspects of organic life and of the evolutionary process. A vision, however, where there is a perceived need for the supplementation, sometimes dramatically, of selection by other processes.

More specifically, in Gould's opinion what one has now (at least, what one needs now) is less a single-level theory—as apparently was true of the synthetic theory—and more something that is hierarchical. The image here is of the Catholic church, with its different levels from the parish priest right up to the pope. Likewise in evolutionary theory, argued Gould, we need a layered perspective, going from bottom to top. Neo-Darwinism is good and right, as far as it goes, but it speaks only to a kind of midlevel to the hierarchy. Beneath natural selection working on individual organisms, one has a microlevel that involves molecular biology. Here, at this molecular level, it is pertinent to note that a number of theoretical biologists, particularly Japanese population biologists, have argued that there is ubiquitous randomness: what came to be known, naturally, as molecular drift. It is a well-known fact that at this molecular level, there is a great deal of redundancy. Different molecules encoding the DNA produce the same cellular products. Hence, there is every reason to think that these differences lie below the forces of natural selection and simply drift from one form or ratio to another. (The classic statement of this thesis can be found in Kimura 1983.)

Then, argued Gould, above the microlevels of individual selection, one has macrolevels involving vast periods of time. Here, other new forces come into play. And here, at this macrolevel, the expertise of the paleontologist comes into its own. One sees that individual selection makes no major difference and that such things as constraints on development start to be the major determining factors. Perhaps some of the ideas raised in the spandrels paper are important here. Initially, a certain *Bauplan* is the all-important constraint on what an organism (or a group of organisms) is and must be. A threshold is reached, and there is a rapid change from one *Bauplan* to another—a change that has nothing to do with natural selection, being rather a shuffling of the internal structure (morphological, biochemical, whatever) of the organism. Then selection comes back into play, refining and elaborating on the new form that has been produced. It is all rather as if a kaleidoscope had been shaken, and a new picture emerges from parts that had been fragmented and reassembled.

Crucial to this whole way of looking at things is the belief that what is going at this upper level simply cannot be explained in terms of the lower levels. Gould (like Lewontin) has long been an ardent critic of what he labels "reductionism": the assumption that the key to understanding the upper levels of reality lies in delving ever more deeply into the lower levels of reality. Gould does not deny that this assumption can be the basis of very fruitful inquiry—in ecology, it may well be the vital method of investigation—but he is adamant that it is very dangerous if taken as an all-determining metaphysical principle. Sometimes one can and should try for an understanding at an emergent level—at a higher hierarchical level. And here the higher simply cannot be reduced to or explained away at the lower level. Specifically with respect to evolution at the macrolevel, one has things happening that cannot be explained at microlevels. Dobzhansky and his fellows were just plain wrong. Genetics, the science of the micro, must be supplemented by paleontology, the science of the macro. To argue otherwise is to slip into the dreadful sins of Panglossianism or the building of "Just So" stories (things encountered in this and the last chapter).

For the last decade or more, Gould has been refining his position, trying to build on and develop his own ideas, while at the same time wearing down the opposition: wearing down the Darwinian opposition, that is. One paper deals with the shapes of snail shells, showing that certain atypical forms of the shells—so-called smokestack shells—are a function of constraints on growth, rather than Darwinian selection as the synthetic theory would argue.

The usual Cerion height to width ratio is less than 3. At extreme sizes (dwarf and giant) Cerion occur with higher height:width ratios. Gould calls these "smokestacks." Gould theorizes that the snails remain in the second stage of shell development, resulting in either more whorls with no change in size (dwarves) or whorls of increased height but no change in number (giants). Gould claims that the snails are stuck in a genetically programmed development phase and that the smokestacks are not themselves an adaptation.

> Evolution is a balance between internal constraint and external pushing to determine whether or not, and how and when, any particular channel of development will be entered. Natural selection is one prominent mode of pushing, but most engendered consequences of any impulse may be complex, nonadaptive sequelae of rules in growth that define a channel. Most changes must then be prescribed by these channels, not by any particular effect of selection. Natural selection does not always determine the evolution of morphology; often it only pushes organisms down a preset, permitted path. (Gould 1984, 191–192)

Another paper, coauthored by Gould, deals with the replacement in the same ecological niche of one organic form by another (Gould and Calloway 1980). Gould's claim is that such a replacement might as well be nonadaptive as anything fueled by selection. We may have "ships that pass in the night." To assume otherwise is simply to make a dogma of Darwinism. And yet a third paper deals with specific forms of nonadaptive characteristics, things that Gould has labeled "exaptations" (Gould and Vrba 1982).

A major contribution to the cause was Gould's *Wonderful Life: The Burgess Shale and the Nature of History,* a book published in 1989. On the surface, this is a work about soft-bodied organisms (dating back to the Cambrian) found fossilized in the Rockies of Western Canada. There are all sorts of strange forms, truly sparking one's imagination and seeming to defy orthodox classification. But the telling of the tale is only one part of what Gould is about. Truly, indeed, this is a work with a mission. Gould uses the Burgess Shale to launch an attack on what he sees as an incorrect picture of the history of life, an incorrect picture that has been brought illicitly into evolutionary studies by enthusiastic Darwinians. A particular bugbear of Gould is the idea of evolutionary progress—our old friend of upward change, from monad to man. He thinks this is a truly false picture of history, which is rather one of randomness and chance and lack of any significant direction. Certainly, humans came last. If they did not, we would not be around now to tell the tale. But we are not the finest culmination of a directed process. Like everything else, we just happened. And the fossils of the Burgess Shale show that this is so. There are all sorts of weird and wonderful forms, all now extinct with

The discoverer of Burgess Shale, Charles Doolittle Walcott

very few exceptions (one of which may be a vertebrate predecessor), and any one of these might have been the progenitor of today's organisms. It was just chance that it all went one way rather than any other. Life has no ultimate meaning and history shows this. Those who think otherwise, Darwinians particularly, are just plain wrong.

A denizen of the Burgess Shale (Marrella)

This continues as Gould's theme, even up to his most recent major book, *Full House: The Spread of Excellence from Plato to Darwin* (1996):

> If one small and odd lineage of fishes had not evolved fins capable of bearing weight on land (though evolved for different reasons in lakes and seas), terrestrial vertebrates would never have arisen. If a large extraterrestrial object—the ultimate random bolt from the blue—had not triggered the extinction of dinosaurs 65 million years ago, mammals would still be small creatures, confined to the nooks and crannies of a dinosaur's world, and incapable of evolving the larger size that brains big enough for self-consciousness require. If a small and tenuous population of protohumans had not survived a hundred slings and arrows of outrageous fortune (and potential extinction) on the savannas of Africa, then *Homo sapiens* would never have emerged to spread throughout the globe. We are glorious accidents of an unpredictable process with no drive to complexity, not the expected results of evolutionary principles that yearn to produce a creature capable of understanding the mode of its own necessary construction. (p. 216)

Taking Things Apart

Whether or not evolution has many different levels or layers, Gould's arguments most certainly do. So let us take them apart and see what we get. At one level, the most basic level, you may say that we have a scientific argument. Is he right that the fossil record is as jerky as he claims and that this is something that a Darwinian cannot handle or explain? Whatever else, one can certainly say that Gould has drawn attention to the question of the rates of evolution. Intense effort has been expended on the path or course of evolution as revealed through the fossil record and on its putative support for the theory of punctuated equilibria. And the answer, I am afraid, is one of extreme ambiguity! Indeed, perhaps by now you might have been expecting that I would say this, because several times before in this book when we have come up to crucial points of decision, I back away and say that the facts cannot decide! Although an exaggeration, there is some truth in this. But I think it probably tells you more about science than it does about me. (Although my father used to complain that I could not open my mouth without telling you something about me.)

The truth is that when scientists hold different positions, it rarely is simply one of the physical facts. Both sides can summon up facts to suit their respective causes: Cuvier points to function, Geoffroy to form; Dobzhansky to heterozygosity, Muller to homozygosity; Stringer to Neanderthal differences, Wolpoff to Neanderthal similarities. The facts are not irrelevant, anything but, yet they are not decisive. And certainly this is the case here. There are cases where evolutionary change seems to have been very rapid indeed—so rapid, that it surely qualifies as sudden or jerky in the terms demanded by punctuated equilibria. It seems likely that the evolution of fish (cichlids) in East African lakes qualifies here—one can show that speciation has been so rapid an event that even if there were fossilization, it would be invisible in the record. (Williamson 1985). There are cases where evolutionary change seems not to have been so very rapid—slow enough, in fact, that the changes do come through in the fossil record. This seems true of the evolution of certain mammals, for instance. And there are cases where, depending on your inclination, you can interpret the record one way or the other. The human fossil trail seems to fall into this camp. It is not that punctuated equilibria theory is wrong and that the Darwinian alternative (what Gould calls "phyletic gradualism") is right, or conversely. Rather it is that the fossil record simply is not decisive.

But this is not the end of the argument, for there are other levels of debate. Just as I am always arguing that the facts are not decisive, so I am also always arguing that philosophical differences really count. I will not disappoint your expectations, for I do think that they are very important here. One thing that may seem important is Marxism. Notoriously, Gould has boasted of his connections to this philosophy—we are told that he learnt it "at his daddy's knee"—and he has certainly drawn attention to the way in which the Marxist view of world history is one of rapid revolutionary change, rather than gradualism (which Gould links with the liberal philosophy that was Charles Darwin's). Also, the antireductionism—seeing different processes at work at different levels is Marxist—is a translation of Engels's law of quantity to quality. (Lewontin, who coauthored the spandrels paper, is an ardent Marxist. See also the comments in Gould and Eldredge 1977.)

But although I am sure that this is important, I doubt that it is all-important, even if we discount the fact that since Stephen Jay Gould became *the* Stephen Jay Gould, he has been rather backtracking his earlier influences and enthusiasms. In line with what we have seen, more pertinent to Gould's thinking, I suspect, is that whole Germanic approach to biology (which was, naturally, shared by Marx and Engels). It is the approach of the *Naturphilosoph,* who thinks that form takes precedence over function, who thinks in terms of hierarchy, whose philosophy of history is one of dialectic, swinging from one pole to another. Add to this a good swig of Herbert Spencer—the very name, "punctuated equilibria," reeks of the old man. More seriously, the obsession with equilibrium is very much a Spencerian concern for evolution (as opposed to Darwinian, where it plays no essential role whatsoever). And certainly in some of his writings Gould has shown a liking for the notion of "homeostasis," an idea developed on Spencerian lines in the 1930s by the physiologist Walter B. Cannon, supposing that organisms get themselves into a kind of balance and have a natural tendency to stay or return to the beginning point.

But the one big problem here with Marxism, *Naturphilosophie,* and especially Spencerianism is the matter of progress. All three of these philosophies are deeply progressive, with humankind as the culmination at the top. Gould has spent twenty years arguing against precisely this. How then can one suppose any significant links? Two points are relevant. First, Gould was not always an antiprogressionist. In fact, up to and including the writing of his *Ontogeny and Phylogeny* in 1977, he was in favor of progress, a process apparently triumphing with our own species. Then Gould swung round against the idea. Which brings in the second point, namely that this was just at the time of the heated human sociobiology debate, a matter on which Gould was as committed negatively as was his colleague Richard Lewontin. Gould—like Lewontin both a Marxist and Jew—saw Wilson's science as being a terrible travesty of the way in which real science should be performed. He saw it as a threat to all that he held sacred and something to be opposed with all his might.

There were various ways that Gould set about opposing biological progress. Some were less than subtle. I have told how Gould fingered Teilhard de Chardin for the Piltdown hoax (Gould 1980b). It is hardly contingent that Teilhard has been one of this century's greatest boosters of biological progress. If Teilhard could be removed from the scene as a hoaxer, then there surely would be a trickle-down effect against progress. More openly in his campaign against biological progress, Gould was led to write a book, *The Mismeasure of Man* (1981), detailing the ways in which biological approaches to humankind have had a long and ugly history of prejudice and bias. One should expect no more from human sociobiology. Another move by Gould was that taken with Lewontin and promoted from thenceforth by Gould. Since sociobiology is so deeply Darwinian, so deeply adaptationist, then a general attack on this is a particular attack on human sociobiology. The spandrels paper and its fellows are attempts to show that there is more to evolution than adaptation, and hence the very project of human sociobiology—so thoroughly adaptationist—is misconceived. And then a third move is the denial of progress. Gould is quite open that he believes in the possibility of social progress and that he sees one of its greatest barriers to be thoughts of biological progress, which latter he takes to be deeply Darwinian and very much part of human sociobiology. He thinks that the thought of and hope for biological progress has been behind many moves to oppress blacks and Jews and women and others. If there is progress, then some have to be higher than others, and this would be a natural and proper state of affairs. But obviously such a conclusion is unacceptable. Hence, the opposition to biological progress and to pure Darwinism that is seen to be part and parcel of the package.

There is one final item that should be added and then the case will be complete. Gould is a paleontologist. In the eyes of the general public, this is what evolution is all about: fossils, dinosaurs, Lucy, and all of that. But as you must now realize, this is not at all the way that professional evolutionists see things. To them, paleontology is just the thing that they have had to escape in order to raise the status of their science. To get out of the museums and away from a quasi-religious system of hypothetical phylogeny building, they have had to turn to tight, mathematical, experimental, causal studies of fast-breeding organisms like fruitflies. I would hardly want to say that dinosaurs are an embarrassment, but even now there are echoes of the past. The past decade or so, for example, has seen a very public and indecisive debate about the origins of the birds (Feduccia 1996)—are they descended from the dinosaurs or from other nondinosaur reptiles? If you cannot answer something as basic as this, what hope of a real quality science?

It is symptomatic of the state of affairs that when, in the early 1980s, Gould began suggesting that one must invoke one-step changes in organisms to account for the fossil record, he was slapped down and into place by the geneticists. Paleontology must do what it is told by the geneticists, rather than conversely. But now, with punctuated equilibria theory, the case is changed. Geneticists must sit up and take notice. Not only does paleontology have its own level or levels of understanding—levels that cannot be eliminated (reduced away) by slick appeals to genetics—but there are dimensions where paleontology can actually tell genetics what it can and cannot do. Equality is now in sight. And anyone who thinks that something like this is not of extreme interest to a person with the ego of Stephen Jay Gould, simply does not know the man. Upgrading his subject has always been high on Gould's list of things to do. No one wants to spend their professional lives in a subject that is regarded with disdain, if not contempt—the sociology of the life sciences. If Gould's program succeeds, if people do accept the need for an expanded Darwinism, then at long last paleontology will come into its own. It can stand shoulder to shoulder with genetics rather than lurk unobtrusively in the background, coming forward only when called. The title of a talk Gould gave back in 1982 tells all: "Irrelevance, Submission and Partnership: The Changing Role of Paleontology in Darwin's Three Centennials, and a Modest Proposal for Macroevolution." (The three centennials were for the birth of Darwin in 1908, the publication of the *Origin* in 1959, and the death of Darwin in 1982. We Darwinians like centennials.)

It is this, as much as anything, that accounts for the bitter note in John Maynard Smith's criticism quoted at the beginning of this chapter. The trouble is that people are starting to take Gould seriously, and that rankles. It rankles also that Gould does not fight his battles just in the professional journals, where only professional scientists would take notice. He gets into the public arena, with his monthly column in *Natural History,* and then in collections and monographs, as well as many other places, notably the influential *New York Review of Books.* For Maynard Smith, geneticist and sociobiologist, this is all the wrong way around. Gould should be judged against the standards set by Maynard Smith and his fellows and should not try to get around difficult points with philosophy and rhetoric. He should be more respectful of and appreciative toward the ideas that have been developed and inherited. And he should not remind the world of the shaky status of so much evolutionary theorizing for so long. It is not just that Gould's ideas are wrong. It is that they are presented as position of reason and tolerance and common sense, and the outside world believes him. That really irritates.

Hierarchy Theory

Let us pull away from the motives and countermotives, charges and countercharges. The really important question is whether Darwinism—an ultra Darwinism, which pushes selection without hesitation or apology—is enough, or whether one really wants and needs more to get a full understanding of the evolutionary process. Start with the level below the physical characteristics (the phenotype), the molecular level. At this level (as Gould noted), it has been hypothesized that selection can have only a minimal effect. Even if natural selection produces the hand and the eye, the molecules making everything up are another matter entirely. Selection may (for instance) decide between a blue eye and a brown eye, but suppose there are two ways (with different molecular patterns) of making a blue eye. Selection could not decide between them. Some biologists, extending this possibility, think that in real life there is a huge amount of molecular redundancy, and it was suggested by a leading Japanese population geneticist, Moto Kimura (1983), that at this level the molecules just drift up or down to total fixation or to elimination. In populations where the effects of selection can make themselves known, drift can operate only on small populations. But where selection is absent or minimalist, drift can (in theory) have major effects on large populations.

But is this true? Well, possibly in some cases. But in other cases it certainly does not hold. Where one is dealing with nonfunctional chunks of DNA (pseudogenes), no doubt drift is the player that counts. But overall the amount of drift at the molecular level has been subject to various experiments, some of which suggest strongly that selection is sifting through the molecules, choosing some and rejecting others. For instance, there has been detailed study of the molecular gene replacements in closely related species of fruitfly (Drosophila). If the genes are drifting up or down, irrespective of selection, then one ought to find the same orders of magnitude of differences between species as one finds within species. Everything is going according to random patterns, so interbreeding and like phenomena should make no difference. In fact, they did make major differences. Between species, one finds significant differences in the molecular genes, but within species although there is some variation, there is far less. This all rather suggests that within the species selection is acting in a positive way to cherish some genes and to eliminate others. A counter to the neutral theory (McDonald and Kreitman 1991).

Move next to the physical level: the phenotype. It is here that selection is supposed to reign supreme. But does it? The Darwinian—the ultra-Darwinian like Richard Dawkins—thinks selection is very, very important. But all important? In fact, no one has ever wanted to claim that selection works in a perfect fashion, forever producing adaptations at their "optimized" peak. One might for instance be dealing with something that had an adaptive function but that now no longer serves such an end. It could be that circumstances have changed, and selection simply has left the feature in place—perhaps selection is unable to reduce the feature. Paradoxically, one ultra-Darwinian has suggested that human sexuality might fall into this category (Williams 1975). Although this is a controversial issue and not all would agree, there are reasons to think that sexuality is really only of adaptive advantage to fast-breeding organisms in unstable environments. For humans, who breed slowly and who stabilize their environments, sexuality may be positively disadvantageous—a single female could do the work herself (as is the case in many mammals and to be candid many human families). But our anatomy and physiology have now become so specialized that we cannot relinquish sexuality. We are stuck with it, for all its problems.

Another reason for nonoptimality of adaptation is relative growth (allometry). Sometimes features are linked together, with one part growing faster than other parts. In fact, this is a well-known and studied phenomenon, and it turns out that the usual relationship is logarithmic—the fast-growing part grows at a very much faster rate than the other parts. It could be that such a fast-growing part is of crucial importance in breeding, but unfortunately it then peaks and goes over into nonadaptive status as the rest of the organism matures and reaches full size. It is thought that possibly the massive horn-growth of the extinct "Irish elk" (actually a deer) could have come through such a process. For early breeding purposes, big horns are a decided advantage. But then the horns just keep growing even though they are maladaptive. Unfortunately by this stage the damage is done—the next generation have the potential for big horns—and so the adult maladaptation is perpetuated.

The Irish elk

Something similar occurs when one has sexual selection working against natural selection. Big tails are sexually desirable in the peacock, but from a natural selection viewpoint—escaping from predators—they are no good at all. Such characteristics are adaptive in one sense and maladaptive in others. And then finally let me mention pleiotropy. Sometimes more that one characteristic is produced by one gene. If the one characteristic is very valuable in the struggle for life, then it can balance other characteristics that are less valuable or even harmful. In a way, this is an individual phenomenon somewhat akin to the group effect that you get with balanced superior heterozygote fitness, where the homozygotes are less fit than the heterozygote—where, indeed, a homozygote may be so unfit as to be absolutely lethal.

So you can see that, although they are all selection connected, the Darwinian certainly sees a place for nonadaptive features. Moreover, turning now to Gould's counterarguments, the Darwinian would challenge many of Gould's supposed examples of nonadaptive characteristics. It would be argued that these features are indeed rooted in adaptive advantage, as brought about by natural selection. Take the key example of vertebrate limb number. Gould suggests that the fact that vertebrates have four limbs rather than six (as insects have) is purely a matter of contingency or constraints on building vertebrates or some such thing. Having four rather than three or five can be explained through adaptive advantage—five legs would be lousy for running, although I suppose the kangaroo, with two legs and a tail, might make one pause about three—but why four rather than six or even eight (like arachnids)? However, Maynard Smith (1981) has seized on this example as precisely one where selection does count! He points out that the early vertebrates were sea creatures, with the need to go up and down rapidly in the water. This, as with airplanes in the air, is best effected by two wings or limbs fore and two wings or limbs aft. In fact, there were vertebrates with other numbers of limbs, but selection favored the four-limbed variety. Today, we live with the relict of this need. It may be that we could get by with a different number—snakes, whales, and chickens obviously do—but that is not to deny the fact that four is rooted in selection, contra Gould's claim. And as we have seen, it is certainly not part of the Darwinian case that all features must have maximum adaptive value right now, and always. The point is that such features are connected to selection in some way, at some point in time.

But there is more than this. Thus far we have been rather defending selection. The strong Darwinian would want to go on the attack. He or she would be aggressive about his or her adaptationism. Even though there may be reasons why full adaptive advantage cannot be reached, until forced to assume otherwise the correct attitude is one of attack. One can and should build "optimality models" showing just how and where selection might have worked. Gould may sneer at these as "Just So" stories, but the fact is that they are valuable and extremely fertile tools for analyzing nature. By way of example, let me draw your attention to a series of papers written by Edward O. Wilson. As you know, he is an expert on the social insects. This series, written at the beginning of the 1980s, focuses specifically on the caste system in certain groups of the ants. Using the metaphor of a division of labor, Wilson was concerned to find why and how it is that the ants have so many different forms. One goes all the way from tiny workers within the nest to large soldier ants outside the nest, protecting their siblings from attackers of all kinds. Wilson worked exclusively on the so-called leaf-cutter ants, a genus known as *Atta*. They send out forgers from the nest looking for vegetation, leaves and the like. Once they have spotted something, they proceed to cut their bounty into small pieces, which they can then carry back into the nest. Another caste now takes over, cutting up the leaves into even smaller pieces and treating them with enzymes on which they grow a kind of fungus. Finally, yet another caste takes the fungus and feeds it to the young. "The fungus-growing ants of the tribe Attini are of exceptional interest because, to cite the familiar metaphor, they alone among the ants have achieved the transition from a hunter-gatherer to an agricultural existence" (Wilson 1980a, 153).

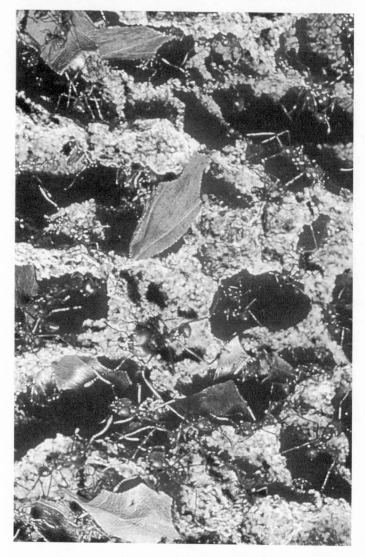

Leaf-cutter ants

Wilson is an ardent Darwinian, so his working assumption was that, from the viewpoint of morphology as well as from behavior, we should find that the ants have been shaped by natural selection. We should find that their body shapes and behavior are about as good (optimized) as is possible to be. Taking this assumption as a tool of research, as much as an established empirical hypothesis, Wilson turned first to the question of the whole overall caste pattern and distribution to be found in *Atta*. Striving to show that there is indeed a division of labor, Wilson's work here was as much descriptive as experimental. First and most obviously, one finds that the soldiers (who take on the roughest work) are bigger and stronger than any of the others: a hundred times bigger than some of their nestmates. Then one finds that those out foraging are in the middle range. Finally, back home in the nest, one finds that here is the place of the most minute and delicate ants.

Why does one have this division? "The elaborate caste system and division of labor that are the hallmark of the genus *Atta* are an essential part of the specialization on fresh vegetation. And, conversely, the utilization of fresh vegetation is the raison d'être of the caste system and division of labor" (Wilson 1980a, 150). And how does this all come about? Wilson was able to show that from a biological point of view, it is done fairly easily. It is a question of relative growth or allometry, combined with a degree of behavioral flexibility. In primitive species, nest members are not differentiated

and anyone can and does do any task. "Most of the monomorphic at-
tines utilize decaying vegetation, insect remains, or insect excre-
ment as substrates, in other words, materials ready made for fungal
growth" (p. 153). In the *Atta,* with specialization, some members of
the nest do some tasks and other members do other tasks. But body
forms are not radically different; rather they are developed proportionately to their ends.

The point is that if one is going to have a kind of specialization
that the *Atta* have developed, namely, the ability to feed on fresh
leaves and to grow fungus on them, one needs much more specialization than one finds with primitive monomorphic forms. But, can one
then show experimentally that there are adaptive reasons behind this?
"Is the colony as efficient in its basic operations as natural selection
can make it, without some basic change in the ground plan of
anatomy and behavior?" (Wilson 1980b, 157). In what way is one to
answer this question?

> The ideal way in which to test the natural selection hypothesis
> and to estimate the degree of optimization is to first write a list
> of all conceivable optimization criteria, deduced a priori from a
> knowledge of the natural history of the species. The next step is
> to conduct experiments to determine which of the criteria has
> been most closely approached, and to what degree. Finally, with
> the results in hand, the theoretician can alter behavioral and
> anatomical parameters in simulations in order to judge whether
> the species is capable of still further optimization by genetic
> evolution. If the approach actually taken by the species cannot
> be significantly improved by the simulations, we are justified in
> concluding that the species has not only been shaped in this par-
> ticular part of its repertory by natural selection, but that it is ac-
> tually on top of an adaptive peak. (p. 158)

One question that interested Wilson centered on the nature of
the ants that would be most efficient for going out foraging, cutting
up leaves, and bringing them back. Why should one find that the middle-range ants do this? Why not bigger ants, who could also act as soldiers, or smaller ants, who could also act as nest tenders? Wilson's hypothesis was that the middle-range ants are the best adapted to their
allotted task: it is they who make optimal use of the energy resources
of the nest. To test this hypothesis, Wilson ran a number of experiments using the so-called pseudomutant strategy. Wilson removed
foragers under certain circumstances and saw whether the other
castes, who were left in the nest, were more efficient at foraging, or

whether the foraging dropped off. For instance, was one better off with smaller foragers or larger foragers? Or was it truly the case that something in between, as one has at the moment, is the best? (Wilson took note of the fact that in natural conditions, the vegetation available to the *Atta* is of a particular kind. In the rain forests, the vegetation is tough. One must therefore recognize that an ant that is good at cutting up rose petals might not function at all well in nature. One needs an ant at least capable of cutting up rhododendron leaves.)

Wilson showed that his hypothesis and research strategy pay off. "What *A. sexdens* has done is to commit the size classes that are energetically the most efficient, by both the criterion of the cost of construction of new workers . . . and the criterion of the cost of maintenance of workers" (p. 164). More than this, Wilson found that the nests are adapted more to the kind of vegetation that they would experience in the wild than to any general range of vegetation. One has natural selection working flat out, most efficiently. The ants are adapted in such a way as to optimize the overall behavior of the nest. In other words, the colony "sits atop an adaptive peak."

You can make up all of the belittling metaphors that you like. You can jeer all you will—although it is interesting to note that critics like Gould seem never to want to focus in on work of this nature—but the fact of the matter is that, through his adaptationism, Wilson was able to throw very considerable light on intricate aspects of hymenopteran social behavior. That ultimately is the answer of the Darwinian to those who would belittle or deny their work and its worth.

Macroquestions

Finally, what about the upper level of the hierarchy? No one is going to deny that you are going to get effects at the macrolevel—that is, over long periods of time—that are more than just microeffects stitched together. I am not sure that there is anything mysterious or "holistic" about this, but the fact is that the course of history over millions of years simply does not follow from the changes in a fruitfly cage. If this is what antireductionism means, then we are all antireductionists. One does not need an Engels to tell us as much. Take, for instance, the whole question of extinction. Not only do individual species go extinct, but sometimes you get a whole range of species going extinct at the same time: "mass extinction." Such events saw out the Devonian, the Permian, and most famously (when the dinosaurs went) the Cretaceous. No one could have inferred these extinctions from microevents, but then no one would ever have pretended to. Clearly, some other factors—possibly random and possible not, possibly extraterrestrial and possibly not—were involved. Everybody knows that the popular hypothesis for the end of the Cretaceous is that an asteroid or some such thing hit the earth, causing a great dust cloud and blocking out of the sun, and that as a consequence there was cooling and paucity of food and that this put paid to the dinosaurs (as opposed to the mammals who were just weedy little runt-sized nocturnal animals). This is not something that could have been predicted by population genetics, but it is something that is part of the causal story of life's history (Alvarez et al. 1980).

It has to be granted then that the macroevolutionist—the paleontologist—will tell us something about evolution as the path that we cannot get from elsewhere. And this will surely lead into discussion of evolution as cause, as one tries to understand and explain the path. The causes might not be directly biological, but they are part of the picture. I am not sure that there is any question of downward causation—of the paleontologist teaching and instructing the geneticist—but there is certainly some measure of autonomy to the macrolevel. But can one go on from here? Are there biological patterns at the macrolevel that would not be expected from the microlevel? Can the macroevolutionist show and explain biologically fueled events that do not appear at smaller levels with shorter times?

In principle there seems no reason why not, and in fact we do find that some workers have tried to provide explanatory models of this nature. By example, let me take a problem that has long puzzled students of life's history, namely the so-called Cambrian explosion. Nearly 600 million years ago, life suddenly started to explode in diversity and number. From fairly sparse numbers and types, at least as revealed in the fossil record, huge numbers and varieties made their appearance, almost overnight as it were. Now there are a number of questions that you can ask—for instance, about why the explosion happened at all. And some of the answers will surely be framed in terms of adaptive advantage. For instance, it may be that the seawater was carrying much more oxygen, thanks to photosynthesis caused by algae, and this then made possible the sustenance of many more and more complex life-forms than previously.

But what about the actual pattern of the explosion? John J. Sepkoski Jr. collected huge amounts of data about the numbers of different kinds of organism that have been recorded as living back then at the time of the explosion. He found, plotting numbers on a graph,

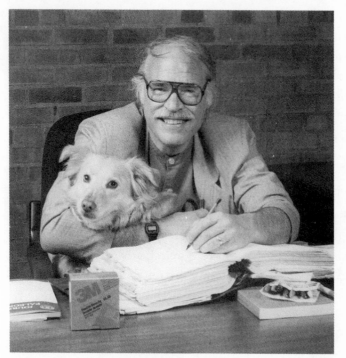

John Sepkoski

that the picture is roughly s-shaped (sigmoidal)—a rapid rise up, and then a flattening out. To explain this, Sepkoski turned to a well-known ecological hypothesis about the colonization of islands by organisms, formulated in the 1960s by Princeton biologist Robert MacArthur and Harvard entomologist Edward O. Wilson (one of Sepkoski's teachers). The island biogeography hypothesis specifies that organism species numbers will reach equilibrium (a function of distance from the mainland and island size) after a period of (exponential) growth—new species arriving on an island will equal the old species leaving or going extinct. Reasoning that colonizing in time is much like colonizing in space, Sepkoski (1976) was able readily to show that one can model the sigmoidal rise of organisms in the Cambrian using the MacArthur-Wilson hypothesis.

The first models produced by Sepkoski were understandably crude, but they were sufficiently promising to stimulate him to further effort. He worked diligently to expand his database—for technical reasons he focused on marine animals—and as the material piled up, he found that he needed to refine his theory. Instead of a nice smooth upward rise, a sigmoidal curve carrying one through the Cambrian and beyond, there is a midlevel break as the growth pauses before picking up again to continue the movement upward (Sepkoski 1979, 235). Tantalizingly, those organisms that seem most successful during the Cambrian reach their peak at the time of this pause, before they start into a long, slow decline.

Tantalizing but suggestive. Surely what is needed is a second set of equations, superimposed on the first, with a second curve therefore taking off on the back of the first. The Cambrian organisms (marine fauna) reach their peak halfway up and then start to decline. But in the meantime, rather like a second-stage rocket that takes over when the first stage is exhausted and is now falling down to the sea, the next batch of organisms has taken over and is rising up through the Paleozoic. "The two-phase kinetic model . . . seems to provide an adequate description of the fundamental patterns observed in the early Phanerozoic diversification of marine metazoan families" (p. 242). This is just a description of what is happening, but the temptation is strong to speculate on causes, and some hypotheses come at

The history of life showing the different phases of the growth of living beings

once to mind. Could the earlier organisms be rather "generalized" in some sense, good for flourishing and increasing when there is lots of empty ecological space, and could the later organisms be rather "specialized" in some sense, good for flourishing and increasing when the ecological space is much more crowded? Are we looking at the replacement of organisms that have "relatively broad feeding and habitat adaptations" by organisms that "might be expected to exhibit lower rates of speciation and extinction and, as a result, lower rates of diversification but higher equilibria" (p. 243)? Are these replacing organisms better at utilizing crowded or restricted environments, so that we end with "more finely divided and stable ecosystems which can be described as having high equilibrial diversities" (p. 243)?

We are not done yet. As more data flowed in, Sepkoski discovered that the new, replacing organisms ran out of steam at some later point, peaking and then going into a slow decline. But now he knew just what to do! A third set of equations yielded a third curve, with a new set of organisms taking off on the back of the second set. After the great extinction at the end of the Permian, life picked up again, increased in diversity, and grew with some force and speed right up to the present (Sepkoski 1984). Humans, of course, are messing things up at the end. The ways in which we are destroying habitats and the denizens thereof has a major impact. But the overall picture of life's history makes good sense.

Moreover, perhaps we can even try our hand at predictions. Humans aside, we seem to be in a bit of a lull right now. Could it be that there is a fourth group of organisms waiting in the wings, ready to take off on the backs of today's animals, ready to scale yet higher peaks? It seems improbable but cannot be discounted entirely. In the plant world, with the arrival of the Cretaceous, we got a new fourth kind of flora, the angiosperms (the flowering plants). Could not the same be true of the animal world? "By analogy to the plant record, we can speculate that one or more unpredictable innovations of importance comparable to angiosperms might appear among future marine animals, leading to major changes in faunal composition and driving diversity to yet higher levels" (p. 264).

The point is made! Sepkoski is certainly not against Darwinism, meaning explaining evolution through selection. Rather, he is interested in somewhat different questions. To be honest, if I were looking for a predecessor, someone in whose shoes he stands, I would opt for Herbert Spencer rather than Darwin. All of the talk about moving up to a plateau and then a period of stability or equilibrium sounds very much like the synthetic philosophy updated. Which would fit in with the influences under which Sepkoski fell. He was a student of Gould as well as Wilson, and both of these men are Spencer-influenced: the fact that they fell out bitterly is almost what you expect from family members. Moreover, since in the case of Sepkoski we have a paleontologist who came into evolution via an intense interest in computers—he never took biology courses as an undergraduate—there is really no reason to seek for strong naturalist influences. I mean that we should not expect to find, nor do we indeed find, influences leading to a fondness for selection.

But however you analyze Sepkoski—on content or on influences—the fact is that he works at a level that is above and beyond that of the Darwinian working on selection-related problems, trying to understand features of and changes in today's organisms. In this sense, Gould is truly right to think of evolutionary theorizing as hierarchical. Darwinism is not the washed-out, inadequate theory he pretends it to be, but there may well be more to the history of life and to our understanding than ultra-Darwinians sometimes claim. In the end, Steve Gould is much like the rest of us. Sometimes he is wrong. And sometimes he is right!

Further Reading

Stephen Jay Gould is a terrific writer, especially of the short essay. For twenty-five years now, in the magazine *Natural History,* he has been writing a monthly column: "This View of Life." Gould ranges over many topics, on and around the life sciences. One month you will get an account of a old volume on natural history he has discovered tucked away in a secondhand book store. The next month you will learn of the mating practices of some tropical bird. But through the diversity of topics and friendly, almost folksy, style, you sense that there is not just a keen intelligence but a burning moral passion. Gould's view of life is fun. Gould's view of life is serious.

The essays have been collected in published volumes. The first and still the best is *Ever Since Darwin* (New York: Norton, 1977). Supplement this with what I think is the best of Gould's full-length books: *Wonderful Life: The Burgess Shale and the Nature of History* (New York: Norton, 1989). Ostensibly this is a discussion of the marvelous finds of soft-bodied, fossilized invertebrates, in a place up in the Rockies between the Canadian provinces of Alberta and British Columbia. These fossils give us an insight into the nature of life just before it exploded up into the rich diversity that marks living beings as we know them, an event that happened during the Cambrian period some half a billion or more years ago. Gould uses the fossils and their interpretation as a vehicle to discourse on the process of scientific discovery and theorizing, as well as on nature and form of life itself, drawing conclusions about the paths and causes of evolution, the status of humankind, and the ways of scientific reasoning. I myself disagree with just about every one of his conclusions, but I have rarely enjoyed a book so much. It is simply science writing at its best—clear, informative, provocative.

Wonderful Life is also, I believe, a rather clever pastiche on books about baseball (Gould's other great passion, along with fossils). See how Gould writes of his characters as if they were managers and players in America's National Pastime. Judge, for instance, how he treats the British paleontologist Simon Conway Morris, brought up from the minors by a manager who saw real talent in his unpolished character and who took him on to win a Cy Young award of science, fellowship in Britain's Royal Society. Conway Morris, I should say, has not entirely appreciated the honor of Gould's analysis and has responded with a somewhat waspish book of his own: *The Crucible of Creation: The Burgess Shale and the Rise of Animals* (Oxford: Oxford University Press, 1998). Unfortunately, although he may be a better scientist, he is nowhere like as good a writer.

Given Gould's engaging prose, what more could you ask of anyone or anything? Well, how about a different perspective, from someone who writes as well yet who is as committed to Darwinism as Gould is questioning. I refer to Richard Dawkins, as English as Gould is American. Dawkins's exposition of the ideas and achievements of Darwinism, *The Blind Watchmaker* (New York: Norton, 1986), is just superb. Dawkins has long been something of a computer buff, and he uses his knowledge and skill to great effect, especially when he is dealing with all of those worries so often expressed about whether so simple a mechanism as natural selection can truly generate the complexity we find distinctive of the living world. How can you generate a line of a Shakespeare play in just a few moves, if all you have is the random processes of nature, akin to a monkey striking randomly on the keys of a typewriter? Read Dawkins and find out. Find out this and much more as you are taken from one dazzling chapter to the next, each one proving that those who think that the scientific study of nature in some way impoverishes our sensibility are themselves the ones truly lacking and blind in spirit.

I have one more recommendation for this chapter. Nearly a century ago, the embryologist E. S. Russell wrote a book trying to trace what he saw as the two conflicting tendencies in biological understanding, between those (like the *Naturphilosophen*) who emphasize the form of organisms and those (like Cuvier) who emphasize the functional nature of organisms. You know that I think this difference is reflected in the opposed thinking of Gould and Dawkins. Read *Form and Function, a Contribution to the History of Animal Morphology* (London: John Murray, 1916) to see that they are the end points of a long tradition of difference. One question I will ask you is whether, given that this divide predates Darwin and continues still today, this means that in some sense the Darwinian revolution is less revolutionary (for all that Darwin established evolution through selection) than many have assumed. Is evolution just a surface dance on philosophical issues of much greater depth? In my *The Darwinian Revolution: Science Red in Tooth and Claw,* 2d ed. (Chicago: University of Chicago Press, 1999), I try to answer this question myself, and I return to it in essays in one of my collections, *The Darwinian Paradigm: Essays on Its History, Philosophy and Religious Implications* (London: Routledge, 1989); but you too might think about it.

Ultimate Questions: Science and Religion

Q: Dr. Ruse, having examined the creationist literature at great length, do you have a professional opinion about whether creation science measures up to the standards and characteristics of science that you have just been describing?

A: Yes, I do. In my opinion, creation science does not have those attributes that distinguish science from other endeavours.

Q: Would you please explain why you think it does not.

A: Most importantly, creation science necessarily looks to the supernatural acts of a Creator. According to creation-science theory, the Creator has intervened in supernatural ways using supernatural forces.

Q: Do you think that creation science is testable?

A: Creation science is neither testable nor tentative. Indeed, an attribute of creation science that distinguishes it quite clearly from science is that it is absolutely certain about all of the answers. And considering the magnitude of the questions it addresses—the origins of man, life, the earth, and the universe—that certainty is all the more revealing. Whatever the contrary evidence, creation science never accepts that its theory is falsified. This is just the opposite of tentativeness and makes a mockery of testing.

Q: Do you find that creation science measures up to the methodological considerations of science?

A: Creation science is woefully lacking in this regard. Most regrettably, I have found innumerable instances of outright dishonesty, deception, and distortion used to advance creation-science arguments.

Q: Dr. Ruse, do you have an opinion to a reasonable degree of professional certainty about whether creation science is science?

A: Yes.

Q: What is your opinion?

A: In my opinion creation science is not science.

Q: What do you think it is?

A: As someone also trained in the philosophy of religion, in my opinion creation science is religion. (Ruse 1988, 304–306)

My moment of glory in Little Rock, Arkansas! It is not often that a philosopher finds himself on national television, and I still dine out on it! It was indeed a moment of glory. In 1981, appearing as an expert witness for the American Civil Liberties Union alongside such evolutionary luminaries as Stephen Jay Gould, I was asked to appear

Michael Ruse

in an attack on the constitutionality of a new law mandating the "balanced treatment" of so-called Creation science with evolution in the publicly financed biology classrooms of the state. And we won! The law was declared unconstitutional, and that was the end of that.

But I should have known better. Court cases, particularly in America, are rarely the end of anything. As we have changed from one millennium to another, the science-religion debate, the evolution-creation debate, rages as never before. Let me bring you up to date, show you where we stand now, and offer a few thoughts of my own.

Creation Science

It really all started with the Russians. In 1957, we were in the depths of the Cold War, and it was then that the Soviet side scored an absolutely massive propaganda victory. Sputnik! They put aloft an unmanned satellite, and then to rub salt in the wounds, they put up another that (they informed the world) was as big as a Cadillac, the epitome of American opulence and success. In fact, looking back, it was not much more than a propaganda victory. Russia was ahead in rocket technology, partly because at the end of World War II they had grabbed more German rocket engineers than the Americans and partly because they needed long-range missiles. America had its nuclear weapons in Turkey, sitting on the Russian border. The Soviets had needs that the Americans did not, and so they had moved to fill them. It was hardly an unbiased question of the superiority of one world system over another. But America certainly perceived itself as lagging behind, not just in rockets but in science and technology generally.

This meant, among other things, that if parity were to be achieved, then education needed to be upgraded dramatically. One way in which this was done (given that education falls under State jurisdiction) was by the Federal government's sponsoring the writing of good quality, new textbooks, which could then be made available to school boards at attractive prices. Naturally enough, although following the chilling effects of the Scopes trial evolution had become something of a nonsubject in high-school biology texts, evolution figured in a major way in the new works. As word of this started to filter out, the provocation set the evangelical literalists moving, and so things start to move toward a new confrontation. Aided, I might say, by something that looks suspiciously like Divine Intervention, for just at the moment of crisis, the men of the hour arrived. John C. Whitcomb, a Bible scholar, and Henry M. Morris, a hydraulic engineer, jointly authored a book, *The Genesis Flood: The Biblical Record and Its Scientific Implications* (1961), which put once again the whole and full case for a literal Genesis-based account of origins. The case was supposedly (in the fashion started by a turn-of-the-century, Canadian, former encyclopaedia salesman turned homegrown geologist, George McCready Price) supported in its entirety by the best quality modern science. Creation that occurred about six thousand years ago, took just a week, and was miraculous, with humans coming last. At some point after all of this had occurred, there was a massive worldwide flood, which wiped out virtually everything except apparently for a few, carefully chosen survivors.

An alternative to evolution was there for all to see and to adopt. Worried about the fossil record? No need to be. The progressiveness of the record is an artifact of the Flood, with the slowest creatures caught at the bottom and more agile creatures getting up to the tops of mountains before perishing. How else do you explain human footprints found down among the dinosaurs? What did lions eat in Eden? A vegetarian diet obviously, since they could hardly have feasted on other animals. Troubled by the age of the earth question? Be assured that you are less troubled than conventional scientists. "Age measurements by radioactivity are not nearly so precise nor so reliable as most writers imply." Indeed, "the great majority of the measurements have had to be rejected as useless for the desired purpose" (p. 343).

Henry Morris, with a group of like-minded thinkers, founded the Institute for Creation Research. Realizing that the situation had changed from the days of the Scopes trial and that no court was going to stand for the elimination or expulsion of evolution, they campaigned rather for the inclusion of their own beliefs. Thus, through the 1970s, Morris and the others (notably Duane T. Gish) wrote and lectured and (very successfully) debated evolutionists on the alternative pictures of origins. At the same time they refined and polished their position—considerable effort had to go into compressing the time-scale down from several billion years to just a few thousand. And also, they took care to see that their position could be presented ostensibly without any reference to biblical matters. In *The Genesis Flood,* for instance, when faced with monstrous-sized human footprints in the fossil record, confident mention is made of the passage in Genesis (6.4) that tells us that there were "giants in the earth in those days" (p. 175). This sort of thing rapidly became unacceptable, at least in "public school editions" of the Creationists' books. The important thing was to offer themselves up as a reasonable, secular alternative to the dominant evolutionarism of the day. Hence, the new name: "Creation science."

One has to say that the Creationists worked hard and succeeded brilliantly in their tactics and aims. They caught evolutionists napping, making them look fools—inarticulate and irrational and prejudiced fools. Working with humor and charm and sincerity, Morris and Gish particularly were masters at the public debate, usually reducing their scientific opponents to choleric rage and intellectual impotence. Moreover, they started to influence state legislatures, and the end result was that early in 1981 Arkansas passed a bill mandating "balanced treatment." I might add that this all happened when Bill Clinton was not in the governor's office, and the bill was signed into law by a man whose unsuitability for the office was equalled only by his surprise at achieving it. And I should say also that the law was a rather unpleasant surprise for many powerful people in the state. The Junior Chamber of Commerce in particular was not happy. It was working flat out to persuade new industry—often high tech, involving electrical engineering or computers—to relocate in the state. The last thing it needed was for a prospective employee, perhaps a newly minted Ph.D. from MIT, to learn that the children would be taught Creationism in the schools. Such a prospective employee would keep on moving until reaching other states—perhaps Arizona, also in the market for the new technology and the people to produce it. Whatever the personal convictions of the leaders of these rival states, they knew enough to maintain a decent hypocrisy of having one set of beliefs for the weekdays and another set for Sundays.

Indeed, the leading Creationists themselves were somewhat torn on the Arkansas law. They knew that once their ideas were made public like this, they would be pilloried in the press and probably defeated in the courts—as indeed they were. Better by far to work at the grassroots level, influencing public opinion, putting pressure on school boards and individual teachers, and like actions. Which is precisely the way the Creationism movement has gone in the past 20 years, again with considerable success. A new round of faces has been recruited, notable for being much more established academically than the earlier Creation scientists. It is true that Morris and Gish have advanced degrees in science—much is made of this point—but now we find supporters of the movement at leading universities, and not just

Phillip Johnson

junior faculty either. Notable are Phillip Johnson, onetime law clerk to former Chief Justice Earl Warren and now professor of law at Berkeley, and Alvin Plantinga, professor at Notre Dame and surely North America's most distinguished philosopher of religion.

With the development of the Creationist side has come, perhaps as a kind of counter in reaction, a development of the evolutionist side. There are still many evolutionists, probably the majority, who want nothing to do with the science/religion conflict. They want to get on with their science and leave matters at that. These days, the majority have probably never heard of Genesis, let alone read it. Among the minority who are interested in religion, one finds Stephen Jay Gould (1999). He has certainly read Genesis: readers of his column will know that he has a biblical knowledge that would challenge any priest or rabbi and will know also that he is prominent among those who think that good fences make good neighbors. Science is science and religion is religion and never the two should meet. He speaks of science and religion as being rival "magisteria"—realms of inquiry and understanding—and advocates what he calls the NOMA principle. Science and religion are *Non-Overlapping Magisteria* and should stay that way.

But many of those interested in the science/religion interface, ardent in their evolutionism—usually ultra-Darwinism—are among those who have really taken a strong and almost personal dislike to Christianity. Richard Dawkins (1996) leads the pack, with the philosopher Dan Dennett (1995) and the historian Will Provine (1988) close behind. They feel very strongly that you cannot serve science and religion at the same time and that Darwinism positively excludes Christianity—not just Creationist Christianity, but any kind. In commenting on a recent letter favorable to evolution, written by Pope John Paul II (1997), Dawkins (1997a) speaks of a "flabbiness of the intellect" affecting those who turn to religion—and if you are prepared to say that about John Paul II, you are prepared to say it about anybody.

What are the pros and cons of the issue? Let us start with the Creationists.

One of the most important things about what I shall call the New Creationism is that it is asymmetric. It tells you what it does not like but is irritatingly (and surely deliberately) silent on what it does like. Phillip Johnson's major book, *Darwin on Trial* (1993), is a paradigm. The new Creationists do not like evolution; especially they do not like Darwinian evolution. So Darwin is put in the dock. But what these Creationists do believe is not specified. Do they believe in a young earth or an old earth? We are not told. Do they believe in a universal Flood or a limited Flood? We are not told. Do they believe that humans necessarily came last (and were Adam and Eve a one-off event or did Eve come later)? We are not told. What we do not know, we cannot criticize— which was a major problem faced by the earlier Creationists. Recently, it is true, some have been arguing that we can say positively that the world shows evidence of a creative intervention (what its supporters call "intelligent design"), but everything is still left so vague that you could plug in just about any belief—so long as it is not Darwinism!

But what about Darwinism? Many of the criticisms are familiar—going back to Cuvier, in fact. Natural selection is a favorite target. It cannot do what is required, it is trivial, probably false, and in any case is simply a redescription of what is going on—it is a "tautology," a necessary truth since it tells you that the fittest survive but then the fittest are defined as those that survive! Mutation is also criticized heavily. It is random, and random means random. You cannot get order from randomness. That is the truth. Organisms need something more—they need something in the intelligence line to put them on the road to being. The fossil record speaks eloquently against evolution. Nor do the so-called missing links help. Consider archaeopteryx, the bird-reptile seized on by Thomas Henry Huxley.

> *Archaeopteryx* is on the whole a point for the Darwinists, but how important is it? Persons who come to the fossil evidence as convinced Darwinists will see a stunning confirmation, but skeptics will see only a lonely exception to consistent pattern of fossil disconfirmation. If we are testing Darwinism rather than merely looking for a confirming example or two, then a single good candidate for ancestor status is not enough to save a theory that posits a worldwide history of continual evolutionary transformation. (Johnson 1993, 81)

The molecular evidence for evolution is found no more convincing. In fact, it is all a little bit of a con job. As far as Johnson is con-

Archaeopteryx, midway between the reptiles and the birds

cerned, it is a classic case of circular argumentation. We start by assuming that the molecules are important and then, backed by this belief, we set out to prove that they are important! "As in other areas, the objective has been to find confirmation of a theory which was conclusively presumed to be true at the start of the investigation" (p. 101). Obviously, although this is very comforting to the true believer, it is a parody of true scientific methodology and understanding. "The true scientific question—Does the molecular evidence as a whole tend to confirm Darwinism when evaluated without a Darwinist bias?—has never been asked" (p. 101). And the same thing holds again and again elsewhere. Indeed, there is little need to go on, for the main thing that remains to be discussed is the origin of life question, and we know on what shaky ground that stands. (As you know, I do not think it is on shaky ground at all. But at the time of writing, Johnson, poor fellow, had not had the opportunity to read this book!)

So much for the scientific evidence. As you might imagine, the critique (and countercritique) continues. I mentioned just above the enthusiasm felt in some Creationist quarters for "intelligent design." One of the biggest boosters in this direction is the Lehigh University biochemist Michael Behe, who argues that the workings of the cell are far too intricate to suppose natural origins. One must suppose that something happened, at some point, out of the usual course of events, to put everything together and to get it to work. As it happens, Behe has been very severely criticized, especially by those biologists whose work he cites in his own defense. Behe argues, for instance, that the blood-clotting mechanism in vertebrates is too complex to have come through evolution, but the world authority on blood clotting (Russell Doolittle of the University of California at San Diego) replies that Behe is just out of date and that the evolution of blood clotting is now well supported. (See Behe 1996 and Miller 1999 for details.)

But, while important, the science is only one part of the Creationist attack. If one stopped here, one would be doing the New Creationists an injustice. There is more to their case, and here we start to move to more philosophical questions. Johnson particularly is strong on this matter. In particular, the New Creationists argue that Darwinism succeeds *faux de mieux,* simply because it is the only game in town. The scene is set so that a position such as Creationism is ruled out of court at the beginning, and then Darwinism is declared the winner! The way in which this is done is through an insistence that science—all science—be naturalistic, that is to say something that works according to unbroken law. Then since this is true of Darwinism and is not true of any theistic position which postulates the action of miracle, Darwinism alone qualifies as a proper answer about origins. It wins by default.

In fact, Johnson's position is a little more forceful than this. Not only does he think that Darwinism wins by sleight of hand, but also he thinks that (for all the Darwinian may say otherwise) the evolutionary position tips one into atheism. In Johnson's opinion, the classic move made by the Darwinian is to distinguish between so-called methodological naturalism and so-called metaphysical naturalism. A methodological naturalist is one who insists that natural explanations can be given for anything, including organic origins. "Hence all events in evolution (before the evolution of intelligence) are assumed to be attributable to unintelligent causes. The question is not *whether* life (genetic information) arose by some combination of chance and chemical laws, to pick one example, but merely *how* it did so" (Johnson 1995, 208).

Johnson is at pains to allow, indeed to stress, that this does not mean to say that methodological naturalists think that all of the crucial scientific problems have now been solved. Indeed they will agree that this is not the case. But their optimism is that through time and effort the unsolved problems will fall away, dissolved and settled by the scientist—the scientist working purely in a naturalistic mode. "Bringing God or intelligent design into the picture is giving up on science by turning to religion (miracle) and invoking a 'God of the gaps.' The Creator belongs to the realm of religion, not scientific investigation" (p. 208).

Metaphysical naturalism, on the other hand, is a philosophical thesis about the nature of reality. Here the assumption is that what you see is what you get is what there is. There is nothing more to existence than basic particles interacting without end, without purpose. "To put it another way, nature is a permanently closed system of material causes and effects that can never be influenced by anything outside of itself—by God, for example. To speak of something as 'supernatural' is therefore to imply that it is imaginary, and belief in powerful imaginary entities is known as superstition" (pp. 37–38). The position here is "metaphysical" because it is making a claim about ultimate reality, in particular that there is no such reality beyond that within the scope of the scientist.

Johnson argues that whatever methodological naturalists may say to the contrary, invariably they find themselves sliding into metaphysical naturalism, and before you know it you have full-blown atheism on your hands. But is this really so? Is it truly the case that, if once you have accepted methodological naturalism, you are on the slippery slope to atheism? I am not at all convinced. I will agree that if you accept methodological naturalism (and I would think of myself as being one who does, incidentally, so you know where I stand), then you are almost certainly going to be an evolutionist. I suppose logically you could think that all of the world's organisms are as old as the universe and that therefore there was no evolution, but we know that this is empirically false. The evidence points to evolution—I myself would say that the evidence points to Darwinism—so it is certainly true that, as things are, descent with modification is a consequence of methodological naturalism.

But does this now mean that the whole god question is ruled out? It would surprise the Pope, and it rather surprises me. Let us suppose, for I do not want to get an easy victory by unfair definition, that you are a Christian and as such you think that the Bible must be true. Obviously if you insist on a literal reading, that is an end to matters. Evolution is out, and you might as well agree at once to a denial of methodological as well as metaphysical naturalism. But—and here I am not making things up but simply reporting fact—it has never been part of orthodox Christianity, Catholic or Protestant, that the

Bible must be taken literally, word for word (McMullin 1985). We have seen already that the most sincere of Christians, people like Cuvier, knew that this is not the way to go. Literalism is a late-nineteenth-century American invention. In fact, as I told you in an earlier chapter, the insistence that the days of creation are of twenty-four-hour duration comes out of that sect known as Seventh Day Adventism who, keen as they were to insist on the Sabbath (Saturday) as the day of rest, wanted the other days to be of the same length so as to reinforce their special beliefs about the seventh day. One could hardly insist on people taking long periods of time off to rest, which would seem the consequence if the six days are understood metaphorically. (George McCready Price, who inspired the authors of *Genesis Flood*, was a Seventh Day Adventist.)

But if Genesis is not literally true, but only metaphorically true, what price God then? Can you be an evolutionist—a genuine one, not the Asa Gray variety who goes in for guided mutations—and yet take in the essential heart of the Bible? The answer of course depends on what you take to be the "essential heart" of the Bible. At a minimum we can say that, to the Christian, this heart speaks of our sinful nature, of God's sacrifice, and of the prospect of ultimate salvation. It speaks of the world as a meaningful creation of God (however caused) and of a foreground drama that takes place within this world. I refer particularly to the original sin, Jesus' life and death, and His resurrection and anything that comes after it. And clearly at once we are plunged into the first of the big problems, namely that of miracles—those of Jesus himself (the turning of water into wine at the marriage at Canna), his return to life on the third day, and (especially if you are a Catholic) such ongoing miracles as transubstantiation and those associated, in response to prayer, with the intervention of saints (Ruse 2000).

The metaphysical naturalist would reject all of these. But what about the methodological naturalist? There are a number of options. You might simply say that such miracles occurred, that they did involve violations of law, but that they are outside your science. People do not usually rise from the dead three days after being crucified, but on one occasion someone did. You cannot explain the event scientifically, but this does not mean that it did not happen. And the same is true of other miracles. They are simply exceptions to the rule. End of argument. A little abrupt, but I am not sure that this is an impossible option. You simply say that God laid the salvation history on top of the normal course of events. The world goes by law, and then Jesus and the saints worked their ways on top of this. In fact, turning an apparent weakness into a strength, you say that what makes the biblical miracles particularly miraculous and wonderful is the fact that they are so uncommon. If miracles happened on a daily basis, the resurrection would be disvalued. Precisely because people do not rise from the dead three days after being crucified, the fact that Jesus did makes it truly significant.

Or you might say that miracles occur but that they are compatible with science, or at least not incompatible. Jesus was in a trance and his rising on the Third Day involved no breaking or lifting of law. Likewise, the cure for cancer after the prayers to Saint Bernadette is according to rare, unknown, but genuine laws. This position is less abrupt, although I will admit that I worry whether it is truly Christian, in letter or in spirit. It seems to me a little bit of a cheat to say that the Jesus taken down from the cross was truly not dead, and the marriage at Canna (when Jesus turned water into wine) starts to sound like outright fraud. Did he bring a barrel of Chardonnay and not tell anybody, or were the guests so drunk that they could not tell what they were drinking? You start stripping away at more and more miracles, downgrading them to regular occurrences blown up and magnified by the Apostles, but in the end this rather defeats the whole purpose.

The third option is simply to refuse to get into the battle at all. You argue that the law/miracle dichotomy is a false one. Miracles are just not the sorts of things that conflict with or confirm natural laws. This is not such a strange or ad hoc suggestion. Christians already accept that some miracles fall into this category. Take for instance transubstantiation—the miracle accepted by Catholics that in the Mass there is a turning of the bread and the wine into the literal body and blood of Christ. This miracle (or if you prefer, this purported miracle) is simply not something open to empirical check. You cannot disconfirm religion or prove science by doing an analysis of the host. Likewise one might say that the same is true even of the resurrection of Jesus. After the Crucifixion, his mortal body was irrelevant. The point was that the disciples, downcast and dispirited, suddenly felt Jesus in their hearts and were thus emboldened to go forth and preach the gospel. Something real happened to them, but it was not a physical reality—nor, for instance, was Paul's conversion a physical event, even though it changed his life and those of countless after him. Today's miracles also are really more a matter of the spirit than the flesh. Does one simply go to Lourdes in hope of a lucky lottery ticket to health or for the comfort that one knows one will get, even if there is no physical cure? Surely the latter at least as much as the former. Miracles are matters of feeling and meaning, not of transgressions of nature. In the words of the philosophers, it is a category mistake to put miracles and laws in the same set.

It seems to me that there are at least these options for the would-be Christian who wants also to be an evolutionist. I myself am not equally keen on each and every one, but there is here surely enough to satisfy the would-be believer. I recognize that not every one would be acceptable to every Christian. Protestants, for instance, do not accept transubstantiation, and although they do not have a shared alternative, many (probably most) think that the Eucharist (the ceremony involving the bread and the wine) simply is symbolic of Jesus' last supper with his disciples. The same is true of the other miracles and their possible explanations. Taking the resurrection metaphorically or in spirit only is certainly not accepted by all or even most Christians. But the point is that these options are all accepted by some Christians, and by no means indifferent or careless believers. Indeed, some of the most passionate and devout go for these alternatives.

Johnson (1995), however, sneers that such options are not "intellectually impressive" (p. 211). He adds: "Makeshift compromises between supernaturalism in religion and naturalism in science may satisfy individuals, but they have little standing in the intellectual world because they are recognized as a forced accommodation of conflicting lines of thought" (p. 212). Which of course is absolutely true. Johnson is right. Makeshift compromises rarely do having much standing in the intellectual world. But are the sort of options I have listed of this nature? To the contrary, the very difficulties I have been discussing—having to take miracles on faith despite the evidence against them or having to admit that there are no physical miracles at all—are taken by some very significant theologians of our age to be the very crux of what it is to be a Christian (Barth [1949] 1959; Bultmann 1958; Gilkey 1985). They believe that, if we can get a guarantee on all of the answers, then commitment is devalued. Faith without difficulty and opposition is not true faith. "As the Danish philosopher Søren Kierkegaard . . . taught us, too much objective certainty deadens the very soul of faith. Genuine piety is possible only in the face of radical uncertainty" (Haught 1995, 59).

Such thinkers, often conservative theologically—revealingly they are known as the "neo-orthodox"—are inspired by the Jewish philosopher Martin Buber (1937) to find God in the center of "I-Thou" personal relationships. For them there is something degrading in the thought of Jesus as a miracle man, a sort of fugitive from the Ed Sullivan show. What happened with the 5,000? Some hocus-pocus over a few loaves and fishes? Was the Redeemer no more than a high-class caterer? Or did Jesus fill the multitude's heart with love, so there was a spontaneous outpouring of generosity and sharing, as everyone in the crowd was fed by the food brought by a few? Surely this is what truly happened. This is what Christianity is really about.

Part of the problem when dealing with matters to do with Christianity and science, evolution in particular, is that so many people believe in so many things. For instance, the position I have just been sketching—that faith is only genuine faith in the face of uncertainty—would be denied by Catholics. They believe that one can in fact prove the existence and nature of God through reason. Although, especially given the Pope's position on evolution, this certainly does not mean that they would now swing round and think that Johnson is right. If anything, Catholics tend to be more opposed to biblical literalism than Protestants. But at this point we can honorably pull back from the details. It is enough to show, and this surely has been shown, that the whole science/religion relationship is more complex than allowed by the New Creationists. More complex, and I would say more interesting and more fruitful.

Darwinian Atheism

Swing around now and look at the other side. Let us focus in on Richard Dawkins and start by quoting a couple of paragraphs from an interview that Dawkins gave a few years ago.

> I am considered by some to be a zealot. This comes partly from a passionate revulsion against fatuous religious prejudices, which I think lead to evil. As far as being a scientist is concerned, my zealotry comes from a deep concern for the truth. I'm extremely hostile towards any sort of obscurantism, pretension. If I think somebody's a fake, if somebody isn't genuinely concerned about what actually is true but is instead doing something for some other motive, if somebody is trying to appear like an intellectual, or trying to appear more profound than he is, or more mysterious than he is, I'm very hostile to that. There's a certain amount of that in religion. The universe is a difficult enough place to understand already without introducing addi-

tional mystical mysteriousness that's not actually there. Another point is esthetic: the universe is genuinely mysterious, grand, beautiful, awe inspiring. The kinds of views of the universe which religious people have traditionally embraced have been puny, pathetic, and measly in comparison to the way the universe actually is. The universe presented by organized religions is a poky little medieval universe, and extremely limited.

I'm a Darwinist because I believe the only alternatives are Lamarckism or God, neither of which does the job as an explanatory principle. Life in the universe is either Darwinian or something else not yet thought of. (Dawkins 1995a, 85–86)

These paragraphs are very revealing, not the least for showing the emotional hostility that Dawkins feels toward religion, including (obviously) Christianity. I am sure the reader will not be surprised to learn that Dawkins has since then characterized his move to atheism from religious belief as a "road to Damascus" experience (Dawkins 1997b). Saint Paul would have recognized a kindred spirit. But my purpose in quoting Dawkins's words here is not so much to pick out the emotion, as to point to the logic of Dawkins's thinking. This comes through particularly in the second paragraph just quoted. It is clear that for Dawkins we have here an exclusive alternation. Either you believe in Darwinism or you believe in God, but *not both*. For Dawkins—as for Phillip Johnson on the other side—there is no place for what philosophers call an inclusive alternation, that is to say either a or b or possibly both. (The third way mentioned is Lamarckism, the inheritance of acquired characteristics. But neither Dawkins nor anybody else today thinks that this is a viable evolutionary mechanism.)

Why not simply slough off Christianity and ignore it? Things are not this simple: as we saw in earlier chapters, Dawkins—like any good Darwinian including Charles Darwin himself—recognizes that the Christian religion poses the important question, namely that of the designlike nature of the world (Dawkins 1986). Moreover, Dawkins believes that until Charles Darwin no one had shown that the God hypothesis, that is to say the God-as-designer hypothesis, is untenable: more particularly, Dawkins argues that until Darwin no one could avoid using the God hypotheses.

In this context, Dawkins is fond of telling a story about a conversation he once had with a well-known philosopher. (Although Dawkins never tells us in print who it is, he himself has told me that it was the late Sir Freddy Ayer, a fellow Oxford professor and a notorious atheist.) Apparently the conversation took place at one of those famed Oxford college feasts, where the food is abundant (in my experience, usually pretty dreadful) and the wine even more abundant (in my experience, always very good). Probably everyone was indulging well if not wisely, and finally the philosopher—in a rather sneering and condescending way—challenged the biologist. Surely he asked, there is nothing in the living world that demands special explanation. All of this nonsense by Christians and biologists alike about the special nature of animals and plants is silly make-believe, pretending that things are more significant and interesting than they really are.

In reply, the aroused biologist demanded an explanation of the complexity that we see around us. Asked Dawkins, does this not require some special understanding? Not at all, replied the philosopher. The living world is as it is and simply exists. That is all and that is enough. But it is not enough, replied Dawkins then, and to this question that continues to haunt him, he still replies sternly. The living world is special. "Paley knew that it needed a special explanation; Darwin knew it, and I suspect that in his heart of hearts my philosopher companion knew it too" (Dawkins 1986, 6).

It is true—as Dawkins concedes—that David Hume, the great Scottish philosopher, made devastating criticisms of the argument from design. In his *Dialogues Concerning Natural Religion* (first published in 1779), he showed, for instance, that the living world might as reasonably have had a team of gods making it as having one unaided creator. He showed that if the world is designed by God or by gods, then it is reasonable to think that there were many previous attempts and trials. There must exist somewhere a whole series of cruder or botched earths, which were the forerunners of our earth—or we must have been formed out of them. Indeed, it may be the case that we ourselves are not living in the final and perfected world. Hume showed in fact that we might as well think that the world is as much like a giant vegetable as like an object of design!

But in Dawkins's opinion there is still a gap requiring a filling. "What Hume did was criticize the logic of using apparent design in nature as *positive* evidence for the existence of a God. He did not offer any *alternative* explanation for apparent design, but left the question open" (Dawkins 1986, 6). Dawkins continues: "An atheist before Darwin could have said, following Hume: 'I have no explanation for complex biological design. All I know is that God isn't a good explanation, so we must wait and hope that somebody comes up with a better one'" (p. 6). But, in Dawkins's opinion, this is not enough. "I can't help feeling that such a position, though logically sound, would have left one feeling pretty unsatisfied, and that although atheism might have been *logically* tenable before Darwin, Darwin made it possible to be an intellectually fulfilled atheist" (Dawkins 1986, 6).

At this point, some of the sorts of questions asked of Johnson start to seem pertinent. Why should we not say that Dawkins is certainly right in stressing the designlike nature of the organic world, but he is wrong in thinking that it is either Darwinism or God, but not both? At least, even if he is not wrong, he has failed to offer an argument for this. Perhaps the designlike nature of the world testifies to God's existence. It is simply that God created through unbroken law. Indeed, as we have seen, people in the past would argue that the very fact that God creates through unbroken law attests to his magnificence. Such a God is much superior to a God who had to act as Paley's watchmaker would have acted, that is, through miracle.

In fairness, I think that at this point Dawkins does have a second argument up his sleeve. It is the venerable argument based on the problem of evil. But for Dawkins it is more than just the traditional argument (which is in itself not particularly evolutionary). What Dawkins would argue is that not only does evolution intensify the problem of evil, but Darwinism in particular makes it an overwhelming barrier to Christian belief. This argument is expressed most clearly in one of Dawkins's recent books: *River out of Eden: A Darwinian View of Life* (1995b). In a chapter entitled "God's Utility Function," he starts by pointing to the fact that many adaptations require that other organisms suffer, sometimes greatly. "A female digger wasp not only lays her egg in a caterpillar (or grasshopper or bee) so that her larva can feed on it but . . . she carefully guides her sting into each ganglion of the prey's central nervous system, so as to paralyze it *but not kill it*. This way, the meat keeps fresh. It is not known whether the paralysis acts as a general anesthetic, or if it is like curare in just freezing the victim's ability to move. If the latter, the prey might be aware of being eaten alive from inside but unable to move a muscle to do anything about it" (p. 95). All of this sounds pretty dreadful and cruel, but Dawkins's conclusion is that speaking of cruelty in such a situation is no better than speaking of beneficence and kindness. "Nature is not cruel, only pitilessly indifferent. This is one of the hardest lessons for humans to learn. We cannot admit that things might be neither good nor evil, neither cruel nor kind but simply callous—indifferent to all suffering, lacking all purpose" (pp. 95–96).

Then, Dawkins goes on to reinforce this point. He talks about organisms being excellent examples of designlike engineering. If we tried to unpack the engineering principles involved in organisms, the problems of pain and evil would come to the fore. Meaning by the notion "utility function" the purpose for which an entity is apparently designed, Dawkins asks about God's Utility Function when it comes to carnivores and their prey. Consider the cheetah, a beautiful piece of design if anything is. We can work backward, "reverse-engineering," trying to ferret out the way it which it was put together and the purposes of its various adaptations. We can probably be fairly successful in our labors, for the problem posed by cheetahs is relatively easy. "They appear to be well designed to kill antelopes. The teeth, claws, eyes, nose, leg muscles, backbone and brain of a cheetah are all precisely what we should expect if God's purpose in designing cheetahs was to maximize deaths among antelopes" (p. 105). The same is true of the cheetah's prey. "If we reverse-engineer an antelope we find equally impressive evidence of design for precisely the opposite end; the survival of antelopes and starvation among cheetahs. It is as though cheetahs had been designed by one deity and antelopes by a rival deity" (p. 105). Or if we want to suppose that there was one designer responsible for both cheetahs and for gazelles, then legitimately we might ask about His intentions. "Is He a sadist who enjoys spectator blood sports? Is He trying to avoid overpopulation in the

Cheetahs

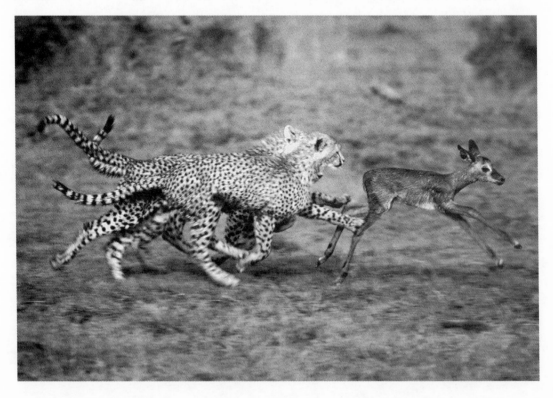

mammals of Africa? Is He maneuvering to maximize David Attenborough's television ratings?" (p. 105).

This is silly of course. No one would draw such a conclusion as this. The point at best seems to be that if there be a God, then He is one who certainly is nothing like the Christian God. He is unkind and unfair or, more likely, totally indifferent. And indeed, this is the point at which Dawkins ends the discussion of this chapter. We simply have to accept that natural selection works by and through pain, pain, and more pain. All of the time, animals are dying: from starvation, from disease, from being eaten by prey, and from many other horrible causes. Things may ease up for a minute or two, but then trouble and pain reappear, even brought on by the pauses. If the predator number is reduced, the prey increase, and then there are more predators in turn and yet more killing than average.

In Dawkins's opinion, this is pointing to an appalling theology, unless we simply stop and realize that our argument is entirely on the wrong track. There is neither a good god nor a bad god. There is simply no god. You ask about human tragedy. Why expect an answer for there is no answer. There is simply nothing.

> In a universe of blind physical forces and genetic replication, some people are going to get hurt, other people are going to get lucky, and you won't find any rhyme or reason in it, nor any justice. The universe we observe has precisely the properties we should expect if there is, at bottom, no design, no purpose, no evil and no good, nothing but blind, pitiless indifference. As that unhappy poet A.E. Houseman put it:
>
> > For Nature, heartless, witless Nature
> > Will neither know nor care.
>
> DNA neither knows nor cares. DNA just is. And we dance to its music. (p. 133)

This is powerful stuff, and whether you agree with Dawkins or disagree, I have no time for anyone who trivializes it. The problem of evil is the biggest of all the obstacles to Christian belief, and Dawkins is absolutely right to point out—to stress—that Darwinism brings it right to the fore. The way in which Darwinian evolution works is through pain and suffering and cruelty and hardship and deprivation and much, much more. You cannot get away from this fact, nor should you pretend to do so. But my suspicion is that Dawkins himself provides the answer! This is a paradox, but true nevertheless. The ardent

Darwinian, the Richard Dawkins (or Michael Ruse for that matter), believes above all that the mark of the organic world is its designlike nature. Animals and plants are adapted. We are with Archdeacon Paley and Georges Cuvier and Charles Darwin on this. But how can one produce this design? If the only way that is possible is through natural selection, then one can argue that God did what He did because He had to. There was no choice. And so the pain follows naturally. It is not God's fault for not preventing it. He could not prevent it. It has always been stressed in Christian theology that God's power—His omnipotence—never meant doing the impossible. God cannot make $2 + 2 = 5$.

Might it not be that, God having decided to create, did then create—perhaps His choice, perhaps not—in an evolutionary fashion? And this being so, might it not be that He was now locked into a path that would necessarily lead to physical evil? It comes with the method employed. The theologian Bruce Reichenbach (1976) makes this objection against the suggestion that God might have used better laws of nature, that is, laws that do not lead to physical evil. At first sight it seems easy for God to have done a better job, making a universe without all of the pain and suffering that we find throughout. But would it really have been all that easy? "For example, what would it entail to alter the natural laws regarding digestion, so that arsenic or other poisons would not negatively affect my constitution? Would not either arsenic or my own physiological composition or both have to be altered such that they would, in effect, be different from the present objects which we now call arsenic or human digestive organs?" (Reichenbach 1976, 185) And this is just the beginning. Think of such everyday things as fire and electricity and the solidity of wood—things that we would be most loathe to relinquish. But in a nonpainful world all sorts of changes would be needed. "Fire would no longer burn or else many things would have to be by nature non-combustible; lightning would have to have a lower voltage or else a consistent repulsion from objects; wood would have to be penetrable so that clubs would not injure" (p. 185).

More than this. Suppose we accept that the world evolved. How could we prevent pain and suffering from occurring? Either we are going to have to change the laws of nature in significant ways, or we are going to have to alter the initial conditions of the universe so that different results come about. Both alternatives raise major problems. If we alter the laws themselves, then at a minimum we will have to alter humans, and this might entail unpleasant or unacceptable theological conclusions. We human are sentient beings, part of nature. That is to say we have a natural physiology, we work according to fixed laws of nature, we see and sense generally because we function like the rest of the world. It comes with the territory that we will encounter unpleasant phenomena—pain and the like—and that we will be conscious of it and not like it. From a biological point of view, if we did not have pain and did not dislike it, we would not function properly. But to alter all of this, we would have to be removed from nature in the sense that we now know it. We would have to be immune from the ways of the world. But if this comes about to be the case, do we now have a being that would be loving and giving (or hating and hurtful)—in other words, would we still have a being of a kind that is supposed to be at the center of God's creation and on which a religion such as Christianity claims that He lavishes so much care and love?

The other alternative suggests that the initial conditions might be altered, thus avoiding unwanted painful conclusions. But what would this mean and entail in fact? If the Big Bang story is right, way back at the near beginning everything was hydrogen. Altering the initial conditions would presumably therefore mean altering the nature and functioning of hydrogen, and probably consequently all of the other elements. But where do you stop and what guarantee do you have that things will now turn out better? In particular, we do not know if humans would have evolved and if they did evolve whether the things that make for pain and so forth would have failed to have evolved alongside. "Whether humans would have evolved but no infectious virus or bacilli, or whether there would have resulted humans with worse and more painful diseases, or whether there would have been no conscious, moral beings at all, cannot be discerned. Given a change in initial conditions, it is possible that this world would not have had any less natural evil while not preserving free moral activity" (Reichenbach 1976, 192–193). All in all therefore, we seem to be in as much trouble after we have made these moves as before. Clearly there is here no devastating argument against the person who believes in a caring and loving God.

I have stressed that the key aspect of organic form is (as we have seen) its adaptedness, and it is this that (as we have also seen) is addressed by natural selection. Physical or natural evil is a result of the causes or a consequence of this selective process. But could one not have got adaptedness by a physical process much nicer than selection? Here we return to the paradox mentioned just above. It is Dawkins (1983) himself who comes to the aid of the theist, for he more than anyone argues strenuously that selection and only selection can do the job. Most putative processes simply do not lead to adaptation: saltationism, evolution by jumps, for instance. Indeed, in Dawkins's opinion, there is a general principle in biology that adaptive complexity always comes through small, gradual processes rather than from big, sudden, incremental changes. And those rivals to selection that address adaptation and that might do things gradually—notably Lamarckism—are known to be false. So it is selection or nothing.

> My general point is that there is one limiting constraint upon all speculations about life in the universe. If a life-form displays adaptive complexity, it must possess an evolutionary mechanism capable of generating adaptive complexity. However diverse evolutionary mechanisms may be, if there is no other generalization that can be made about life all around the Universe, I am betting that it will always be recognizable as Darwinian life. The Darwinian Law . . . may be as universal as the great laws of physics. (p. 423)

God had no choice but to take the option that He chose.

In the end therefore, for all that so many people think that a true Christian could never be an evolutionist—certainly could never be a Darwinian—it turns out that, for the Christian, Darwinism is to be welcomed positively at this point. Physical evil exists, and Darwinism explains why God had no choice but to allow it to occur. He wanted to produce designlike effects—without producing these He would not have organisms, including humankind—and natural selection is the only option open. Natural selection has costs—physical pain—but these are costs that must be paid. What more need be said?

Darwinian Religion

I am arguing what history has shown: there is really no reason why a Christian should not be a Darwinian, and there is really no reason why a Darwinian should not be a Christian. I am not saying that you should be a Christian, and I am not really saying that you should be a Darwinian, but I am saying that the one does not preclude the other. But is this not all a bit redundant? We have seen that Darwinism has been used as a kind of secular religion—Religion without Revelation. Should we not all be going that way now? You can be a Christian if you want to be, but better by far to be a Darwinian in religion as well as science. We have an upward rise to humankind, yielding moral prescriptions—telling us what we should do—and all other religious commitments are pushed out or seen to be secondary.

Some people, some biologists in particular, think that this route of secular religion (often known as "humanism") is precisely the one we should choose. Let me quote one thinker whom we have met many times before. Edward O. Wilson (1978) writes:

> But make no mistake about the power of scientific materialism. It presents the human mind with an alternative mythology that until now has always, point for point in zones of conflict, defeated traditional religion. Its narrative form is the epic: the evolution of the universe from the big bang of fifteen billion years ago through the origin of the elements and celestial bodies to the beginnings of life on earth. The evolutionary epic is mythology in the sense that the laws it adduces here and now are believed but can never be definitively proved to form a cause-and-effect continuum from physics to the social sciences, from this world to all other worlds in the visible universe, and backward through time to the beginning of the universe. (p. 192)

And, in fact, Wilson goes even further than this. He thinks that biology now can explain religion, as something that is needed for group cohesion or some such thing. This means that religion is on the way out. In the future, it will at best be seen as a consequence of a more powerful, more adequate, world picture. Theology will no longer survive as an autonomous subject.

Well, perhaps! But what if you do not share Wilson's vision of progress up to humankind? What if you think that any progress you see in evolution is something that you have read into the process rather than found and read out? What if you do not share Wilson's materialism? What if, like me, you think that (in this quantum age) materialism is slightly silly and that even if you extend your understanding of the term, it still does not follow that evolutionism equals materialism? What if you think you can be an evolutionist and a nonmaterialist? And most particularly, what if you think that whether or not evolution can explain religion, nothing is said about what if anything is more important or basic? After all, if evolution be true, at some level everything we know or understand has to come from evolution. But does this tell us about ontological status or importance? I feel hunger and there are good evolutionary reasons. Does this make my hunger any less real? I feel sexual pangs and there are good evolutionary reasons. Does this make my love any less real or genuine or worthwhile? Does this reduce to nothing all of the poetry that has been written? If evolution be true, I fully expect there to be good evolutionary reasons for religion. But does this mean that God does not exist?

Well, I am sure you know by now how I am going to answer these questions. Frankly if God is going to create in an evolutionary fashion, He would be tempting fate if He then made all belief in Him and all religious practice into things that went against our evolved nature. The fact that religion may have an evolutionary base—a selective base even—tells us nothing about the nonreality of religion. For that, we need an additional argument that there are reasons, perhaps evolutionary reasons, to think that our biology is deceiving us over the religion matter. These may exist, but they are not forthcoming. As it is, one has a feeling here that the Creationist Johnson may have a good point. The philosophy is being fed in at the beginning of the paragraph, and then triumphantly at the end of the paragraph it is being produced as proven. There is no secret to success in hide and seek that beats first hiding the prizes yourself.

None of what I have just said is to stop someone making a religion of evolution if they so wish. Edward O. Wilson is my friend, and I am proud to acknowledge our relationship. He is a good and gentle man, generous to a fault, with a real moral concern for the world's ills, for problems of biodiversity and ecological preservation in particular. If he wants to take evolution as the new myth, something replacing Christianity, I am happy for him to do so. I see only good coming from this move. Those who condemn the man because they do not share his beliefs are bigots and worse. But I do not see that his fellow evolutionists have to follow him into making a religion of our shared science. This has nothing to do with whether or not we want to opt for some other religion, Christianity for instance, or if we have no religion at all—if perhaps we find no ultimate meaning to life, other than that of everyday living and the joys and troubles with that. The point is that just as being an evolutionist neither compels nor denies Christian belief, so also being an evolutionist neither forces one into nor, for that matter, prevents one from being a member of the Church of Darwin. And that is my final (well, almost final) word on the subject.

Further Reading

Spearheading the New Creationist attack on evolution are Phillip Johnson's *Darwin on Trial,* 2d ed. (Downers Grove, Ill.: InterVarsity Press, 1993) and his follow-up work *Reason in the Balance: The Case against Naturalism in Science, Law and Education* (Downers Grove, Ill.: InterVarsity Press, 1995). Although his mentor, Supreme Court Chief Justice Earl Warren, must now be revolving in his grave at his protégé's behavior, Johnson is a brilliant man and these are clever and skilfully written books. I hope you are not convinced by them but do not underestimate them. Note that Johnson pursues a dual strategy. He wants to discredit evolution as science but also he wants to argue that the real issues are not scientific but philosophical: a battle between rank materialism or naturalism and a theistic world picture, the latter being one that has place for designing intervention by the Creator. My own feeling (and I suspect that Johnson would agree) is that the real battle is being fought at the philosophical level and that we evolutionists do not always do as good a job as we might.

Johnson has had terrific supporting help from biochemist Michael Behe. The latter's *Darwin's Black Box: The Biochemical Challenge to Evolution* (New York: Free Press, 1996) is indeed a great read and very persuasive. I am sure that Behe is a wonderful teacher, for he has a great ability to make a difficult point clear through a simple but appropriate example. I think he is wrong, wrong, wrong, but do not take my word for it. Rather turn to *Finding Darwin's God: A Scientist's Search for Common Ground between God and Evolution* (New York: Cliff Street Books, 1999) by biologist (and practicing Christian) Kenneth Miller. He knocks down both Johnson and Behe with great skill, drawing on a deep and profound understanding of modern biology, both evolutionary and those parts more directed toward physiology and the molecular realm. Miller like Behe has the ability to pick on the right and illuminating example, and I am sure he is also a great teacher. Some people just have a gift for communication.

Judged as a scientist, Edward O. Wilson is today's leading evolutionist. He is also the leading spokesman for a religious-type reading of evolutionary thought. If you have already looked at *On Human Nature* (Cambridge: Harvard University Press, 1978), you will know what I mean. His most recent book, *Consilience: The Unity of Knowledge* (New York: Knopf, 1998), continues the theme. My experience is that there are very few who react to this latter book with indifference. Either you love it or you hate it. You have to approach it with a certain open mindedness. If you are looking for subtle technical philosophy, you will come away empty handed. But if you want inspiration and a push to thinking about things in new ways, then you may well find that there is much to stir you. I would also recommend Wilson's autobiography, *Naturalist* (Washington, D.C.: Island Press/Shearwater Books, 1994). It too is inspirational, but the bit I like best is about how miffed was Jim Watson of double helix fame, when he was beaten by Wilson in the race to get tenure at Harvard. More seriously, Wilson's book gives great insight into the life and mind of a scientist—the dedication necessary for real success is rather frightening.

And so as I come toward the end, let me recommend two more of my own books. *But Is It Science? The Philosophical Question in the Creation / Evolution Controversy*

(Buffalo, N.Y.: Prometheus, 1988) is an edited volume that brings together many different readings on and about the debate over evolution and Creationism. There is some historical material as well as a good selection dealing with the 1981 Arkansas creation trial, not to mention criticisms from my fellow philosophers over the kind of performance I gave in the witness box. My most recent book, *Can a Darwinian Be a Christian? The Relationship between Science and Religion* (Cambridge: Cambridge University Press, 2000) tries to look seriously not only at the pertinent science in the evolution/creation debate but also at the relevant theology. Too often people on both sides are arguing from a very inadequate conception of what constitutes Christian belief—at least, what constitutes traditional Christian belief, for let there be no pretence that Creationism is other than a particular form of American evangelical religion.

I think that indeed a Darwinian can be a Christian. You may disagree, but before you do disagree, you owe it to both sides (as well as to yourself) to learn what people want to claim. If nothing else, in my book I try to give you the pertinent information and the tools to start on the job yourself—the job, that is, of deciding where you stand on evolution and how if at all this affects your beliefs in other domains, particularly those of religion. Since this is a book that deals primarily with the ideas of others, trying deliberately to put my own convictions into the background, and since you do have the right to know where I myself stand, I would direct you to my *Taking Darwin Seriously: A Naturalistic Approach to Philosophy*, 2d ed. (Buffalo: Prometheus, 1998). There I do try to sketch out the philosophy of a Darwinian, one who takes the science as seriously and as importantly as does someone like Edward O. Wilson, but who does not (unlike Wilson) turn the science into a form of secular religion. Wilson rather thinks that this makes my enterprise somewhat sterile. I think he is chasing dreams from his past. You judge.

Epilogue

Charles Darwin's body lies a mould'ring in his grave, but his soul goes marching on. And it marches in the form of the wonderful theory that he bequeathed to us. I am not now particularly interested in whether you are a Darwinian or even an evolutionist—although I do hope that you are certainly the latter if not the former. I am certainly not interested in whether or not you are a Christian or a subscriber to some other faith. I do not mean to be rude: I expect and hope that your religious beliefs are important to you as mine are to me. But I am not about to convert you to or from Christianity or any other religion. That is for you to decide.

What I do care is that, at the very least, you are now able to stand back and appreciate what Darwin and his fellow evolutionists did and still do. I want you to recognize that these were magnificent achievements, even if in the end you decide you cannot accept them. I want you to see that the work of these scientists is the real miracle of life. That grubby little primates should be able to work out all of these things is something one should respect and admire. If God exists—certainly if the Christian God exists—then we have in this life an intellectual challenge as much as a moral challenge. It is our job to discern and understand this wonderful creation, and to give thanks and praise. That is what the evolutionists have been doing. And whether you are a Christian or not, history and logic dictate that you can accept evolution—Darwinism even—for what it is. A wonderful scientific theory—no more but certainly no less.

And with this I come to an end. I have spent a lot of my life working on and around Darwin and his achievements. I have had a lot of fun doing so. If I have passed on to you some of my enthusiasm, then this is a good reason for my having written this book and for your having read it. What more can either of us say?

References

Agassiz, E. C., ed. 1885. *Louis Agassiz: His Life and Correspondence*. Boston: Houghton Mifflin.

Agassiz, L. 1859. *Essay on Classification*. London: Longman, Brown, Green, Longmans, and Roberts and Trubner.

Allan, J. M. 1869. On the real difference in the minds of men and women. *Journal of Anthropology* 7: cxcv–ccxix.

Allen, G. E. 1978. *Thomas Hunt Morgan: The Man and His Science*. Princeton, N.J.: Princeton University Press.

Allen, L. et al. 1977. Sociobiology: A new biological determinism. In Science for the People, ed., *Biology as a Social Weapon*. Minneapolis, Minn.: Burgess Publishing.

Alvarez, L. W., W. Alvarez, F. Asaro, and H. V. Michel. 1980. Extraterrestrial cause for the Cretaceous-Tertiary extinction. *Science* 208: 1095–1098.

Appel, T. A. 1987. *The Cuvier-Geoffroy Debate: French Biology in the Decades before Darwin*. New York: Oxford University Press.

Arensburg, B., L. A. Schepartz, A. M. Tillier, B. Vandermeersch, and Y. Rak. 1990. A reappraisal of the anatomical basis for speech in Middle Palaeolithic hominids. *American Journal of Physical Anthropology* 83: 137–146.

Arensburg, B., A. M. Tillier, B. Vandermeersch, H. Duday, L. A. Schepartz, and Y. Rak. 1989. A Middle Palaeolithic human hyoid bone. *Nature* 338: 758–760.

Bannister, R. 1979. *Social Darwinism: Science and Myth in Anglo-American Social Thought*. Philadelphia: Temple University Press.

Barth, K. [1949] 1959. *Dogmatics in Outline*. New York: Harper and Brothers.

Beatty, J. 1987. Weighing the risks: Stalemate in the classical/balance controversy. *Journal of the History of Biology* 20: 289–319.

Beecher, H. W. 1885. *Evolution and Religion*. Part I: *Eight Sermons Discussing the Bearings of the Evolutionary Philosophy on the Fundamental Doctrines of Evangelical Christianity;* Part II: *Eighteen Sermons Discussing the Application of the Evolutionary Principles and Theories to the Practical Aspects of Religious Life*. New York: Fords, Howard, & Hulbert.

Behe, M. 1996. *Darwin's Black Box: The Biochemical Challenge to Evolution*. New York: Free Press.

Bowler, P. J. 1976. *Fossils and Progress*. New York: Science History Publications.

————. 1996. *Life's Splendid Drama*. Chicago: Chicago University Press.

Brewster, D. 1844. Review of "Vestiges." *North British Review* 3: 470–515.

Browne, J. 1995. *Charles Darwin: Voyaging. Volume 1 of a Biography*. New York: Knopf.

Buber, M. 1937. *I and Thou*. Edinburgh: T. and T. Clark.

Bullock, A. 1991. *Hitler and Stalin: Parallel Lives*. London: HarperCollins.

Bultmann, R. 1958. *Jesus Christ and Mythology*. New York: Scribner.

Burchfield, J. D. 1975. *Lord Kelvin and the Age of the Earth*. New York: Science History Publications.

Burdach, C. F. 1832. *Die Physiologie als Erfahrungswissenschaft*. Leipzig.

Burkhardt, R. W. 1995. *The Spirit of System: Lamarck and Evolutionary Biology (with a New Forword by the Author)*. Cambridge: Harvard University Press.

Cain, J. A. 1993. Common problems and cooperative solutions: Organizational activity in evolutionary studies, 1936–1947. *Isis* 84: 1–25.

Cairns-Smith, A. G., and H. Hartman, eds. 1986. *Clay Minerals and the Origin of Life*. Cambridge: Cambridge University Press.

Calvin, J. 1847–1850. *Commentaries on the First Book of Moses Called Genesis*. Translated by J. King. Edinburgh: Calvin Translation Society.

Cann, R. L., M. Stoneking, and A. C. Wilson. 1987. Mitochondrial DNA and human evolution. *Nature 25:* 31–36.

Chambers, R. 1844. *Vestiges of the Natural History of Creation*. London: Churchill.

————. 1846. *Explanations: A Sequel to "Vestiges of the Natural History of Creation."* London: Churchill.

Coleman, W. 1964. *Georges Cuvier Zoologist. A Study in the History of Evolutionary Thought*. Cambridge: Harvard University Press.

Crook, D. P. 1994. *Darwinism, War, and History: The Debate over the Biology of War from the "Origin of Species" to the First World War*. Cambridge: Cambridge University Press.

Cuvier, G. 1810. *Rapport Historique sur les progrés des sciences naturelles*. Paris.

Darwin, C. 1859. *On the Origin of Species*. London: John Murray.

————. 1871. *The Descent of Man*. London: John Murray.

Darwin, E. 1794–1796. *Zoonomia: or, The Laws of Organic Life*. London: J. Johnson.

————. 1801. *Zoonomia: or, The Laws of Organic Life*. 3d ed. London: J. Johnson.

————. 1803. *The Temple of Nature*. London: J. Johnson.

Darwin, F. 1887. *The Life and Letters of Charles Darwin, Including an Autobiographical Chapter*. London: John Murray.

Darwin, F., and A. C. Seward, eds. 1903. *More Letters of Charles Darwin*. 2 vols. London: John Murray.

Davies, N. 1992. *Dunnock Behaviour and Social Evolution*. Oxford: Oxford University Press.

Dawkins, R. 1976. *The Selfish Gene*. Oxford: Oxford University Press.

————. 1983. Universal Darwinism. In D. S. Bendall, ed., *Evolution from Molecules to Men*, 403–425. Cambridge: Cambridge University Press.

————. 1986. *The Blind Watchmaker*. New York: Norton.

————. 1995a. Richard Dawkins: "A survival machine." In J. Brockman, ed., *The Third Culture*, 74–95. New York: Simon and Schuster.

————. 1995b. *River out of Eden: A Darwinian View of Life*. New York: Basic Books.

————. 1996. *Climbing Mount Improbable*. New York: Norton.

————. 1997a. Obscurantism to the rescue. *Quarterly Review of Biology* 72: 397–399.

————. 1997b. Religion is a virus. *Mother Jones*. November/December, 60–61.

Dennett, D. C. 1995. *Darwin's Dangerous Idea: Evolution and the Meanings of Life*. New York: Simon and Schuster.

Desmond, A. 1994. *Huxley, the Devil's Disciple*. London: Michael Joseph.

————. 1997. *Huxley, Evolution's High Priest*. London: Michael Joseph.

Desmond, A., and J. Moore. 1992. *Darwin: The Life of a Tormented Evolutionist*. New York: Warner.

Dobzhansky, T. 1937. *Genetics and the Origin of Species*. New York: Columbia University Press.

Dobzhansky, T., and H. Levene. 1955. Developmental homeostasis in natural populations of *Drosophila pseudoobscura*. *Genetics* 40: 797–808.

Dobzhansky, T., and B. Wallace. 1953. The genetics of homeostasis in *Drosophila*. *Proceedings of the National Academy of Sciences* 39: 162–171.

Duarte, C., J. Mauricio, P. B. Pettitt, P. Souto, E. Trinkaus, H. van der Plicht, and J. Zilhao. 1999. The early Upper Paleolithic human skeleton from the Abrigo do Lagar Velho (Portugal) and modern human emergence in Iberia. *Proceedings of the National Academy of Sciences* 96: 7604–7609.

Duncan, D., ed. 1908. *Life and Letters of Herbert Spencer*. London: Williams and Norgate.

Dupree, A. H. 1959. *Asa Gray 1810–1888*. Cambridge: Harvard University Press.

Eldredge, N., and S. J. Gould. 1972. Punctuated equilibria: An alternative to phyletic gradualism. In T. J. M. Schopf, ed, *Models in Paleobiology*, 82–115. San Francisco: Freeman, Cooper.

Ellegård, A. 1958. *Darwin and the General Reader.* Goteborg: Goteborgs Universitets Arsskrift.

Elliot Smith, G. 1924. *Evolution of Man.* London: Oxford University Press.

Erskine, F. 1995. "The Origin of Species" and the science of female inferiority. In D Amigoni and J. Wallace, eds., *Charles Darwin's "The Origin of Species": New Interdisciplinary Essays,* 95–121. Manchester: Manchester University Press.

Farley, J. 1977. *The Spontaneous Generation Controversy from Descartes to Oparin.* Baltimore: Johns Hopkins University Press.

Feduccia, A. 1996. *The Origin and Evolution of Birds.* New Haven: Yale University Press.

Fodor, J. A. 1983. *Modularity of Mind: An Essay on Faculty Psychology.* Cambridge: MIT Press.

Fox, S. W. 1988. *The Emergence of Life: Darwinian Evolution from the Inside.* New York: Basic Books.

Friedlander, S. 1997. *Nazi Germany and the Jews: The Years of Persecution 1933–1939.* London: Weidenfeld and Nicolson.

Gasman, D. 1971. *The Scientific Origins of National Socialism: Social Darwinism in Ernst Haeckel and the German Monist League.* New York: American Elsevier.

Gilkey, L. B. 1985. *Creationism on Trial: Evolution and God at Little Rock.* Minneapolis, Minn.: Winston Press.

Gould, S. J. 1966. Allometry and size in ontogeny and phylogeny. *Biological Reviews of the Cambridge Philosophical Society* 41: 587–640.

———. 1969. An evolutionary microcosm: Pleistocene and Recent history of the land snail 'P. (Poecilozonites)' in Bermuda. *Bulletin of the Museum of Comparative Zoology* 138: 407–532.

———. 1977a. *Ever since Darwin.* New York: Norton.

———. 1977b. *Ontogeny and Phylogeny.* Cambridge: Belknap Press of Harvard University Press.

———. 1980a. Is a new and general theory of evolution emerging? *Paleobiology* 6: 119–130.

———. 1980b. The Piltdown conspiricy. *Natural History* 89(August): 8–28.

———. 1981. *The Mismeasure of Man.* New York: Norton.

———. 1982. Darwinism and the expansion of evolutionary theory. *Science* 216: 380–387.

———. 1983. Irrelevance, submission, and partnership: The changing role of paleontology in Darwin's three centennials, and a modest proposal for macroevolution. In D. S. Bendall, ed., *Evolution from Molecules to Men,* 347–66. Cambridge: Cambridge University Press.

———. 1984. Morphological channeling by structural constraint: Convergence in styles of dwarfing and gigantism in *Cerion,* with a description of two new fossil species and a report on the discovery of the largest *Cerion. Paleobiology* 10: 172–194.

————. 1989. *Wonderful Life: The Burgess Shale and the Nature of History.* New York: W. W. Norton.

————. 1996. *Full House: The Spread of Excellence from Plato to Darwin.* New York: Harmony Books.

————. 1997. Darwinian fundamentalism. *The New York Review* 44(10): 34–37.

————. 1999. *Rocks of Ages.* New York: Norton.

Gould, S. J., and C. B. Calloway. 1980. Clams and brachipods—ships that pass in the night. *Paleobiology* 6: 383–396.

Gould, S. J., and N. Eldredge. 1977. Punctuated equilibria: The tempo and mode of evolution reconsidered. *Paleobiology* 3: 115–151.

Gould, S. J., and R. C. Lewontin. 1979. The spandrels of San Marco and the Panglossian paradigm: A critique of the adaptationist program. *Proceedings of the Royal Society of London, B: Biological Sciences* 205: 581–598.

Gould, S. J., and E. S. Vrba. 1982. Exaptation—a missing term in the science of form. *Paleobiology* 8: 4–15.

Graham, L. R. 1987. *Science, Philosophy, and Human Behavior in the Soviet Union.* 2d ed. New York: Columbia University Press.

Grant, M. 1916. *The Passing of the Great Race: Or the Racial Basis of European History.* New York: Charles Scribner's Sons.

Gray, A. 1876. *Darwiniana.* New York: Appleton.

Gray, J. L. 1894. *Letters of Asa Gray.* Boston: Houghton Mifflin.

Haeckel, E. 1866. *Generelle Morphologie der Organismen.* Berlin: Reimer.

————. 1898. *The Last Link: Our Present Knowledge of the Descent of Man.* 2d ed. London: A. & C. Black.

Haldane, J. B. S. 1929. The origin of life. *The Rationalist Annual for the Year 1929,* 1–10. London: Watts.

Hamilton, W. D. 1964a. The genetical evolution of social behaviour I. *Journal of Theoretical Biology* 7: 1–16.

————. 1964b. The genetical evolution of social behaviour II. *Journal of Theoretical Biology* 7: 17–32.

Haught, J. F. 1995. *Science and Religion: From Conflict to Conversation.* New York: Paulist Press.

Hodge, C. 1872. *Systematic Theology.* London and Edinburgh: Nelson.

Hrdy, S. 1981. *The Woman That Never Evolved.* Cambridge: Harvard University Press.

Hume, D. 1779 [1947]. *Dialogues Concerning Natural Religion,* Chaps. 10–20. Indianapolis: Bobbs-Merrill.

Huxley, J. S. 1927. *Religion without Revelation.* London: Ernest Benn.

————. 1942. *Evolution: The Modern Synthesis*. London: Allen and Unwin.

Huxley, L. 1900. *The Life and Letters of Thomas Henry Huxley*. London: Macmillan.

Huxley, T . H. 1863. *Evidence as to Man's Place in Nature*. London: Williams and Norgate.

————. 1868a. On the animals which are most nearly intermediate between birds and reptiles. *Geological Magazine* 5: 357–365.

————. 1868b. On some organisms living at great depths in the North Atlantic Ocean. *Quarterly Journal of Microscopical Science* 8: 203–212.

————. [1871] 1893. Administrative nihilism. *Methods and Results*, 251–289. London: Macmillan.

————. 1893. Evolution and ethics. In *Evolution and Ethics and Other Essays*, 46–116. London: Macmillan.

Hyatt, A. 1857. *Alpheus Hyatt's Travel Book*. Hyatt Papers, Syracuse University Library, Syracuse, New York.

————. 1889. *Genesis of the Arietidae*. Washington, D.C.: Smithsonian Institution.

John Paul II. 1997. The Pope's message on evolution. *Quarterly Review of Biology* 72: 377–383.

Johnson, P. E. 1993. *Darwin on Trial*. 2d ed. Downers Grove, Ill.: InterVarsity Press.

————. 1995. *Reason in the Balance: The Case against Naturalism in Science, Law and Education*. Downers Grove, Ill.: InterVarsity Press.

Jones, G. 1980. *Social Darwinism and English Thought*. Brighton: Harvester.

Joravsky, D. 1970. *The Lysenko Affair*. Cambridge: Harvard University Press.

Kelly, A. 1981. *The Descent of Darwin: The Popularization of Darwinism in Germany, 1860–1914*. Chapel Hill: University of North Carolina Press.

Kimura, M. 1983. *Neutral Theory of Molecular Evolution*. Cambridge: Cambridge University Press.

King-Hele, D., ed. 1981. *The Letters of Erasmus Darwin*. Cambridge: Cambridge University Press.

Krings, M., A. Stone, R. W. Schmitz, H. Krainitzki, M. Stoneking, and S. Pääbo. 1997. Neanderthal DNA sequences and the origin of modern humans. *Cell* 90: 19–30.

Kropotkin, P. [1902] 1955. *Mutual Aid*. Boston: Extending Horizon Books.

Lamarck, J. B. 1809. *Philosophie zoologique*. Paris: Dentu.

Landau, M. 1991. *Narratives of Human Evolution*. New Haven: Yale University Press.

Larson, E. J. 1997. *Summer for the Gods: The Scopes Trial and America's Continuing Debate over Science and Religion*. New York: Basic Books.

LeVay, S. 1996. *Queer Science: The Use and Abuse of Research into Homosexuality*. Cambridge: MIT Press.

Lewin, R. 1989. *Human Evolution: An Illustrated Introduction*. 2d ed. Oxford: Blackwell Scientific.

Lewontin, R. C. 1982. *Human Diversity*. New York: Scientific American Library.

Lewontin, R. C., and J. L. Hubby. 1966. A molecular approach to the study of genic heterozygosity in natural populations. II: Amount of variation and degree of heterozygosity in natural populations of *Drosophila pseudoobscura*. *Genetics* 54: 595–609.

Lewontin, R. C., J. A. Moore, W. B. Provine, and B. Wallace, eds. 1981. *Dobzhansky's Genetics of Natural Populations I-XLIII*. New York: Columbia University Press.

Lieberman, P. 1994. Hyoid bone position and speech: Reply to Dr. Arensburg et al. (1990). *American Journal of Physical Anthropology* 94: 275–278.

———. 1999. Silver-tongued Neanderthals? *Science* 283:175.

Lovejoy, A. O. 1936. *The Great Chain of Being*. Cambridge: Harvard University Press.

Lurie, E. 1960. *Louis Agassiz: A Life in Science*. Chicago: University of Chicago Press.

Lyell, C. 1830–1833. *Principles of Geology*. London: John Murray.

———. 1863. *Antiquity of Man*. London: John Murray.

Malthus, T. R. [1826] 1914. *An Essay on the Principle of Population*. 6th ed. London: Everyman.

Marchant, J., ed. 1916. *Alfred Russel Wallace: Letters and Reminiscences*. London: Cassell.

Margulis, L. 1982. *Early Life*. Boston: Science Books International.

Maynard Smith, J. 1981. Did Darwin get it right? *London Review of Books* 3(11): 10–11.

———. 1982. *Evolution and the Theory of Games*. Cambridge: Cambridge University Press.

———. 1995. Genes, memes, and minds. *New York Review of Books* 42(19): 46–48.

Mayr, E. 1942. *Systematics and the Origin of Species*. New York: Columbia University Press.

———. 1954. Change of genetic environment and evolution. In J. Huxley, A. C. Hardy, and E. B. Ford, eds., *Evolution as a Process*, 157–180. London: Allen and Unwin.

McCosh, J. 1882. *The Method of Divine Government, Physical and Moral*. London: Macmillan.

McDonald, J., and M. Kreitman. 1991. Adaptive protein evolution at the Adh locus in Drosophila. *Nature* 351: 652–654.

McMullin, E., ed. 1985. *Evolution and Creation*. South Bend, Ind.: University of Notre Dame Press.

McNeil, M. 1987. *Under the Banner of Science: Erasmus Darwin and His Age*. Manchester: Manchester University Press.

Medawar, P. 1961. Review of *The Phenomenon of Man*. *Mind* 70: 99–106.

Miller, K. 1999. *Finding Darwin's God: A Scientist's Search for Common Ground between God and Evolution*. New York: Cliff Street Books.

Mitman, G. 1992. *The State of Nature: Ecology, Community, and American Social Thought, 1900–1950.* Chicago: University of Chicago Press.

Moore, A. L. 1890. The Christian doctrine of God. In C. Gore, ed, *Lux Mundi: A Series of Studies in the Religion of the Incarnation,* 10th ed. 41–81. London: John Murray.

Moore, J. R. 1979. *The Post-Darwinian Controversies: A Study of the Protestant Struggle to Come to Terms with Darwin in Great Britain and America, 1870–1900.* Cambridge: Cambridge University Press.

Muller, H. J. 1949. The Darwinian and modern conceptions of natural selection. *Proceedings of the American Philosophical Society* 93: 459–470.

Numbers, R. L. 1992. *The Creationists.* New York: A. A. Knopf.

Oakley, K. P. 1964. The problem of man's antiquity. *Bulletin of the British Museum (Natural History), Geological Series* 9(5).

O'Brien, C. F. 1970. Eozoon canadense: The dawn animal of Canada. *ISIS* 61: 206–223.

Oparin, A. [1924] 1967. The origin of life. Translated by A. Synge. In J. D. Bernal, ed., *The Origin of Life,* 199–234. Cleveland: World. Originally published as *Proishkhozhdenie zhizni* (1928).

Osborn, H. F. 1931. *Cope: Master Naturalist: The Life and Writings of Edward Drinker Cope.* Princeton, N.J.: Princeton University Press.

Paley, W. [1802] 1819. *Natural Theology (Collected Works: IV).* London: Rivington.

Parker, G. A. 1970. The reproductive behaviour and the nature of the sexual selection in *Scatophaga stercovaria* L. (Diptera: Scatophagidae)—VIII. The origin and evolution of the passive phase. *Evolution* 24: 744–788.

————. 1978. Searching for mates. Review article in J. R. Krebs and N. B. Davies, eds., *Behavioural Ecology: An Evolutionary Approach.* Oxford: Blackwell.

Pasteur, L. [1883] 1922. La dissymétrie moléculaire. In Pasteur Vallery-Radot, ed., *Oeuvres de Pasteur,* Vol. 1. Paris: Masson.

Pilbeam, D. 1984. The descent of hominoids and hominids. *Scientific American* 250(3): 84–97.

Pinker, S. 1994. *The Language Instinct: How the Mind Creates Language.* New York: William Morrow.

————. 1997. *How the Mind Works.* New York: Norton.

Pittenger, M. 1993. *American Socialists and Evolutionary Thought, 1870–1920.* Madison: University of Wisconsin Press.

Plantinga, A. 1991. When faith and reason clash: Evolution and the Bible. *Christian Scholar's Review* 21(1): 8–32.

Provine, W. B. 1971. *The Origins of Theoretical Population Genetics.* Chicago: University of Chicago Press.

————. 1988. Progress in evolution and meaning in life. In M. N. Nitecki, ed., *Evolutionary Progress,* 49–74. Chicago: University of Chicago Press.

Rainger, R. 1991. *An Agenda for Antiquity: Henry Fairfield Osborn and Vertebrate Paleontology at the American Museum of Natural History, 1890–1935.* Tuscaloosa: University of Alabama Press.

Redi, F. [1688] 1909. *Experiments on the Generation of Insects.* Translated by M. Bigelow. Chicago: Open Court Publishing.

Reichenbach, B. R. 1976. Natural evils and natural laws: A theodicy for natural evils. *International Philosophical Quarterly* 16: 179–196.

Richards, R. J. 1987. *Darwin and the Emergence of Evolutionary Theories of Mind and Behavior.* Chicago: University of Chicago Press.

————. 1992. *The Meaning of Evolution: The Morphological Construction and Ideological Reconstruction of Darwin's Theory.* Chicago: University of Chicago Press.

Roger, J. 1997. *Buffon: A Life of Natural History.* Ithaca, N.Y.: Cornell University Press.

Ruse, M. 1979. *The Darwinian Revolution: Science Red in Tooth and Claw.* Chicago: University of Chicago Press.

————. 1980. Charles Darwin and group selection. *Annals of Science* 37: 615–630.

————. 1982. *Darwinism Defended: A Guide to the Evolution Controversies.* Reading, Mass.: Addison-Wesley.

————. 1984. Is there a limit to our knowledge of evolution? *BioScience* 34(2): 100–104.

————. 1986. *Taking Darwin Seriously: A Naturalistic Approach to Philosophy.* Oxford: Blackwell.

————. 1996. *Monad to Man: The Concept of Progress in Evolutionary Biology.* Cambridge: Harvard University Press.

————. 1999. *Mystery of Mysteries: Is Evolution a Social Construction?* Cambridge: Harvard University Press.

————. 2001. *Can a Darwinian Be a Christian? The Relationship between Science and Religion.* Cambridge: Cambridge University Press.

Ruse, M., ed. 1988. *But Is It Science? The Philosophical Question in the Creation/Evolution Controversy.* Buffalo, N.Y.: Prometheus.

Russell, E. S. 1916. *Form and Function, a Contribution to the History of Animal Morphology.* London: John Murray.

Russett, C. E. 1976. *Darwin in America: The Intellectual Response, 1865–1912.* San Francisco: Freeman.

Sahlins, M. 1976. *The Use and Abuse of Biology.* Ann Arbor: University of Michigan Press.

Schiff, M., and R. C. Lewontin. 1986. *Education and Class: The Irrelevance of IQ Studies.* Oxford: Oxford University Press.

Sepkoski, J. J., Jr. 1976. Species diversity in the Phanerozoic—species-area effects. *Paleobiology* 2: 298–303.

————. 1979. The kinetic model of Phanerozoic taxonomic diversity. II: Early Phanerozoic families and multiple equilibria. *Paleobiology* 5: 222–251.

————. 1984. A kinetic model of Phanerozoic taxonomic diversity. III: Post-Paleozoic families and mass extinctions. *Paleobiology* 10: 246–267.

Settle, M. L. 1972. *The Scopes Trial: The State of Tennesse v. John Thomas Scopes.* New York: Franklin Watts.

Shor, E. N. 1974. *The Fossil Feud between E. D. Cope and O. C. Marsh.* Hicksville, N.Y.: Exposition Press.

Simpson, G. G. 1944. *Tempo and Mode in Evolution.* New York: Columbia University Press.

————. 1949. *The Meaning of Evolution: A Study of the History of Life and of Its Significance for Man.* New Haven: Yale University Press.

————. 1953. *The Major Features of Evolution.* New York: Columbia University Press.

————. 1969. Biology and ethics. In G. G. Simpson, ed., *Biology and Man,* 130–148. New York: Harcourt Brace and World.

Spencer, H. 1851. *Social Statics: Or, the Conditions Essential to Human Happiness Specified, and the First of Them Developed.* London: J. Chapman.

————. 1852. A theory of population, deduced from the general law of animal fertility. *Westminster Review* n.s. 1: 468–501.

————. 1857. Progress: Its law and cause. *Westminster Review* 67: 244–267.

————. 1864. *Principles of Biology.* London: Williams and Norgate.

————. [1873] 1961. *Study of Sociology.* Ann Arbor: University of Michigan Press.

————. 1904. *Autobiography.* London: Williams and Norgate.

Stebbins, G. L. 1950. *Variation and Evolution in Plants.* New York: Columbia Unversity Press.

————. 1969. *The Basis of Progressive Evolution.* Chapel Hill: University of North Carolina Press.

Stebbins, G. L., and F. J. Ayala. 1981. Is a new evolutionary synthesis necessary? *Science* 213: 967–971.

Sulloway, F. J. 1979. *Freud, Biologist of the Mind: Beyond the Psychoanalytic Legend.* New York: Basic Books.

Sumner, W. G. 1914. *The Challenge of Facts and Other Essays.* New Haven: Yale University Press.

Teilhard de Chardin, P. 1959. *The Phenomenon of Man.* London: Collins.

Trivers, R. 1971. The evolution of reciprocal altruism. *Quarterly Review of Biology* 46: 35–57.

Wallace, A. R. 1900. *Studies: Scientific and Social*. London: Macmillan.

—. 1905. *My Life: A Record of Events and Opinions*. London: Chapman and Hall.

Westfall, R. S. 1980. *Never at Rest: A Biography of Isaac Newton*. Cambridge: Cambridge University Press.

Whitcomb, J. C., and H. M. Morris. 1961. *The Genesis Flood: The Biblical Record and its Scientific Implications*. Philadelphia: Presbyterian and Reformed Publishing.

Williams, G. C. 1966. *Adaptation and Natural Selection*. Princeton, N.J.: Princeton University Press.

—. 1975. *Sex and Evolution*. Princeton, N.J.: Princeton University Press.

Williamson, P. 1985. Punctuated equilibrium, morphological stasis and the paleontological documentation of speciation. *Biological Journal of the Linnaean Society of London* 26: 307–324.

Wilson, E. O. 1975. *Sociobiology: The New Synthesis*. Cambridge: Harvard University Press.

—. 1978. *On Human Nature*. Cambridge: Harvard University Press.

—. 1980a. Caste and division of labor in leaf-cutter ants (Hymenoptera: Formicidae: *Atta*). I: The overall pattern in *A. sexdens*. *Behavioral Ecology and Sociobiology* 7: 143–156.

—. 1980b. Caste and division of labor in leaf-cutter ants (Hymenoptera: Formicidae: *Atta*). II: The ergonomic optimization of leaf cutting. *Behavioral Ecology and Sociobiology* 7: 157–165.

—. 1994. *Naturalist*. Washington, D.C.: Island Press/Shearwater Books.

Winsor, M. P. 1991. *Reading the Shape of Nature: Comparative Zoology at the Agassiz Museum*. Chicago: University of Chicago Press.

Wolpoff, M., and R. Caspari. 1997. *Race and Human Evolution: A Fatal Attraction*. New York: Simon and Schuster.

Wright, G. F. 1882. *Studies in Science and Religion*. Andover, Mass.: W. F. Draper.

Wright, S. 1932. The roles of mutation, inbreeding, crossbreeding and selection in evolution. *Proceedings of the Sixth International Congress of Genetics* 1: 356–366.

Wynne-Edwards, V. C. 1962. *Animal Dispersion in Relation to Social Behaviour*. Edinburgh: Oliver and Boyd.

Young, R. M. 1985. *Darwin's Metaphor: Nature's Place in Victorian Culture*. Cambridge: Cambridge University Press.

Zihlman, A. 1981. Women as shapers of the human adaptation. In F. Dahlberg, ed., *Woman the Gatherer*, 75–120. New Haven: Yale University Press.

Chronology

1862	Louis Pasteur crushes idea of spontaneous generation	1939–1945	World War II
1863	*Evidence as to Man's Place in Nature* by Thomas Henry Huxley	1942	*Evolution: The Modern Synthesis* by Julian Huxley
1869	Founding of American Museum of Natural History		*Systematics and the Origin of Species* by Ernst Mayr
1871	*The Descent of Man* by Charles Darwin	1944	*Tempo and Mode in Evolution* by George Gaylord Simpson
1876	Thomas Henry Huxley announces fossil history of horse	1947	Journal *Evolution* founded
	Darwiniana by Asa Gray	1950	*Variation and Evolution in Plants* by G. Ledyard Stebbins
1879	*Principles of Ethics* by Herbert Spencer	1953	Stanley Miller and Harold Urey simulate the natural production of amino acids
1881	Opening of British Museum (Natural History)		Discovery of double helix by James Watson and Francis Crick
1882	Death of Charles Darwin	1955	*Le phénomène humaine* by Pierre Teilhard de Chardin
1900	Mendel's laws rediscovered		
	Albert Einstein discovers relativity theory	1957	Sputnik
1901	*Mutual Aid* by Prince Petr Kropotkin		*Syntactic Structures* by Noam Chomsky
	Death of Queen Victoria	1961	*Genesis Flood* by John Whitmore and Henry Morris
1908	Hardy-Weinberg law	1964	W. D. Hamilton proposes theory of kin selection
1912	Discovery of Piltdown Man	1966	Richard Lewontin uses gel electrophoretic techniques
1914–1918	World War I		
1925	Raymond Dart discovers Taung baby	1972	Niles Eldredge and Stephen Jay Gould announce punctuated equilibria theory
	Scopes Monkey Trial	1974	Don Johanson discovers Lucy
1928	*The Origin of Life* by A. I. Oparin	1975	*Sociobiology: The New Synthesis* by Edward O. Wilson
1929	*The Origin of Life* by J. B. S. Haldane	1976	*The Selfish Gene* by Richard Dawkins
	Wall Street crash		
1930	*The Genetical Theory of Natural Selection* by R. A. Fisher	1981	Arkansas Creation Trial
1932	Adaptive landscape metaphor of Sewall Wright	1997	John Paul II issues papal letter accepting modern evolutionary theory
1933	Adolf Hitler comes to power		
1937	*Genetics and the Origin of Species* by Theodosius Dobzhansky	1999	State of Kansas takes evolution out of school curricula.

ABIOGENESIS: the natural development of life from nonliving materials (traditionally applied to the spontaneous generation of life from inorganic, never-living material).

ADAPTATION: any characteristic that aids its possessor to survive and reproduce.

ADAPTATIONISM: the belief that all organic characters are indeed adaptive.

ADAPTIVE LANDSCAPE: a metaphor introduced by Sewall Wright claiming that the fitness of organisms can be mapped as if on a hilly terrain.

AGNOSTICISM: the belief that one cannot know whether or not God exists.

ALLELE: any one of a number of forms of a gene that can occupy the same place (locus) on a chromosome.

ALLOMETRY: the study of the relative growth of some parts of an organism in comparison with other parts or the whole.

ALTRUISM: help given by one organism to another, at some biological cost, for the donor's long-term reproductive advantage.

AMINO ACID: the major complex organic molecules that serve as the building blocks of proteins.

ARACHNID: an arthropod (an invertebrate with segmented body, jointed limbs, and an external skeleton) with eight legs; includes spiders and scorpions.

ARCHAEOPTERYX: an ancient bird with many reptilian features (a missing link).

ARCHETYPE: the basic building plan of a group of animals such as vertebrates.

ASTROLOGY: the system claiming that our destinies are controlled by the configurations of the heavens.

ATHEISM: the belief that God does not exist.

AUSTRALOPITHECUS: a genus that gave rise immediately to our genus, *Homo,* consisting of animals intermediate between apelike forms and humans.

AUTOCATALYTIC: becoming ever more powerful thanks to positive feedback mechanisms.

BALANCE HYPOTHESIS: the belief that natural selection holds many different alleles or genes in a balance or equilibrium within a population.

BALANCED SUPERIOR HETEROZYGOTE FITNESS: the claim that natural selection keeps different alleles in balance or equilibrium within a population, because the heterozygote is fitter than either homozygote.

BAUPLAN: an archetype.

BIOGENESIS: the natural development of life from living materials (as in normal generation).

BIOGENETIC LAW: the claim that ontogeny, individual development, recapitulates phylogeny, the evolution of the group.

BIOGEOGRAPHY: the study of the distribution of organisms.

BIOMETRICS/BIOMETRICIAN: the school of biologists at the beginning of the century committed to the belief that evolution could be studied through detailed quantification and statistical analysis.

CALVINISM: a branch of Protestantism which follows the sixteenth-century reformer John Calvin, marked especially by a belief in predestination (that is, that our fates are known to God from the first).

CAMBRIAN EXPLOSION: the rapid evolution and diversification of life forms in the Cambrian period, occurring between 500 and 600 million years ago.

CASTE SYSTEM: the different forms or morphs found within species in the hymenoptera.

CATASTROPHISM: a geological theory of periodic, violent, earth upheavals, endorsed by Georges Cuvier, believed responsible for major geological formations.

CELL: the building blocks of complex organisms, being membrane-contained units containing within them the genes and other components necessary for life.

CENANCESTOR: the most recent, jointly shared ancestor of all living things.

CHAIN OF BEING: a doctrine popular in the Middle Ages claiming that all organisms can be put in a continuous line from the very simple to the very complex.

CHROMOSOME: a string-like entity, in the cell, that carries the genes.

CLASSICAL HYPOTHESIS: the claim that there is little genetic variation within a population thanks to the cleansing effect of natural selection.

CONSTRAINTS: the physical factors that keep an organism developing along certain fixed limited paths.

CREATION SCIENCE: the claim that the early chapters of Genesis can be given good scientific backing.

CREATIONIST: one who believes in the literal truth of the Bible, especially the early chapters of Genesis.

CYTOLOGY: the systematic study of cell structure.

DARWINIAN: a person who accepts Charles Darwin's theory of evolution through natural selection.

DARWINISM: the theory of evolution through natural selection.

DEEP STRUCTURE: the claim, first made by Noam Chomsky, that all languages have a fundamental, underlying similarity.

DEISM: the belief in God as unmoved mover.

DEOXYRIBONUCLEIC ACID (DNA): the macromolecule that transmits genetic information (the molecular gene).

DIVISION OF LABOR: the process of breaking down an activity into specialized tasks so that it can be performed much more efficiently.

DOMINANT: a gene (allele) whose effects mask the effects of its paired opposite at the same locus.

DROSOPHILA: a fruitfly, a popular organism for study by geneticists.

DUALIST/DUALISM: the belief that the mind is a substance (usually thought of as a thinking or spiritual substance) corresponding to the substance of the body (usually thought of as material substance).

DYNAMIC EQUILIBRIUM: a balance between opposing forces that is in constant motion, usually upwards.

ÉLAN VITAL/VITALISM: the force the vitalists believe animates living bodies.

EMBRANCHEMENT: one of the four divisions of the animal kingdom as supposed by Georges Cuvier (vertebrates, mollusks, articulata, and radiata).

EMBRYOLOGY: the study of developing multicellular organisms, particularly those in early development.

EUGENICS: the movement that aimed to improve humankind through selective breeding.

EUKARYOTE: an organism (like a mammal) where the DNA is on the chromosomes within the nucleus (see also, prokaryote).

EVANGELICAL CHRISTIANITY: a form of Protestantism that puts a major emphasis on personal commitment to Jesus as Lord, stressing the significance of the Bible taken fairly literally.

EVOLUTION (CAUSE): the mechanism or force behind evolutionary change.

EVOLUTION (FACT): the belief that organisms living and dead are the end results of a natural process of development from one or a few forms.

EVOLUTION (PATH): the particular track that organisms have taken through history, usually known as phylogeny.

EVOLUTIONARILY STABLE STRATEGY (ESS): a genetically programmed path or strategy taken by members of a group such that no other strategy or path can dislodge it.

EXAPTATION: an organic feature without adaptive function.

FITNESS: the comparative ability of an organism to survive and reproduce and pass on its genes.

FORM: the particular shape of features of an organism, usually distinguished from function.

FOUNDER PRINCIPLE: the claim by the systematist Ernst Mayr that the making of a new species involves just a few organisms isolated from the main group.

FUNCTION: a thing that a characteristic does, believed by Darwinians to have been produced by natural selection and hence adaptive.

FUNDAMENTALIST/FUNDAMENTALISM: a form of American Protestant evangelicalism popular at the beginning of the last century, committed to the literal truth of the Bible.

GEL ELECTROPHORESIS: a technique for detecting variations in molecules (and hence genes) by tracking their progress through a gel under the influence of an electric field.

GENE: the ultimate unit of heredity, believed today to consist of ribonucleic acid (usually DNA).

GENE POOL: the collective genes of a population of organisms.

GENETIC DRIFT: the claim by Sewall Wright that sometimes characteristics and their genes have so little effect that they escape the effects of natural selection and hence (proportionately) move randomly up or down within a population.

GENETICS: the theory of heredity dating back to the nineteenth-century monk Gregor Mendel.

GENOTYPE: the collective genes of any particular organism, to be contrasted with the phenotype, which refers to the collective physical characteristics of an organism.

GROUP SELECTION: the belief that sometimes natural selection can work for the benefit of the group against the interests of some individuals.

HETEROGENESIS: the belief that life is created naturally from non-organic matter (traditionally applied to the spontaneous generation of life from formerly living materials, as maggots from meat).

HETEROSIS: superior heterozygote fitness.

HETEROZYGOTE: an organism that has different alleles occupying the paired loci of some chromosome.

HIERARCHY THEORY: the claim that supposes that different evolutionary forces operate at different levels, from microevolution through to macroevolution (hence, anti-reductionistic).

HOLISTIC/HOLISM: the belief that in order to understand the individual one must look at the whole functioning organism and not just at the parts; to be contrasted with reductionism, the claim that higher level macrofunctions of organisms can be explained fully in terms of lower level microprocesses.

HOMEOSTASIS: the state of balance or equilibrium brought on by different forces within an individual or a population.

HOMINIDS: humans and their relatively recent ancestors and relations (technically, members of the family Hominidae).

HOMO: the genus of organisms that includes *Homo sapiens* as one of its members.

HOMOLOGY/HOMOLOGIES: the isomorphisms between the parts of different organisms of different species, believed today to be a result of common ancestry.

HOMOZYGOTE: an organism with two identical alleles at some locus on the chromosomes.

HYMENOPTERA: the ants, the bees, and the wasps.

IDENTITY THEORY: the philosophy of mind that takes mind and brain to be different aspects of the same substance.

INDIVIDUAL SELECTION: the claim that selection works for, and only for, the individual as opposed to the group.

INTELLIGENT DESIGN: a contemporary position arguing that the complexity of organisms is such that the only possible causal explanation is some form of intelligence.

"JUST SO" STORIES: fantastical adaptive scenarios that could not possibly be true (a term borrowed from the tales of Rudyard Kipling).

KIN SELECTION: the action of a selective force that promotes the well being of close relatives as well as of the individual; a form of biological altruism.

KINETIC MODEL: a theory about movement.

LAISSEZ-FAIRE: a sociopolitical economic doctrine which claims that, for the most good to be achieved, there should be no state interference in the workings of individual firms or businesses.

LAMARCKISM: the belief that acquired characters can be inherited.

LAW: a statement about some particular regularity of nature as in Mendel's Laws or the Hardy-Weinberg Law.

LOCUS: a particular point on a chromosome matched by a corresponding point on the other member of paired chromosomes; can be occupied by the various different alleles in the set peculiar to a species.

MACROEVOLUTION: organic change over large quantities of time.

MENDELIAN(ISM): the genetical theory going back to Gregor Mendel, and stressing that the units of heredity, the genes, remain unchanged from generation to generation (a particulate theory).

METAPHYSICAL NATURALISM: the belief that physical nature is all that exists, hence excluding any deity or other supernatural entity.

METHODOLOGICAL NATURALISM: the assumption that physical nature is all that exists, taken in order to promote the success of science, but not making any ultimate metaphysical assumptions.

MICROEVOLUTION: organic change over short periods of time.

MIRACLE: a special intervention by the Deity.

MITOCHONDRIA/MITOCHONDRIAL DNA: bodies within the cell, that are separate from the central nucleus, but carry genes.

MITOCHONDRIAL EVE: the female supposedly ancestral to all living humans, dating from about 140,000 years ago.

MODERNIST/MODERNISM: a movement, particularly in religion at about the beginning of the last century, which tried to accommodate and harmonize with the science of the day.

MOLECULAR CODE: the pattern governing the ordering of the DNA molecule's components, yielding information needed in the production of other cellular parts, notably amino acids.

MOLECULAR DRIFT: the belief that there is random change in populations at the molecular level, below the forces of natural selection (see also genetic drift, of which molecular drift is a special case).

MONAD: the most primitive (and presumably earliest) of all organisms.

MONOMORPHIC: having only one particular form, as opposed to dimorphic or polymorphic.

MORPHOLOGY: the study of the structure or form of organisms.

MUTATION: the spontaneous change in a gene, which may lead to new variations or characteristics.

NATURAL RELIGION/THEOLOGY: the belief that through reason one can get to know of the existence and nature of the Deity.

NATURAL SELECTION: Charles Darwin's mechanism of evolution supposing that, since more organisms are born than can survive and reproduce, fitter organisms will be successful in the struggle for existence and hence permanent change will be effected.

NATURALISM: a belief that one can explain everything through unbroken law.

NATURPHILOSOPHIE/NATURPHILOSOPHEN: an idealistic German morphological movement at the beginning of the nineteenth century that stressed form over function.

NEANDERTHAL: a subspecies of *Homo sapiens* that lived about 100,000 years ago.

NUCLEIC ACID: a chainlike, macromolecule found in cells, either DNA which (usually) carries the information of heredity, or RNA which reads the information from the DNA (in some viruses, RNA carries the information).

NUCLEUS/NUCLEI: the central part of a cell that contains chromosomes carrying genes.

ONTOGENY: the individual development of an organism.

ORIGINAL SIN: a Christian doctrine claiming that all humans are tainted with sin, due in some sense to the original misdeeds of Adam and Eve.

PALEOANTHROPOLOGY: the study of the fossil evidence for and evolutionary history of humankind.

PALEONTOLOGY: the study of life's past as revealed particularly through the fossils.

PANGLOSS(IAN): an extreme form of adaptationism, unjustifiably seeing selection as having produced everything (usually acting exclusively for the good of the individual), often marked by undue use of "just so" stories (taken from Voltaire's *Candide,* where the Leibnizian philosopher Dr. Pangloss sees value in everything, including the most extreme disasters).

PHENOTYPE: the collective physical characteristics of an organism, contrasted with the genotype.

PHOTOSYNTHESIS: the capture of the sun's energy by green plants for use in the manufacture of carbon compounds.

PHRENOLOGY: a nineteenth-century pseudoscience which claimed that you could read character from the shape of the skull.

PHYLETIC GRADUALISM: Stephen Jay Gould's term for the Darwinian commitment to the gradual (as opposed to jerky) nature of the evolutionary process.

PHYLOGENY: the path of evolution.

PLEIOTROPY: the production of different characteristics by the same gene.

POLYMORPHIC: having different forms within the same population.

POPULATION GENETICS: the generalization of Mendelian genetics to deal with groups, thus focussing on the spread and change of genetic variation within a population.

PROBLEM OF EVIL: the theological problem of reconciling a Christian God who is both all good and all powerful with the existence of evil and pain in the world.

PROGRESS (BIOLOGICAL): the belief that the course of evolution has been from the simple to the complex.

PROGRESS (CULTURAL): the belief that humankind can and is improving its lot.

PROKARYOTE: a single-celled organism, like a bacterium, where the DNA floats free (as opposed to a eukaryote).

PROTEIN: a chain-like macromolecule, one of the building blocks of the cell, composed of amino acids as produced by the translation of the RNA coded from the DNA.

PROVIDENTIALISM: the belief that God is always present (immanent) in the universe and ready to intercede on our behalf.

PUNCTUATED EQUILIBRIA: the theory of Stephen Jay Gould that the course of evolution is one of lack of action (stasis) followed by short, sharp breaks or jumps.

RECESSIVE: an allele whose effects are masked by the other allele at the same locus (see dominant).

RECIPROCAL ALTRUISM: help given by one organism to another with the expectation that such help will be returned.

REDUCTIONISTIC/REDUCTIONISM: the claim that higher-level macrofunctions of organisms can be explained fully in terms of lower-level microprocesses (to be contrasted with holism).

REVEALED RELIGION/THEOLOGY: that area of religious belief dealing with claims based on faith or dogma or authority; to be contrasted with natural theology.

REVERSE ENGINEERING: the process of discovering an organism's function by working backwards from its form to its possible design.

RIBONUCLEIC ACID (RNA): the kind of nucleic acid that carries the information from the DNA to produce the amino acids that make up proteins; in some viruses, RNA acts as the carrier of information itself.

RIBOSOMES: particles within cells where RNA makes proteins.

SALTATION(ISM): the evolutionary theory claiming that organic change goes in jumps, usually through macromutations.

SELFISH GENE: a metaphor produced by Richard Dawkins to emphasize that selection ultimately works for the benefit of the individual, perhaps even the individual gene, rather than the group.

SEXUAL SELECTION: a secondary form of selection posited by Charles Darwin and involving competition within species for mates.

SOCIOBIOLOGY: that area of evolutionary thought dealing with the development and maintenance of social behavior.

SPANDRELS: Stephen Gould's term for the nonfunctional byproducts of adaptations, taken from the triangular mosaic-covered areas to be seen at the tops of the columns in St. Mark's in Venice (properly these areas are called pendentives).

SPONTANEOUS GENERATION: the belief that life appears in one move, naturally from nonliving matter.

STASIS: the periods of stability or nonchange as supposed in Stephen Gould's theory of punctuated equilibria.

SYNTHETIC THEORY: the theory of evolution combining Darwinian selection with Mendelian or post-Mendelian particulate genetics; sometimes called neo-Darwinism.

SYSTEMATICS: the area of evolutionary biology that classifies and explains the relationships between organisms.

TELEOLOGY: the claim that entities, organisms particularly, can and should be considered with reference to their ends or functions and not simply by the causes that brought them about.

THEIST: someone who believes in an immanent god, that is a god who is prepared to intervene in the creation: a term traditionally reserved for the followers of the major religions of Judaism, Christianity, and Islam.

TRANSUBSTANTIATION: the Catholic miracle where the bread and wine in the mass are changed into the body and blood of Jesus Christ.

UNIFORMITARIANISM: the geological theory associated with Charles Lyell that supposes that all geological change takes place gradually through causes now in operation; to be contrasted with catastrophism.

UNITY OF TYPE: the fact that organisms exhibit significant isomorphisms or homologies, linked through the sharing of a common archetype or Bauplan, and explained by evolutionists as a function of shared descent.

UTILITY FUNCTION: the purpose or function supposed by engineers of machines and thus the thing to be discovered through reverse engineering.

Documents

Chapter 1

My first document is an extract from Jean Baptiste de Lamarck's Philosophie Zoologique *(1809). Lamarck's theory of evolution was of an upwardly moving escalator, with organisms somehow being carried along rising ever higher until they evolved into humans. On this, as a kind of secondary mechanism, he laid the force for which he is best known, the inheritance of acquired characteristics: through use and disuse features are developed or lost, and then they are passed on to future genera-tions. This is the idea presented in the first document. Note how Lamarck presents his thinking in the form of laws of nature. He wants to come across as a genuine scientist like those in the physical sciences. Georges Cuvier of course would have none of this, and in the second document he takes on Lamarck's ideas, arguing that they are simply not borne out in nature: breeders can never cross the species barrier and the forms of ancient animals are those that live today. Note that whatever his ideological objections to evolutionism, Cuvier the scientist knew that he had to oppose it on scien-tific grounds.*

The third document of this chapter contains extracts from In Memoriam *by the English poet Alfred Tennyson. Published in 1850, the poem is dedicated to the memory of a friend (Arthur Hallam) who had died some twenty years before. It was long in conception and writing as Tennyson wrestled with his troublesome and worrying thoughts, especially those sparked by his reading of the science of his day. As a young man Tennyson had read Charles Lyell's* Principles of Geology, *and in the first of the two extracts we see the poet recoiling from the endless, mindless picture of nature that the geologist sketched. Nothing seems to make sense. Later Tennyson read a detailed review of Robert Chambers's* Vestiges of the Natural History of Creation, *that evolutionary work pub-lished in 1844. It claimed that all of nature is in an evolutionary progression up to humankind and perhaps beyond to something superior. In the second of the extracts, Tennyson recovers hope by sup-posing that Hallam was a precursor of this future "crowning race," doomed because he was born be-fore his time. I offer these passages here to show that even before the* Origin *people were getting ready for evolutionary ideas.* In Memoriam, *with its evolutionary underpinnings offering hope for the future, was the Victorians' favorite poem. It was a source of great comfort to many, not the least Queen Victoria after her husband, Prince Albert, died suddenly in 1860. You should not think that evolutionary ideas faced simple, complete, absolute opposition. The story is more complex than that, with many having more reasons to accept some form of evolution than to reject it outright.*

Zoological Philosophy

J. B. Lamarck

In every locality where animals can live, the conditions constituting any one order of things remain the same for long periods: indeed they alter so slowly that man cannot di-rectly observe it. It is only by an inspection of ancient monuments that he becomes con-vinced that in each of these localities the order of things which he now finds has not al-ways been existent; he may thence infer that it will go on changing.

Races of animals living in any of these localities must then retain their habits equally long: hence the apparent constancy of the races that we call species,—a con-stancy which has raised in us the belief that these races are as old as nature.

But in the various habitable parts of the earth's surface, the character and situation of places and climates constitute both for animals and plants environmental influences of extreme variability. The animals living in these various localities must therefore differ among themselves, not only by reason of the state of complexity of organisation attained in each race, but also by reason of the habits which each race is forced to acquire; thus when the observing naturalist travels over large portions of the earth's surface and sees conspicuous changes occurring in the environment, he invariably finds that the characters of species undergo a corresponding change.

Now the true principle to be noted in all this is as follows:

1. Every fairly considerable and permanent alteration in the environment of any race of animals works a real alteration in the needs of that race.

2. Every change in the needs of animals necessitates new activities on their part for the satisfaction of those needs, and hence new habits

3. Every new need, necessitating new activities for its satisfaction, requires the animal, either to make more frequent use of some of its parts which it previously used less, and thus greatly to develop and enlarge them; or else to make use of entirely new parts, to which the needs have imperceptibly given birth by efforts of its inner feeling; this I shall shortly prove by means of known facts.

Thus to obtain a knowledge of the true causes of that great diversity of shapes and habits found in the various known animals, we must reflect that the infinitely diversified but slowly changing environment in which the animals of each race have successively been placed, has involved each of them in new needs and corresponding alterations in their habits. This is a truth which, once recognised, cannot be disputed. Now we shall easily discern how the new needs may have been satisfied, and the new habits acquired, if we pay attention to the two following laws of nature, which are always verified by observation.

First Law.

In every animal which has not passed the limit of its development, a more frequent and continuous use of any organ gradually strengthens, develops and enlarges that organ, and gives it a power proportional to the length of time it has been so used; while the permanent disuse of any organ imperceptibly weakens and deteriorates it, and progressively diminishes its functional capacity, until it finally disappears.

Second Law.

All the acquisitions or losses wrought by nature on individuals, through the influence of the environment in which their race has long been placed, and hence through the influence of the predominant use or permanent disuse of any organ; all these are preserved by reproduction to the new individuals which arise, provided that the acquired modifications are common to both sexes, or at least to the individuals which produce the young.

Here we have two permanent truths, which can only be doubted by those who have never observed or followed the operations of nature, or by those who have allowed themselves to be drawn into the error which I shall now proceed to combat.

Naturalists have remarked that the structure of animals is always in perfect adaptation to their functions, and have inferred that the shape and condition of their parts have determined the use of them. Now this is a mistake: for it may be easily proved by observation that it is on the contrary the needs and uses of the parts which have caused the development of these same parts, which have even given birth to them when they did not exist, and which consequently have given rise to the condition that we find in each animal.

If this were not so, nature would have had to create as many different kinds of structure in animals, as there are different kinds of environment in which they have to live; and neither structure nor environment would ever have varied.

This is indeed far from the true order of things. If things were really so, we should not have race-horses shaped like those in England; we should not have big draught-horses so heavy and so different from the former, for none such are produced in nature; in the same way we should not have basset-hounds with crooked legs, nor grey-hounds so fleet of foot, nor water-spaniels, etc.; we should not have fowls without tails, fantail pigeons, etc.; finally, we should be able to cultivate wild plants as long as we liked in the rich and fertile soil of our gardens, without the fear of seeing them change under long cultivation.

A feeling of the truth in this respect has long existed; since the following maxim has passed into a proverb and is known by all, *Habits form a second nature.*

Assuredly if the habits and nature of each animal could never vary, the proverb would have been false and would not have come into existence, nor been preserved in the event of any one suggesting it.

If we seriously reflect upon all that I have just set forth, it will be seen that I was entirely justified when in my work entitled *Recherches sur les corps vivants* (p. 50), I established the following proposition:

"It is not the organs, that is to say, the nature and shape of the parts of an animal's body, that have given rise to its special habits and faculties; but it is, on the contrary, its habits, mode of life and environment that have in course of time controlled the shape of its body, the number and state of its organs and, lastly, the faculties which it possesses."

If this proposition is carefully weighed and compared with all the observations that nature and circumstances are incessantly throwing in our way, we shall see that its importance and accuracy are substantiated in the highest degree.

Time and a favourable environment are as I have already said nature's two chief methods of bringing all her productions into existence: for her, time has no limits and can be drawn upon to any extent.

As to the various factors which she has required and still constantly uses for introducing variations in everything that she produces, they may be described as practically inexhaustible.

The principal factors consist in the influence of climate, of the varying temperatures of the atmosphere and the whole environment of the variety of localities and their situation, of habits, the commonest movements, the most frequent activities, and, lastly, of the means of self-preservation, the mode of life and the methods of defence and multiplication.

Now as a result of these various influences, the faculties become extended and strengthened by use, and diversified by new habits that are long kept up. The conformation, consistency and, in short, the character and state of the parts, as well as of the organs, are imperceptibly affected by these influences and are preserved and propagated by reproduction.

These truths, which are merely effects of the two natural laws stated above, receive in every instance striking confirmation from facts; for the facts afford a clear indication of nature's procedure in the diversity of her productions.

Theory of the Earth

G. Cuvier

Nature appears also to have guarded against the alterations of species which might proceed from mixture of breeds, by influencing the various species of animals with mutual aversion from each other. Hence all the cunning and all the force that man is able to exert is necessary to accomplish such unions, even between species that have the nearest resemblances. And when the mule-breeds that are thus produced by these forced conjunctions happen to be fruitful, which is seldom the case, this fecundity never continues beyond a few generations, and would not probably proceed so far, without a continuance of the same cares which excited it at first. Thus we never see in a wild state intermediate productions between the hare and the rabbit, between the stag and the doe, or between the marten and the weasel. But the power of man changes this established order, and contrives to produce all these intermixtures of which the various species are susceptible, but which they would never produce if left to themselves.

The degrees of these variations are proportional to the intensity of the causes that produce them, namely, the slavery or subjection under which those animals are to man. They do not proceed far in half-domesticated species. In the cat, for example, a softer or harsher fur, more brilliant or more varied colours, greater or less size—these form the whole extent of the varieties in the species; the skeleton of the cat of Angora differs in no regular and constant circumstances from the wild cat of Europe.

In the domesticated herbivorous quadrupeds, which man transports into all kinds of climates, and subjects to various kinds of management, both in regard to labour and nourishment, he procures certainly more considerable variations, but still they are all merely superficial. Greater or less size; longer or shorter horns, or even the want of these entirely; a hump of fat, larger or smaller, on the shoulder; these form the chief differences among particular races of the *bos taurus,* or domestic black cattle; and these differences continue long in such breeds as have been transported to great distances from the countries in which they were originally produced, when proper care is taken to prevent crossing.

The innumerable varieties in the breeds of the *ovis aries,* or common sheep, are of a similar nature, and chiefly consist in differences of their fleeces, as the wool which they produce is a very important object of attention. These varieties, though not quite so perceptible, are yet sufficiently marked among horses. In general the forms of the bones are very little changed; their connections and articulations, and the form and structure of the large grinding teeth, are invariably the same. The small size of the tusks in the domesticated hog, compared with the wild boar, of which it is only a cultivated variety, and the junction of its cloven hoofs into one solid hoof, observable in some races, form the extreme point of the differences which man has been able to produce among herbivorous domesticated quadrupeds.

The most remarkable effects of the influence of man are produced upon that animal which he has reduced most completely under subjection. Dogs have been transported by mankind into every part of the world, and have submitted their actions to his entire direction. Regulated in their sexual unions by the pleasure or caprice of their masters, the almost endless variety of dogs differ from each other in colour; in length and abundance of hair, which is sometimes entirely wanting; in their natural instincts; in size, which varies in measure as one to five, amounting, in some instances, to more than an hundred fold in bulk; in the forms of their ears, noses, and tails; in the relative length of their legs; in the progressive development of the brain in several of the domesticated varieties, occasioning alterations, even in the form of the head; some of them having long slender muzzles with a flat forehead; others having short muzzles, with the forehead convex, &c. insomuch that the apparent differences between a mastiff and a water spaniel, and between a greyhound and a pug-dog, are even more striking than between almost any of the wild species of a genus. Finally, and this may be considered as the maximum of known variation in the animal kingdom, some races of dogs have an additional claw on each hind foot, with corresponding bones of the tarsus; as there sometimes occur in the human species some families that have six fingers on each hand. Yet, in all these varieties, the relations of the bones with each other remain essentially the same, and the form of the teeth never changes in any perceptible degree, except that in some individuals one additional false grinder occasionally appears, sometimes on the one side, and sometimes on the other. *

It follows from these observations, that animals have certain fixed and natural characters, which resist the effect of every kind of influence, whether proceeding from natural causes or human interference; and we have not the smallest reason to suspect that time has any more effect upon them than climate.

I am well aware that some naturalists lay prodigious stress on the thousands of years which they can call into action by a dash of their pens. In such matters, however, our only way of judging as to the effects which may be produced by a long period of time, is by multiplying, as it were, such as are produced by a shorter known time. With this view I have endeavoured to collect all the ancient documents respecting the forms of animals; and there are none equal to those furnished by the Egyptians, both in regard to their antiquity and abundance. They have not only left us representations of animals, but even their identical bodies embalmed and preserved in the catacombs.

I have examined with the greatest attention the engraved figures of quadrupeds and birds upon the numerous obelisks brought from Egypt to ancient Rome; and all these figures, one with another, have a perfect resemblance to their intended objects, such as they still are in our days. On examining the copies made by Kirker and Zoega, we find that, without preserving every trait of the original in its utmost purity, they have yet given us figures which are easily recognised. We readily distinguish the ibis, the vulture, the owl, the falcon, the Egyptian goose, the lapwing, the land rail, the asp, the cerastes, the Egyptian hare with its long ears, even the hippopotamus; and among the numerous remains engraved in the great work on Egypt, we sometimes observe the rarest animals, the algazel, for example, which was known in Europe only a few years ago.

My learned colleague, M. Geoffroy Saint Hilaire, convinced of the importance of this research, carefully collected in the tombs and temples of Upper and Lower Egypt as many mummies of animals as he could procure. He has brought home the mummies of cats, ibises, birds of prey, dogs, monkies, crocodiles, and the head of a bull; and after the most attentive and detailed examination, not the smallest difference is to be per-

---

* See, in the Annals of the Museum, XVIII. 338., a memoir by my brother on the varieties of dogs, which he drew up at my request, from a series of skeletons of all the varieties of dogs, prepared by me expressly on purpose.

ceived between these animals and those of the same species which we now see, any more than between the human mummies and the skeletons of men of the present day. Some slight differences are discoverable between ibis and ibis, for example, just as we now find differences in the descriptions of naturalists; but I have removed all doubts on that subject, in a Memoir on the Ibis of the ancient Egyptians, in which I have clearly shewn that this bird is precisely the same in all respects at present that it was in the days of the Pharaohs.* I am aware that in these I only cite the monuments of two or three thousand years back: but this is the most remote antiquity to which we can resort in such a case.

From all these well-established facts, there does not seem to be the smallest foundation supposing, that the new genera which I have discovered or established among extraneous fossils, such as the *palæotherium, anoplotherium, megalonyx, mastodon, pterodactylis,* &c. have ever been the sources of any of our present animals, which only differ so far as they are influenced by time or climate. Even if it should prove true, which I am far from believing to be the case, that the fossil elephants, rhinoceroses, elks, and bears do not differ farther from the presently existing species of the same genera, than the present races of dogs differ among themselves, this would by no means be a sufficient reason to conclude that they were of the same species; since the races or varieties of dogs have been influenced by the trammels of domesticity, which these other animals never did, and indeed never could experience.

In Memoriam

A. Tennyson

54.
Oh yet we trust that somehow good
   Will be the final goal of ill,
   To pangs of nature, sins of will,
Defects of doubt, and taints of blood;

That nothing walks with aimless feet;
   That not one life shall be destroy'd,
   Or cast as rubbish to the void,
When God hath made the pile complete,

That not a worm is cloven in vain;
   That not a moth with vain desire
   Is shrivell'd in a fruitless fire,
Or but subserves another's gain.

---

* In that dissertation, the ibis of the ancient Egyptians is shewn to be a species of *numenius,* or curlew, denominated by Cuvier *numenius ibis:* the same bird described in Bruce's Travels under the name of *abu-hannes*—Transl.

Behold, we know not anything;
　　I can but trust that good shall fall
　　At last—far off—at last, to all,
And every winter change to spring.

So runs my dream: but what am I?
　　An infant crying in the night:
　　An infant crying for the light:
And with no language but a cry.

55.
The wish, that of the living whole
　　No life may fail beyond the grave,
　　Derives it not from what we have
The likest God within the soul?

Are God and Nature then at strife,
　　That Nature lends such evil dreams?
　　So careful of the type she seems,
So careless of the single life;

That I, considering everywhere
　　Her secret meaning in her deeds,
　　And finding that of fifty seeds
She often brings but one to bear,

I falter where I firmly trod,
　　And falling with my weight of cares
　　Upon the great world's altar-stairs
That slope thro' darkness up to God,

I stretch lame hands of faith, and grope,
　　And gather dust and chaff, and call
　　To what I feel is Lord of all,
And faintly trust the larger hope.

56.
'So careful of the type?' but no.
　　From scarped cliff and quarried stone
　　She cries, 'A thousand types are gone:
I care for nothing, all shall go.

'Thou makest thine appeal to me:
　　I bring to life, I bring to death:
　　The spirit does but mean the breath:
I know no more.' And he, shall he,

Man, her last work, who seem'd so fair,
   Such splendid purpose in his eyes,
   Who roll'd the psalm to wintry skies,
Who built him fanes of fruitless prayer,

Who trusted God was love indeed
   And love Creation's final law—
   Tho' Nature, red in tooth and claw
With ravine, shriek'd against his creed—

Who loved, who suffer'd countless ills,
   Who battled for the True, the Just.
   Be blown about the desert dust,
Or seal'd within the iron hills?

No more? A monster then, a dream,
   A discord. Dragons of the prime,
   That tare each other in their slime,
Were mellow music match'd with him.

O life as futile, then, as frail!
   O for thy voice to soothe and bless!
   What hope of answer, or redress?
Behind the veil, behind the veil.

...
   A soul shall draw from out the vast
And strike his being into bounds,

And, moved thro' life of lower phase,
   Result in man, be born and think,
   And act and love, a closer link
Betwixt us and the crowning race

Of those that, eye to eye, shall look
   On knowledge; under whose command
   Is Earth and Earth's, and in their hand
Is Nature like an open book;

No longer half-akin to brute,
   For all we thought and loved and did,
   And hoped, and suffer'd, is but seed
Of what in them is flower and fruit;

Whereof the man, that with me trod
   This planet, was a noble type

Appearing ere the times were ripe,
That friend of mine who lives in God,

That God, which ever lives and loves,
    One God, one law, one element,
    And one far-off divine event,
To which the whole creation moves.

Chapter 2

Chapter 2 is on Darwin. I could have given you something from the Origin *itself but have decided, for the sake of completeness and to suggest a sense of the excitement of the event, to present as the sole document for this chapter the extracts given by Darwin, read for him by his friends Lyell and the botanist Joseph Hooker, at the Linnaean Society meeting in London in early July 1858. The first extract is from a book Darwin was writing (and never published) on natural selection when the arrival of Wallace's essay interrupted his plans. It deals with the struggle for existence leading to natural selection, which gives rise to adaptation. The second extract, talking of something that Darwin called his "principle of divergence," is from a letter that Darwin sent to his American friend the botanist Asa Gray. Here Darwin is trying to show how his evolutionary ideas lead to branching, to new species, and ultimately to a tree of life.*

On the Variation of Organic Beings in a State of Nature: On the Natural Means of Selection; on the Comparison of Domestic Races and True Species

Charles Darwin

De Candolle, in an eloquent passage, has declared that all nature is at war, one organism with another, or with external nature. Seeing the contented face of nature, this may at first well be doubted: but reflection will inevitably prove it to be true. The war, however, is not constant, but recurrent in a slight degree at short periods, and more severely at occasional more distant periods; and hence its effects are easily overlooked. It is the doctrine of Malthus applied in most cases with tenfold force. As in every climate there are seasons, for each of its inhabitants, of greater and less abundance, so all annually breed; and the moral restraint which in some small degree cheeks the increase of mankind is entirely lost. Even slow-breeding mankind has doubled in twenty-five years; and if he could increase his food with greater ease, he would double in less time. But for animals without artificial means, the amount of food for each species must, *on an average,* be constant, whereas the increase of all organisms tends to be geometrical, and in a vast majority of cases at an enormous ratio. Suppose in a certain spot there are eight pairs of birds, and that *only* four pairs of them annually (including double hatches) rear only four young, and that these go on rearing their young at the same rate, then at the end of seven years (a short life, excluding violent deaths, for any bird) there will be 2048 birds, instead of the original sixteen. As this increase is quite impossible, we must conclude either that birds do not rear nearly half their young, or that the average life of a bird is, from accident, not nearly seven years. Both cheeks probably concur. The same kind of calculation applied to all plants and animals affords results more or less striking, but in very few instances more striking than in man.

Many practical illustrations of this rapid tendency to increase are on record, among which, during peculiar seasons, are the extraordinary numbers of certain animals; for instance, during the years 1826 to 1828, in La Plata, when from drought some millions of cattle perished, the whole country actually *swarmed* with mice. Now I think it cannot be doubted that during the breeding season all the mice (with the exception of a few males or females in excess) ordinarily pair, and therefore that this astounding increase during three years must be attributed to a greater number than usual surviving the first year, and then breeding, and so on till the third year, when their numbers were brought down to their usual limits on the return of wet weather. Where man has introduced plants and animals into a new and favourable country, there are many accounts in how surprisingly few years the whole country has become stocked with them. This increase would necessarily stop as soon as the country was fully stocked; and yet we have every reason to believe, from what is known of wild animals, that *all* would pair in the spring. In the majority of cases it is most difficult to imagine where the check falls— though generally, no doubt, on the seeds, eggs, and young; but when we remember how impossible, even in mankind (so much better known than any other animal), it is to infer from repeated casual observations what the average duration of life is, or to discover the different percentage of deaths to births in different countries, we ought to feel no surprise at our being unable to discover where the check falls in any animal or plant. It should always be remembered, that in most cases the checks are recurrent yearly in a small, regular degree, and in an extreme degree during unusually cold, hot, dry, or wet years, according to the constitution of the being in question. Lighten any check in the least degree, and the geometrical powers of increase in every organism will almost instantly increase the average number of the favoured species. Nature may be compared to a surface on which rest ten thousand sharp wedges touching each other and driven inwards by incessant blows. Fully to realize these views much reflection is requisite. Malthus on man should be studied; and all such cases as those of the mice in La Plata, of the cattle and horses when first turned out in South America, of the birds by our calculation, etc., should be well considered. Reflect on the enormous multiplying power *inherent and annually in action* in all animals; reflect on the countless seeds scattered by a hundred ingenious contrivances, year after year, over the whole face of the land; and yet we have every reason to suppose that the average percentage of each of the inhabitants of a country usually remains constant. Finally, let it be borne in mind that this average number of individuals (the external conditions remaining the same) in each country is kept up by recurrent struggles against other species or against external nature (as on the borders of the arctic regions, where the cold checks life), and that ordinarily each individual of every species holds its place, either by its own struggle and capacity of acquiring nourishment in some period of its life, from the egg upwards; or by the struggle of its parents (in short-lived organisms, when the main check occurs at longer intervals) with other individuals of the *same* or *different* species.

But let the external conditions of a country alter. If in a small degree, the relative proportions of the inhabitants will in most cases simply be slightly changed; but let the number of inhabitants be small, as on an island, and free access to it from other countries be circumscribed, and let the change of conditions continue progressing (forming new stations), in such a case the original inhabitants must cease to be as perfectly adapted to the changed conditions as they were originally. It has been shown in a former part of this work, that such changes of external conditions would, from their acting on the reproductive system, probably cause the organization of those beings which were most affected to become, as under domestication, plastic. Now, can it be doubted, from the struggle each individual has to obtain subsistence, that any minute variation in structure, habits, or instincts, adapting that individual better to the new conditions, would tell upon its vigour and health? In the struggle it would have a better *chance* of surviving; and those of its offspring which inherited the variation, be it ever so slight, would also have a better *chance*. Yearly more are bred than can survive; the smallest grain in the balance, in the long run, must tell on which death shall fall, and which shall survive. Let this work of selection on the one hand, and death on the other, go on for a thousand generations, who will pretend to affirm that it would produce no effect, when we remember what, in a few years, Bakewell effected in cattle, and Western in sheep, by this identical principle of selection?

To give an imaginary example from changes in progress on an island: let the organization of a canine animal which preyed chiefly on rabbits, but sometimes on hares, become slightly plastic; let these same changes cause the number of rabbits very slowly to decrease, and the number of hares to increase; the effect of this would be that the fox or dog would be driven to try to catch more hares: his organization, however, being slightly plastic, those individuals with the lightest forms, longest limbs, and best eyesight, let the difference be ever so small, would be slightly favoured, and would tend to live longer, and to survive during that time of the year when food was scarcest; they would also rear more young, which would tend to inherit these slight peculiarities. The less fleet ones would be rigidly destroyed. I can see no more reason to doubt that these causes in a thousand generations would produce a marked effect, and adapt the form of the fox or dog to the catching of hares instead of rabbits, than that greyhounds can be improved by selection and careful breeding. So would it be with plants under similar circumstances. If the number of individuals of a species with plumed seeds could be increased by greater powers of dissemination within its own area (that is, if the check to increase fell chiefly on the seeds), those seeds which were provided with ever so little more down, would in the long run be most disseminated; hence a greater number of seeds thus formed would germinate, and would tend to produce plants inheriting the slightly better-adapted down.[1]

Besides this natural means of selection, by which those individuals are preserved, whether in their egg, or larval, or mature state, which are best adapted to the place they fill in nature, there is a second agency at work in most unisexual animals, tending to produce the same effect, namely the struggle of the males for the females. These struggles are generally decided by the law of battle, but in the case of birds, apparently, by the charms of their song, by their beauty or their power of courtship, as in the dancing rock-thrush of Guiana. The most vigorous and healthy males, implying perfect adaptation, must generally gain the victory in their contests. This kind of selection, however, is less

---

[1] I can see no more difficulty in this, than in the planter improving his varieties of the cotton plant. C. D. 1858.

rigorous than the other; it does not require the death of the less successful, but gives to them fewer descendants. The struggle falls, moreover, at a time of year when food is generally abundant, and perhaps the effect chiefly produced would be the modification of the secondary sexual characters, which are not related to the power of obtaining food, or to defence from enemies, but to fighting with or rivalling other males. The result of this struggle amongst the males may be compared in some respects to that produced by those agriculturists, who pay less attention to the careful selection of all their young animals, and more to the occasional use of a choice male.

Abstract of a Letter from Charles Darwin to Professor Asa Gray of Boston, U.S.A., Dated Down, 5 September 1857

It is wonderful what the principle of Selection by Man, that is the picking out of individuals with any desired quality, and breeding from them, and again picking out, can do. Even Breeders have been astounded at their own results. They can act on differences inappreciable to an uneducated eye. Selection has been *methodically* followed in *Europe* for only the last half century; but it was occasionally, and even in some degree methodically, followed in the most ancient times. There must have been also a kind of unconscious selection from a remote period, namely in the preservation of the individual animals (without any thought of their offspring) most useful to each race of man in his particular circumstances. The 'roguing', as nurserymen call the destroying of varieties which depart from their type, is a kind of selection. I am convinced that intentional and occasional selection has been the main agent in the production of our domestic races; but however this may be, its great power of modification has been indisputably shown in later times. Selection acts only by the accumulation of slight or greater variations, caused by external conditions, or by the mere fact that in generation the child is not absolutely similar to its parent. Man, by this power of accumulating variations, adapts living beings to his wants—may be said to make the wool of one sheep good for carpets, of another for cloth, etc.

Now suppose there were a being who did not judge by mere external appearances, but who could study the whole internal organization, who was never capricious, and should go on selecting for one object during millions of generations; who will say what he might not effect? In nature we have some *slight* variation occasionally in all parts; and I think it can be shown that changed conditions of existence is the main cause of the child not exactly resembling its parents; and in nature geology shows us what changes have taken place, and are taking place. We have almost unlimited time; no one but a practical geologist can fully appreciate this. Think of the Glacial period, during the whole of which the same species at least of shells have existed; there must have been during this period millions on millions of generations.

I think it can be shown that there is such an unerring power at work in *Natural Selection* (the title of my book), which selects exclusively for the good of each organic being. The elder de Candolle, W. Herbert, and Lyell have written excellently on the struggle for life; but even they have not written strongly enough. Reflect that every being (even the elephant) breeds at such a rate, that in a few years, or at most a few centuries, the surface of the earth would not hold the progeny of one pair. I have found it hard constantly to bear in mind that the increase of every single species is checked during some part of its life, or during some shortly recurrent generation. Only a few of those annually born can live to propagate their kind. What a trifling difference must often determine which shall survive, and which perish!

Now take the case of a country undergoing some change. This will tend to cause some of its inhabitants to vary slightly—not but that I believe most beings vary at all times enough for selection to act on them. Some of its inhabitants will be exterminated; and the remainder will be exposed to the mutual action of a different set of inhabitants, which I believe to be far more important to the life of each being than mere climate. Considering the infinitely various methods which living beings follow to obtain food by struggling with other organisms, to escape danger at various times of life, to have their eggs or seeds disseminated, etc., I cannot doubt that during millions of generations individuals of a species will be occasionally born with some slight variation, profitable to some part of their economy. Such individuals will have a better chance of surviving, and of propagating their new and slightly different structure; and the modification may be slowly increased by the accumulative action of natural selection to any profitable extent. The variety thus formed will either coexist with, or more commonly will exterminate its parent form. An organic being like the woodpecker or mistletoe, may thus come to be adapted to a score of contingencies: natural selection accumulating those slight variations in all parts of its structure, which are in any way useful to it during any part of its life.

Multiform difficulties will occur to every one, with respect to this theory. Many can, I think, be satisfactorily answered. *Natura non facit saltum* answers some of the most obvious. The slowness of the change, and only a very few individuals undergoing change at any one time, answers others. The extreme imperfection of our geological records answers others.

Another principle, which may be called the principle of divergence, plays, I believe, an important part in the origin of species. The same spot will support more life if occupied by very diverse forms. We see this in the many generic forms in a square yard of turf, and in the plants or insects on any little uniform islet, belonging almost invariably to as many genera and families as species. We can understand the meaning of this fact amongst the higher animals, whose habits we understand. We know that it has been experimentally shown that a plot of land will yield a greater weight if sown with several species and genera of grasses, than if sown with only two or three species. Now, every organic being, by propagating so rapidly, may be said to be striving its utmost to increase in numbers. So it will be with the offspring of any species after it has become diversified into varieties, or subspecies, or true species. And it follows, I think, from the foregoing facts, that the varying offspring of each species will try (only few will succeed) to seize on as many and as diverse places in the economy of nature as possible. Each new variety or species, when formed, will generally take the place of, and thus exterminate its less well-fitted parent. This I believe to be the origin of the classification and affinities of organic beings at all times; for organic beings always *seem* to branch and sub-branch like the limbs of a tree from a common trunk, the flourishing and diverging twigs destroying the less vigorous—the dead and lost branches rudely representing extinct genera and families.

This sketch is *most* imperfect; but in so short a space I cannot make it better. Your imagination must fill up very wide blanks.

Chapter 3

Chapter 3 has three documents. The first two are by Thomas Henry Huxley, beginning with an autobiographical account of Huxley's becoming an evolutionist. See how he too had read Vestiges *(not just a review like Tennyson!). But far from being convinced of the truth of evolution, as a professional biologist Huxley was simply irritated. Nor did the pleading of Huxley's friend Herbert Spencer change his mind. Yet the idea would not go away, and under the influence of Charles Darwin not only was Huxley introduced to a more scientifically plausible evolutionism but eventually he was convinced of its truth. Note however that Huxley was less enthusiastic about natural selection, thinking that it was not yet proven by experimental study, and this was a skepticism that he carried to his grave.*

The second document is from a letter Huxley wrote to the Reverend Charles Kingsley, known to us today as the author of the popular children's book The Water Babies. *Huxley's son had just died of scarlet fever at the age of 4, and Kingsley wrote to Huxley offering Christian consolation. This spurred Huxley to respond, stating his own unbelief—he was to coin the term* agnostic—*but also his belief in the power and worth of science. Note how explicitly Huxley is regarding this as a religious experience: to "sit down before the fact as a little child" is akin to surrendering to the will of God. It is small wonder that a man like this should have seized on evolutionism and regarded and treated it as a form of secular religion: something to give meaning to life.*

The third document is from an essay by Alfred Russel Wallace, and it expands on the theme mentioned in my main text, the argument that future human progress depends on the power and influence of women. Supposedly they will select the best men and thus there will be a general increase in human well-being and happiness. Note that Wallace's is a biological position—explicitly he downplays the lasting potential and influence of education. Note also how Wallace's position belies the traditional view of Social Darwinism—that is, a view of a philosophy that endorses a stern pro-capitalism economics, with success to the strong and failure for the weak. Wallace's socialism comes through strongly in our extracts. A major influence in the later part of Wallace's life, from which this document is taken, was the novel Looking Backward *by the American author Edward Bellamy. It is a tale of a man who sleeps until he awakens in the future, about now, in fact. All of the stuff about female power is lifted directly from the novel. The paradox is that generally Wallace was (as we saw) uncomfortable with the notion that selection involves female choice: he thought the peahen is relatively dowdy not because she has chosen bright males but because she needs camouflage. There really is no paradox, for, as I pointed out also, Wallace's real worry was that Darwin was making the peahen with her choice seem more humanlike. Humans are already humanlike, so they can do all the choosing that they will!*

The Reception of the "Origin of Species"

T. H. Huxley

I think I must have read the V*estiges* before I left England in 1846; but, if I did, the book made very little impression upon me, and I was not brought into serious contact with the "Species" question until after 1850. At that time, I had long done with the Pentateuchal cosmogony, which had been impressed upon my childish understanding as Divine truth, with all the authority of parents and instructors, and from which it had cost me many a struggle to get free. But my mind was unbiassed in respect of any doctrine which presented itself, if it professed to be based on purely philosophical and scientific reasoning. It seemed to me then (as it does now) that "creation," in the ordinary sense of the word, is perfectly conceivable. I find no difficulty in conceiving that, at some former period, this universe was not in existence; and that it made its appearance in six days (or instantaneously, if that is preferred), in consequence of the volition of some pre-existing Being. Then, as now, the so-called *a priori* arguments against Theism, and, given a Deity, against the possibility of creative acts, appeared to me to be devoid of reasonable foundation. I had not then, and I have not now, the smallest *a priori* objection to raise to the account of the creation of animals and plants given in *Paradise Lost,* in which Milton so vividly embodies the natural sense of Genesis. Far be it from me to say that it is untrue because it is impossible. I confine myself to what must be regarded as a modest and reasonable request for some particle of evidence that the existing species of animals and plants did originate in that way, as a condition of my belief in a statement which appears to me to be highly improbable.

And, by way of being perfectly fair, I had exactly the same answer to give to the evolutionists of 1851–58. Within the ranks of the biologists, at that time, I met with nobody, except Dr. Grant of University College, who had a word to say for Evolution—and his advocacy was not calculated to advance the cause. Outside these ranks, the only person known to me whose knowledge and capacity compelled respect, and who was, at the same time, a thorough-going evolutionist, was Mr. Herbert Spencer, whose acquaintance I made, I think, in 1852, and then entered into the bonds of a friendship which, I am happy to think, has known no interruption. Many and prolonged were the battles we fought on this topic. But even my friend's rare dialectic skill and copiousness of apt illustration could not drive me from my agnostic position. I took my stand upon two grounds:—Firstly, that up to that time, the evidence in favour of transmutation was wholly insufficient; and secondly, that no suggestion respecting the causes of transmutation assumed, which had been made, was in any way adequate to explain the phenomena. Looking back at the state of knowledge at that time, I really do not see that any other conclusion was justifiable.

In those days I had never even heard of Treviranus' *Biologie*. However, I had studied Lamarck attentively, and I had read the *Vestiges* with due care; but neither of them afforded me any good ground for changing my negative and critical attitude. As for the *Vestiges,* I confess that the book simply irritated me by the prodigious ignorance and thoroughly unscientific habit of mind manifested by the writer. If it had any influence on me at all, it set me against Evolution; and the only review I ever have qualms of conscience about, on the ground of needless savagery, is one I wrote on the *Vestiges* while under that influence. ...

But, by a curious irony of fate, the same influence which led me to put as little faith in modern speculations on this subject as in the venerable traditions recorded in the first two chapters of Genesis, was perhaps more potent than any other in keeping alive a sort of pious conviction that Evolution, after all, would turn out true. I have recently read afresh the first edition of the *Principles of Geology;* and when I consider that this remarkable book had been nearly thirty years in everybody's hands, and that it brings home to any reader of ordinary intelligence a great principle and a great fact,—the principle that the past must be explained by the present, unless good cause be shown to the contrary; and the fact that so far as our knowledge of the past history of life on our globe goes, no such cause can be shown,—I cannot but believe that Lyell, for others, as for myself, was the chief agent in smoothing the road for Darwin. For consistent uniformitarianism postulates Evolution as much in the organic as in the inorganic world. The origin of a new species by other than ordinary agencies would be a vastly greater "catastrophe" than any of those which Lyell succesfully eliminated from sober geological speculation.

Thus, looking back into the past, it seems to me that my own position of critical expectancy was just and reasonable, and must have been taken up, on the same grounds, by many other persons. If Agassiz told me that the forms of life which have successively tenanted the globe were the incarnations of successive thoughts of the Deity, and that He had wiped out one set of these embodiments by an appalling geological catastrophe as soon as His ideas took a more advanced shape, I found myself not only unable to admit the accuracy of the deductions from the facts of paleontology, upon which this astounding hypothesis was founded, but I had to confess my want of any means of testing the correctness of his explanation of them. And besides that, I could by no means see what the explanation explained. Neither did it help me to be told by an eminent anatomist that species had succeeded one another in time, in virtue of "a continuously operative creational law." That seemed to me to be no more than saying that species had succeeded one another in the form of a vote-catching resolution, with "law" to catch the man of science, and "creational" to draw the orthodox. So I took refuge in that "thätige Skepsis" which Goethe has so well defined; and, reversing the apostolic precept to be all things to all men, I usually defended the tenability of the received doctrines when I had to do with the transmutationists; and stood up for the possibility of transmutation among the orthodox—thereby, no doubt, increasing an already current, but quite undeserved, reputation for needless combativeness.

I remember, in the course of my first interview with Mr. Darwin, expressing my belief in the sharpness of the lines of demarcation between natural groups and in the absence of transitional forms, with all the confidence of youth and imperfect knowledge. I was not aware, at that time, that he had then been many years brooding over the species-question; and the humorous smile which accompanied his gentle answer, that such was not altogether his view, long haunted and puzzled me. But it would seem that four or five years' hard work had enabled me to understand what it meant; for Lyell, writing to Sir Charles Bunbury (under date of April 30, 1856), says :—

"When Huxley, Hooker and Wollaston were at Darwin's last week, they (all four of them) ran atilt against species—further, I believe, than they are prepared to go."

I recollect nothing of this beyond the fact of meeting Mr. Wollaston; and except for Sir Charles's distinct assurance as to "all four," I should have thought my *outrecuidance* was probably a counterblast to Wollaston's conservatism. With regard to Hooker, he was already, like Voltaire's Habbakuk, *capable de tout* in the way of advocating Evolution.

As I have already said, I imagine that most of those of my contemporaries who thought seriously about the matter were very much in my own state of mind—inclined to say to both Mosaists and Evolutionists, "a plague on both your houses!" and disposed to turn aside from an interminable and apparently fruitless discussion, to labour in the fertile fields of ascertainable fact. And I may therefore suppose that the publication of the Darwin and Wallace paper in 1858, and still more that of the "Origin" in 1859, had the effect upon them of the flash of light which, to a man who has lost himself on a dark night, suddenly reveals a road which, whether it takes him straight home or not, certainly goes his way. That which we were looking for, and could not find, was a hypothesis respecting the origin of known organic forms which assumed the operation of no causes but such as could be proved to be actually at work. We wanted, not to pin our faith to that or any other speculation, but to get hold of clear and definite conceptions which could be brought face to face with facts and have their validity tested. The "Origin" provided us with the working hypothesis we sought. Moreover, it did the immense service of freeing us for ever from the dilemma—Refuse to accept the creation hypothesis, and what have you to propose that can be accepted by any cautious reasoner? In 1857 I had no answer ready, and I do not think that anyone else had. A year later we reproached ourselves with dulness for being perplexed with such an inquiry. My reflection, when I first made myself master of the central idea of the "Origin" was, "How extremely stupid not to have thought of that!" I suppose that Columbus' companions said much the same when he made the egg stand on end. The facts of variability, of the struggle for existence, of adaptation to conditions, were notorious enough; but none of us had suspected that the road to the heart of the species problem lay through them, until Darwin and Wallace dispelled the darkness, and the beacon-fire of the "Origin" guided the benighted.

Whether the particular shape which the doctrine of Evolution, as applied to the organic world, took in Darwin's hands, would prove to be final or not, was to me a matter of indifference. In my earliest criticisms of the "Origin" I ventured to point out that its logical foundation was insecure so long as experiments in selective breeding had not produced varieties which were more or less infertile; and that insecurity remains up to the present time. But, with any and every critical doubt which my sceptical ingenuity could suggest, the Darwinian hypothesis remained incomparably more probable than the creation hypothesis. And if we had none of us been able to discern the paramount significance of some of the most patent and notorious of natural facts, until they were, so to speak, thrust under our noses, what force remained in the dilemma—creation or nothing? It was obvious that hereafter the probability would be immensely greater, that the links of natural causation were hidden from our purblind eyes, than that natural causation should be incompetent to produce all the phenomena of nature. The only rational course for those who had no other object than the attainment of truth was to accept "Darwinism" as a working hypothesis and see what could be made of it. Either it would prove its capacity to elucidate the facts of organic life, or it would break down under the strain. This was surely the dictate of common sense, and, for once, common sense carried the day.

Letter to Charles Kingsley

T. H. Huxley

14 Waverley Place, Sept. 23, 1860.

My Dear Kingsley—I cannot sufficiently thank you, both on my wife's account and my own, for your long and frank letter, and for all the hearty sympathy which it exhibits—and Mrs. Kingsley will, I hope, believe that we are no less sensible of her kind thought of us. To myself your letter was especially valuable, as it touched upon what I thought even more than upon what I said in my letter to you. My convictions, positive and negative, on all the matters of which you speak, are of long and slow growth and are firmly rooted. But the great blow which fell upon me seemed to stir them to their foundation, and had I lived a couple of centuries earlier I could have fancied a devil scoffing at me and them—and asking me what profit it was to have stripped myself of the hopes and consolations of the mass of mankind? To which my only reply was and is—Oh devil! truth is better than much profit. I have searched over the grounds of my belief, and if wife and child and name and fame were all to be lost to me one after the other as the penalty, still I will not lie.

And now I feel that it is due to you to speak as frankly as you have done to me. An old and worthy friend of mine tried some three or four years ago to bring us together— because, as he said, you were the only man who would do me any good. Your letter leads me to think he was right, though not perhaps in the sense he attached to his own words.

To begin with the great doctrine you discuss. I neither deny nor affirm the immortality of man. I see no reason for believing in it, but, on the other hand, I have no means of disproving it.

Pray understand that I have no *a priori* objections to the doctrine. No man who has to deal daily and hourly with nature can trouble himself about *a priori* difficulties. Give me such evidence as would justify me in believing anything else, and I will believe that. Why should I not ? It is not half so wonderful as the conservation of force, or the indestructibility of matter. Whoso clearly appreciates all that is implied in the failing of a stone can have no difficulty about any doctrine simply on account of its marvellousness. But the longer I live, the more obvious it is to me that the most sacred act of a man's life is to say and to feel, "I believe such and such to be true." All the greatest rewards and all the heaviest penalties of existence cling about that act. The universe is one and the same throughout; and if the condition of my success in unravelling some little difficulty of anatomy or physiology is that I shall rigorously refuse to put faith in that which does not rest on sufficient evidence, I cannot believe that the great mysteries of existence will be laid open to me on other terms. It is no use to talk to me of analogies and probabilities. I know what I mean when I say I believe in the law of the inverse squares, and I will not rest my life and my hopes upon weaker convictions. I dare not if I would.

Measured by this standard, what becomes of the doctrine of immortality?

You rest in your strong conviction of your personal existence, and in the instinct of the persistence of that existence which is so strong in you as in most men.

To me this is as nothing. That my personality is the surest thing I know—may be true. But the attempt to conceive what it is leads me into mere verbal subtleties. I have champed up all that chaff about the ego and the non-ego, about noumena and phenomena, and all the rest of it, too often not to know that in attempting even to think of these questions, the human intellect flounders at once out of its depth.

It must be twenty years since, a boy, I read Hamilton's essay on the unconditioned, and from that time to this, ontological speculation has been a folly to me. When Mansel took up Hamilton's argument on the side of orthodoxy(!) I said he reminded me of nothing so much as the man who is sawing off the sign on which he is sitting, in Hogarth's picture. But this by the way.

I cannot conceive of my personality as a thing apart from the phenomena of my life. When I try to form such a conception I discover that, as Coleridge would have said, I only hypostatise a word, and it alters nothing if, with Fichte, I suppose the universe to be nothing but a manifestation of my personality. I am neither more nor less eternal than I was before.

Nor does the infinite difference between myself and the animals alter the case. I do not know whether the animals persist after they disappear or not. I do not even know whether the infinite difference between us and them may not be compensated by *their* persistence and *my* cessation after apparent death, just as the humble bulb of an annual lives, while the glorious flowers it has put forth die away.

Surely it must be plain that an ingenious man could speculate without end on both sides, and find analogies for all his dreams. Nor does it help me to tell me that the aspirations of mankind—that my own highest aspirations even—lead me towards the doctrine of immortality. I doubt the fact, to begin with, but if it be so even, what is this but in grand words asking me to believe a thing because I like it.

Science has taught to me the opposite lesson. She warns me to be careful how I adopt a view which jumps with my pre-conceptions, and to require stronger evidence for such belief than for one to which I was previously hostile.

My business is to teach my aspirations to conform themselves to fact, not to try and make facts harmonise with my aspirations.

Science seems to me to teach in the highest and strongest manner the great truth which is embodied in the Christian conception of entire surrender to the will of God. Sit down before fact as a little child, be prepared to give up every preconceived notion, follow humbly wherever and to whatever abysses nature leads, or you shall learn nothing. I have only begun to learn content and peace of mind since I have resolved at all risks to do this.

There are, however, other arguments commonly brought forward in favour of the immortality of man, which are to my mind not only delusive but mischievous. The one is the notion that the moral government of the world is imperfect without a system of future rewards and punishments. The other is: that such a system is indispensable to practical morality. I believe that both these dogmas are very mischievous lies.

With respect to the first, I am no optimist, but I have the firmest belief that the Divine Government (if we may use such a phrase to express the sum of the "customs of matter") is wholly just. The more I know intimately of the lives of other men (to say nothing of my own), the more obvious it is to me that the wicked does *not* flourish nor is the righteous punished. But for this to be clear we must bear in mind what almost all forget, that the rewards of life are contingent upon obedience to the *whole* law—physical as well as moral—and that moral obedience will not atone for physical sin, or *vice versa*.

The ledger of the Almighty is strictly kept, and every one of us has the balance of his operations paid over to him at the end of every minute of his existence.

Life cannot exist without a certain conformity to the surrounding universe—that conformity involves a certain amount of happiness in excess of pain. In short, as we live we are paid for living.

And it is to be recollected in view of the apparent discrepancy between men's acts and their rewards that Nature is juster than we. She takes into account what a man brings with him into the world, which human justice cannot do. If I, born a bloodthirsty and savage brute, inheriting these qualities from others, kill you, my fellow-men will very justly hang me, but I shall not be visited with the horrible remorse which would be my real punishment if, my nature being higher, I had done the same thing.

The absolute justice of the system of things is as clear to me as any scientific fact. The gravitation of sin to sorrow is as certain as that of the earth to the sun, and more so—for experimental proof of the fact is within reach of us all—nay, is before us all in our own lives, if we had but the eyes to see it.

Not only, then, do I disbelieve in the need for compensation, but I believe that the seeking for rewards and punishments out of this life leads men to a ruinous ignorance of the fact that their inevitable rewards and punishments are here.

If the expectation of hell hereafter can keep me from evildoing, surely *a fortiori* the certainty of hell now will do so? If a man could be firmly impressed with the belief that stealing damaged him as much as swallowing arsenic would do (and it does), would not the dissuasive force of that belief be greater than that of any based on mere future expectations?

And this leads me to my other point.

As I stood behind the coffin of my little son the other day, with my mind bent on anything but disputation, the officiating minister read, as a part of his duty, the words, "If the dead rise not again, let us eat and drink, for to-morrow we die." I cannot tell you how inexpressibly they shocked me. Paul had neither wife nor child, or he must have known that his alternative involved a blasphemy against all that was best and noblest in human nature. I could have laughed with scorn. What! because I am face to face with ir-reparable loss, because I have given back to the source from whence it came, the cause of a great happiness, still retaining through all my life the blessings which have sprung and will spring from that cause, I am to renounce my manhood, and, howling, grovel in bes-tiality? Why, the very apes know better, and if you shoot their young, the poor brutes grieve their grief out and do not immediately seek distraction in a gorge.

Kicked into the world a boy without guide or training, or with worse than none, I confess to my shame that few men have drunk deeper of all kinds of sin than I. Happily, my course was arrested in time—before I had earned absolute destruction—and for long years I have been slowly and painfully climbing, with many a fall, towards better things. And when I look back, what do I find to have been the agents of my redemption? The hope of immortality or of future reward? I can honestly say that for these fourteen years such a consideration has not entered my head. No, I can tell you exactly what has been at work. *Sartor Resartus* led me to know that a deep sense of religion was compatible with the entire absence of theology. Secondly, science and her methods gave me a rest-ing-place independent of authority and tradition. Thirdly, love opened up to me a view of the sanctity of human nature, and impressed me with a deep sense of responsibility.

If at this moment I am not a worn-out, debauched, useless carcass of a man, if it has been or will be my fate to advance the cause of science, if I feel that I have a shadow of a claim on the love of those about me, if in the supreme moment when I looked down into my boy's grave my sorrow was full of submission and without bitterness, it is be-cause these agencies have worked upon me, and not because I have ever cared whether my poor personality shall remain distinct for ever from the All from whence it came and whither it goes.

And thus, my dear Kingsley, you will understand what my position is. I may be quite wrong, and in that case I know I shall have to pay the penalty for being wrong. But I can only say with Luther, "Gott helfe mir, Ich kann nichts anders."

I know right well that 99 out of 100 of my fellows would call me atheist, infidel, and all the other usual hard names. As our laws stand, if the lowest thief steals my coat, my evidence (my opinions being known) would not be received against him.[1]

---

[1] The law with respect to oaths was reformed in 1869.

But I cannot help it. One thing people shall not call me with justice and that is—a liar. As you say of yourself, I too feel that I lack courage; but if ever the occasion arises when I am bound to speak, I will not shame my boy.

I have spoken more openly and distinctly to you than I ever have to any human being except my wife.

If you can show me that I err in premises or conclusion, I am ready to give up these as I would any other theories. But at any rate you will do me the justice to believe that I have not reached my conclusions without the care befitting the momentous nature of the problems involved.

And I write this the more readily to you, because it is clear to me that if that great and powerful instrument for good or evil, the Church of England, is to be saved from being shivered into fragments by the advancing tide of science—an event I should be very sorry to witness, but which will infallibly occur if men like Samuel of Oxford are to have the guidance of her destinies—it must be by the efforts of men who, like yourself, see your way to the combination of the practice of the Church with the spirit of science. Understand that all the younger men of science whom I know intimately are *essentially* of my way of thinking. (I know not a scoffer or an irreligious or an immoral man among them, but they all regard orthodoxy as you do Brahmanism.) Understand that this new school of the prophets is the only one that can work miracles, the only one that can constantly appeal to nature for evidence that it is right, and you will comprehend that it is of no use to try to barricade us with shovel hats and aprons, or to talk about our doctrines being "shocking."

I don't profess to understand the logic of yourself, Maurice, and the rest of your school, but I have always said I would swear by your truthfulness and sincerity, and that good must come of your efforts. The more plain this was to me, however, the more obvious the necessity to let you see where the men of science are driving, and it has often been in my mind to write to you before.

If I have spoken too plainly anywhere, or too abruptly, pardon me, and do the like to me.

My wife thanks you very much for your volume of sermons.—Ever yours very faithfully,

T. H. Huxley.

Studies, Scientific and Social

A. R. Wallace

Social Advance Will Result in Improvement of Character.

It is my firm conviction, for reasons which I shall state presently, that, when we have cleansed the Augean stable of our existing social organization, and have made such arrangements that *all* shall contribute their share of either physical or mental labour, and that all workers shall reap the *full* and equal reward of their work, the future of the race will be ensured by those laws of human development that have led to the slow but continuous advance in the higher qualities of human nature. When men and women are alike free to follow their best impulses; when idleness and vicious or useless luxury on the one hand, oppressive labour and starvation on the other, are alike unknown; when all receive the best and most thorough education that the state of civilization and knowledge at the time will admit; when the standard of public opinion is set by the wisest and the best, and that standard is systematically inculcated on the young; then we shall find that a system of selection will come spontaneously into action which will steadily tend to eliminate the lower and more degraded types of man, and thus continuously raise the average standard of the race. I therefore strongly protest against any attempt to deal with this great question by legal enactments in our present state of unfitness and ignorance, or by endeavouring to modify public opinion as to the beneficial character of monogamy and permanence in marriage. ...

We have now to consider what would be the probable effect of a condition of social advancement, the essential characteristics of which have been already hinted at, on the two great problems—the increase of population, and the continuous improvement of the race by some form of selection which we have reason to believe is the only method available. In order to make this clear, however, and in order that we may fully realize the forces that would come into play in a just and rational state of society, such as may certainly be realized in the not distant future, it will be necessary to have a clear conception of its main characteristics. For this purpose, and without committing myself in any way to an approval of all the details of his scheme, I shall make use of Mr. Bellamy's clear and forcible picture of the society of the future, as he supposes it may exist in America in little more than a century hence.[2]

The essential principle on which society is supposed to be founded is that of a great family. As in a well-regulated modern family the elders, those who have experience of the labours, the duties, and the responsibilities of life, determine the general mode of living and working, with the fullest consideration for the convenience and real well-being of the younger members, and with a recognition of their essential independence. As in a family, the same comforts and enjoyments are secured to all, and the very idea of making any difference in this respect to those who from mental or physical disability are unable to do so much as others, never occurs to any one, since it is opposed to the essential principles on which a true society of human brotherhood is held to rest. As regards education all have the same advantages, and all receive the fullest and best training, both intellectual and physical; every one is encouraged to follow out those studies or pursuits for which they are best fitted, or for which they exhibit the strongest inclination. This education, the complete and thorough training for a life of usefulness and enjoyment, continues in both sexes till the age of twenty-one (or thereabouts), when all alike, men and women, take their place in the lower ranks of the industrial army in which they serve for three years. During the latter years of their education, and during the succeeding three years of industrial service, every opportunity is given them to see and understand every kind of work that is carried on by the community, so that at the end of the term of probation they can choose what department of the public service they prefer to enter. As every one—men, women, and children alike—receives the same amount of public credit—their equal share of the products of the labour of the community, the attractiveness of various pursuits is equalized by differences in the hours of labour, in holidays, or in special privileges attached to the more disagreeable kinds of necessary work, and these are so modified from time to time that the volunteers for every occupation are

---

[2] *Looking Backward.* See specially chapters vii., ix., xii., and xxv.

always about equal to its requirements. The only other essential feature that it is necessary to notice for our present purpose is the system of grades, by which good conduct, perseverance, and intelligence in every department of industry and occupation are fully recognized, and lead to appointments as foremen, superintendents, or general managers, and ultimately to the highest offices of the state. Every one of these grades and appointments is made public; and as they constitute the only honours and the only differences of rank, with corresponding insignia and privileges, in an otherwise equal body of citizens, they are highly esteemed, and serve as ample inducements to industry and zeal in the public service.

At first sight it may appear that in any state of society whose essential features were at all like those here briefly outlined, all the usual restraints to early marriage as they now exist would be removed, and that a rate of increase of the population unexampled in any previous era would be the result, leading in a few generations to a difficulty in obtaining subsistence, which Malthus has shown to be the inevitable result of the normal rate of increase of mankind when all the positive as well as the preventive checks are removed. As the positive checks—which may be briefly summarized as war, pestilence, and famine—are supposed to be non-existent, what, it may be asked, are the preventive checks which are suggested as being capable of reducing the rate of increase within manageable limits? This very reasonable question I will now endeavour to answer.

Natural Checks to Rapid Increase.

The first and most important of the checks upon a too rapid increase of population will be the comparatively late average period of marriage, which will be the natural result of the very conditions of society, and will besides be inculcated during the period of education, and still further enforced by public opinion. As the period of systematic education is supposed to extend to the age of twenty-one, up to which time both the mental and physical powers will be trained and exercised to their fullest capacity, the idea of marriage during this period will rarely be entertained. During the last year of education, however, the subject of marriage will be dwelt upon, in its bearing on individual happiness and on social well-being, in relation to the welfare of the next generation and to the continuous development of the race. The most careful and deliberate choice of partners for life will be inculcated as the highest social duty; while the young women will be so trained as to look with scorn and loathing on all men who in any way wilfully fail in their duty to society—on idlers and malingerers, on drunkards and liars, on the selfish, the cruel, or the vicious. They will be taught that the happiness of their whole lives will depend on the care and deliberation with which they choose their husbands, and they will be urged to accept no suitor till he has proved himself to be worthy of respect by the place he holds and the character he bears among his fellow-labourers in the public service.

Under social conditions which render every woman absolutely independent, so far as the necessaries and comforts of existence are concerned, surrounded by the charms of family life and the pleasures of society, which will be far greater than anything we now realize when *all* will possess the refinements derived from the best possible education, and all will be relieved from sordid cares and the struggle for mere existence, is it not in the highest degree probable that marriage will rarely take place till the woman has had three or four years' experience of the world after leaving college—that is, till the age of 25, while it will very frequently be delayed till 30 or upwards? Now Mr. Galton has shown, from the best statistics available, that if we compare women married at 20 with those married at 29, the proportionate fertility is about as 8 to 5. But this difference, large as it is, only represents a portion of the effect on the rate of increase of population caused by a delay in the average period of marriage. For when the age of marriage is delayed the time between successive generations is correspondingly lengthened; while a still further effect is produced by the fact that the greater the average age of marriage the fewer generations are alive at the same time, and it is the combined effect of these three factors that determines the actual rate of increase of the population.[3]

But there is yet another factor tending to check the increase of population that would come into play in a society such as we have been considering. In a remarkable essay on the *Theory of Population* Herbert Spencer has shown, by an elaborate discussion of the phenomena presented by the whole animal kingdom, that the maintenance of the individual and the propagation of the race vary inversely, those species and groups which have the shortest and most uncertain lives producing the greatest number of offspring; in other words, individuation and reproduction are antagonistic. But individuation depends almost entirely on the development and specialization of the nervous system, through which, not only are the several activities and co-ordinations of the various organs carried on, but all advance in instinct, emotion, and intellect is rendered possible. The actual rate of increase in man has been determined by the necessities of the savage state, in which, as in most animal species, it has usually been only just sufficient to maintain a limited average population. But with civilization the average duration of life increases, and the possible increase of population under favourable conditions becomes very great, because fertility is greater than is needed under the new conditions. The advance in civilization as regards the preservation of life has in recent times become so rapid, and the increased development of the nervous system has been limited to so small a portion of the whole population, that no general diminution in fertility has yet occurred. That the facts do, however, accord with the theory is indicated by the common observation that highly intellectual parents do not as a rule have large families, while the most rapid increase occurs in those classes which are engaged in the simpler kinds of manual labour. But in a state of society in which all will have their higher faculties fully cultivated and fully exercised throughout life, a slight general diminution of fertility would at once arise, and this diminution added to that caused by the later average period of marriage would at once bring the rate of increase of population within manageable limits. The same general principle enables us to look forward to that distant future when the world will be fully peopled, in perfect confidence that an equilibrium between the birth and death rates will then be brought about by a combination of physical and social agencies, and the bugbear of over-population become finally extinct.[4]

How Natural Selection Will Improve the Race.

There now only remains for consideration the means by which, in such a society, a continuous improvement of the race could be brought about, on the assumption that for this

---

[3] See *Inquiries into Human Faculty and Its Development,* p. 321; and *Hereditary Genius,* p. 353.

[4] *A theory of Population deduced from the General Law of Animal Fertility.* Republished from the *Westminster Review* for April, 1852.

purpose education is powerless as a direct agency, since its effects are not hereditary, and that some form of selection is an absolute necessity. This improvement I believe will certainly be effected through the agency of female choice in marriage. Let us, therefore, consider how this would probably act.

It will be generally admitted that, although many women now remain unmarried from necessity rather than from choice, there are always a considerable number who feel no strong inclination to marriage, and who accept husbands to secure a subsistence or a home of their own rather than from personal affection or sexual emotion. In a society in which women were all pecuniarily independent, were all fully occupied with public duties and intellectual or social enjoyments, and had nothing to gain by marriage as regards material well-being, we may be sure that the number of the unmarried from choice would largely increase. It would probably come to be considered a degradation for any woman to marry a man she could not both love and esteem, and this feeling would supply ample reasons for either abstaining from marriage altogether or delaying it till a worthy and sympathetic husband was encountered. In man, on the other hand, the passion of love is more general, and usually stronger; and as in such a society as is here postulated there would be no way of gratifying this passion but by marriage, almost every woman would receive offers, and thus a powerful selective agency would rest with the female sex. Under the system of education and of public opinion here suggested there can be no doubt how this selection would be exercised. The idle and the selfish would be almost universally rejected. The diseased or the weak in intellect would also usually remain unmarried; while those who exhibited any tendency to insanity or to hereditary disease, or who possessed any congenital deformity would in hardly any case find partners, because it would be considered an offence against society to be the means of perpetuating such diseases or imperfections. ...

When we allow ourselves to be guided by reason, justice, and public spirit in our dealings with our fellow-men, and determine to abolish poverty by recognizing the equal rights of all the citizens of our common land to an equal share of the wealth which all combine to produce—when we have thus solved the lesser problem of a rational social organization adapted to secure the equal well-being of all, then we may safely leave the far greater and deeper problem of the improvement of the race to the cultivated minds and pure instincts of the men, and especially of the Women of the Future.

Chapter 4

Chapter 4 has two documents. The first is an exchange of ideas between Asa Gray and Charles Darwin. Gray, an evangelical Christian, thought that selection could not bring on all of the adaptation—which he took as evidence of God's design—that he saw about him in nature. Hence, in the essay from which the extract printed here is taken he argued that God must have been standing directly behind evolutionary change. For Darwin this would be to make selection redundant, and he expressed himself forcefully in the extracts given here, taken from his letters to Gray. Darwin thought science should keep direction and purpose and unseen forces out of its explanations. He wanted no truck with guided variations bringing on adaptation. For him it was natural selection or nothing. Note however that Darwin himself seems to allow that there is (or may be) something or someone behind everything. It is just that he does not want this guiding force to be active on the job. Rather, leave it all to normal unguided laws of nature.

The second document comes from the end of our period. Henry Fairfield Osborn, whom we met in the text as director of the American Museum of Natural History in New York, had been chosen as one of the expert witnesses to appear on behalf of evolution in the Scopes trial. As it happened, the judge would not admit any of this evidence, but this did not stop Osborn—a sincere Christian—from expressing his views in print. In the passage given, from a collection revealingly entitled The Earth Speaks to Bryan, Osborn is affirming his faith in evolution. He is also giving his decided opinion on the primitive nature of Neanderthals, whom he contrasts unfavorably with a later race, Cro-Magnon man. Osborn was treading a careful line here, intending his history to convey a powerful message for the present. As a modern, enlightened American from the North, he thought that all humans—including Negroes—were the same species. His mother had read Uncle Tom's Cabin to him as a child. But as an upper-class, white, Protestant American he had no doubt that not all humans were equal and that blacks, Jews, Catholics, Irish, Italians—in short, anyone not of his own narrow group—were inferior to good old Anglo-Saxon stock. Osborn was keen on eugenics and supported moves to restrict immigration to the United States. The Neanderthal/Cro-Magnon divide was intended to show that these differences were a natural state of affairs among humans and that his own position was merely a reflection and appreciation of the history of humankind. Neanderthals were human but, like recent immigrants, not the best kinds of humans. As I have argued in the text, evolution was proving a perfect vehicle for social messages of a kind that one associates with an ideology or religion.

Darwiniana

A. Gray

However that may be, it is undeniable that Mr. Darwin has purposely been silent upon the philosophical and theological applications of his theory. This reticence, under the circumstances, argues design, and raises inquiry as to the final cause or reason why. Here, as in higher instances, confident as we are that there is a final cause, we must not be over-confident that we can infer the particular or true one. Perhaps the author is more familiar with natural-historical than philosophical inquiries, and, not having decided which particular theory about efficient cause is best founded, he meanwhile argues the scientific questions concerned—all that relates to secondary causes—upon purely scientific grounds, as he must do in any case. Perhaps, confident, as he evidently is, that his view will finally be adopted, he may enjoy a sort of satisfaction in hearing it denounced as sheer atheism by the inconsiderate, and afterward, when it takes its place with the nebular hypothesis and the like, see this judgment reversed, as we suppose it would be in such event.

Whatever Mr. Darwin's philosophy may be, or whether he has any, is a matter of no consequence at all, compared with the important questions, whether a theory to account for the origination and diversification of animal and vegetable forms through the operation of secondary causes does or does not exclude design; and whether the establishment by adequate evidence of Darwin's particular theory of diversification through variation and natural selection would essentially alter the present scientific and philosophical grounds for theistic views of Nature. The unqualified affirmative judgment rendered by the two Boston reviewers, evidently able and practised reasoners, "must give us pause." We hesitate to advance our conclusions in opposition to theirs. But, after full and serious consideration we are constrained to say that, in our opinion, the adoption of a derivative hypothesis, and of Darwin's particular hypothesis, if we understand it, would leave the doctrines of final causes, utility, and special design, just where they were before. We do not pretend that the subject is not environed with difficulties. Every view is so environed; and every shifting of the view is likely, if it removes some difficulties, to bring others into prominence. But we cannot perceive that Darwin's theory brings in any new kind of scientific difficulty, that is, any with which philosophical naturalists were not already familiar.

Since natural science deals only with secondary or natural causes, the scientific terms of a theory of derivation of species—no less than of a theory of dynamics—must needs be the same to the theist as to the atheist. The difference appears only when the inquiry is earned up to the question of primary cause—a question which belongs to philosophy. Wherefore, Darwin's reticence about efficient cause does not disturb us. He considers only the scientific questions. As already stated, we think that a theistic view of Nature is implied in his book, and we must charitably refrain from suggesting the contrary until the contrary is logically deduced from his premises. If, however, he anywhere maintains that the natural causes through which species are diversified operate without an ordaining and directing intelligence, and that the orderly arrangements and admirable adaptations we see all around us are fortuitous or blind, undesigned results—that the eye, though it came to see, was not designed for seeing, nor the hand for handling—then, we suppose, he is justly chargeable with denying, and very needlessly denying, all design in organic Nature; otherwise, we suppose not. Why, if Darwin's well-known passage about the eye[1]—equivocal though some of the language be—does not imply ordaining and directing intelligence, then he refutes his own theory as effectually as any of his opponents are likely to do. He asks:

"May we not believe that [under variation proceeding long enough, generation multiplying the better variations times enough, and natural selection securing the improvements] a living optical instrument might be thus formed as superior to one of glass as the works of the Creator are to those of man?"

This must mean one of two things: either that the living instrument was made and perfected under (which is the same thing as by) an intelligent First Cause, or that it was not. If it was, then theism is asserted; and as to the mode of operation, how do we know, and why must we believe, that, fitting precedent forms being in existence, a living instrument (so different from a lifeless manufacture) would be originated and perfected in any other way, or that this is not the fitting way? If it means that it was not, if he so misuses words that by the Creator he intends an unintelligent power, undirected force, or necessity, then he has put his case so as to invite disbelief in it. For then blind forces have produced not only manifest adaptations of means to specific ends—which is absurd enough—but better adjusted and more perfect instruments or machines than intellect (that is, human intellect) can contrive and human skill execute—which no sane person will believe.

On the other hand, if Darwin even admits—we will not say adopts—the theistic view, he may save himself much needless trouble in the endeavor to account for the absence of every sort of intermediate form. Those in the line between one species and another supposed to be derived from it he may be bound to provide; but as to "an infinite number of other varieties not intermediate, gross, rude, and purposeless, the unmeaning creations of an unconscious cause," born only to perish, which a relentless reviewer has imposed upon his theory—rightly enough upon the atheistic alternative—the theistic view

---

[1] Page 188, English edition.

rids him at once of this "scum of creation." For, as species do not now vary at all times and places and in all directions, nor produce crude, vague, imperfect, and useless forms, there is no reason for supposing that they ever did. Good-for-nothing monstrosities, failures of purpose rather than purposeless, indeed, sometimes occur; but these are just as anomalous and unlikely upon Darwin's theory as upon any other. For his particular theory is based, and even over-strictly insists, upon the most universal of physiological laws, namely, that successive generations shall differ only slightly, if at all, from their parents; and this effectively excludes crude and impotent forms. Wherefore, if we believe that the species were designed, and that natural propagation was designed, how can we say that the actual varieties of the species were not equally designed? Have we not similar grounds for inferring design in the supposed varieties of species, that we have in the case of the supposed species of a genus? When a naturalist comes to regard as three closely-related species what he before took to be so many varieties of one species, how has he thereby strengthened our conviction that the three forms are designed to have the differences which they actually exhibit? Wherefore, so long as gradatory, orderly, and adapted forms in Nature argue design, and at least while the physical cause of variation is utterly unknown and mysterious, we should advise Mr. Darwin to assume, in the philosophy of his hypothesis, that variation has been led along certain beneficial lines. Streams flowing over a sloping plain by gravitation (here the counterpart of natural selection) may have worn their actual channels as they flowed; yet their particular courses may have been assigned; and where we see them forming definite and useful lines of irrigation, after a manner unaccountable on the laws of gravitation and dynamics, we should believe that the distribution was designed.

To insist, therefore, that the new hypothesis of the derivative origin of the actual species is incompatible with final causes and design, is to take a position which we must consider philosophically untenable. We must also regard it as highly unwise and dangerous, in the present state and present prospects of physical and physiological science. We should expect the philosophical atheist or skeptic to take this ground; also, until better informed, the unlearned and unphilosophical believer; but we should think that the thoughtful theistic philosopher would take the other side. Not to do so seems to concede that only supernatural events can be shown to be designed, which no theist can admit— seems also to misconceive the scope and meaning of all ordinary arguments for design in Nature. This misconception is shared both by the reviewers and the reviewed. At least, Mr. Darwin uses expressions which imply that the natural forms which surround us, because they have a history or natural sequence, could have been only generally, but not particularly designed—a view at once superficial and contradictory; whereas his true line should be, that his hypothesis concerns the *order* and not the *cause,* the *how* and not the *why* of the phenomena, and so leaves the question of design just where it was before.

Letters to Asa Gray

Charles Darwin

3 July 1860

One word more on "designed laws" & "undesigned results." I see a bird which I want for food, take my gun & kill it, I do this *designedly.*——An innocent & good man stands under tree & is killed by flash of lightning. Do you believe (& I really sh^d like to hear) that God *designedly* killed this man? Many or most persons do believe this; I can't & don't.——If you believe so, do you believe that when a swallow snaps up a gnat that God designed that that particular swallow sh^d snap up that particular gnat at that particular instant? I believe that the man & the gnat are in same predicament.—If the death of neither man or gnat are designed, I see no good reason to believe that their *first* birth or production sh^d be necessarily designed. Yet, as I said before, I cannot persuade myself that electricity acts, that the tree grows, that man aspires to loftiest conceptions all from blind, brute force.

26 November 1860

Yesterday I read over with care the third Article; & it seems to me, as before, *admirable*. But I grieve to say that I cannot honestly go as far as you do about Design. I am conscious that I am in an utterly hopeless muddle. I cannot think that the world, as we see it, is the result of chance; & yet I cannot look at each separate thing as the result of Design.—To take a crucial example, you lead me to infer (p. 414) that you believe "that variation has been led along certain beneficial lines."—I cannot believe this; & I think you would have to believe, that the tail of the Fan-tail was led to vary in the number & direction of its feathers in order to gratify the caprice of a few men. Yet if the fan-tail had been a wild bird & had used its abnormal tail for some special end, as to sail before the wind, unlike other birds, everyone would have said what beautiful & designed adaptation. Again I say I am, & shall ever remain, in a hopeless muddle.—

17 September 1861

Your question what would convince me of Design is a poser. If I saw an angel come down to teach us good, & I was convinced, from others seeing him, that I was not mad, I sh^d believe in design.—If I could be convinced thoroughily that life & mind was in an unknown way a function of other imponderable forces, I sh^d be convinced.—If man was made of brass or iron & in no way connected with any other organism which had ever lived, I sh^d perhaps be convinced. But this is childish writing.—

I have lately been corresponding with Lyell, who, I think, adopts your idea of the stream of variation having been led or designed. I have asked him (& he say he will hereafter reflect & answer me) whether he believes that the shape of my nose was designed. If he does, I have nothing more to say. If not, seeing what Fanciers have done by selecting individual differences in the nasal bones of Pigeons, I must think that it is illogical to suppose that the variations, which Nat. Selection, preserves for the good of any being, have been designed.

The Earth Speaks to Bryan

H. F. Osborn

The Testimony of the Rocks

> "Day unto day uttereth speech, and night unto night sheweth knowledge." (Psalm 19:2.)

The Earth Speaks, clearly, distinctly, and, in many of the realms of Nature, loudly, to William Jennings Bryan, but *he fails to hear a single sound*. The earth speaks from the remotest periods in its wonderful life history in the Archæozoic Age, when it reveals only a few tissues of its primitive plants. Fifty million years ago it begins to speak as "the waters bring forth abundantly the moving creatures that hath life." In successive eons of time the various kinds of animals leave their remains in the rocks which compose the deeper layers of the earth, and when the rocks are laid bare by wind, frost, and storm we find wondrous lines of ascent invariably following the principles of *creative evolution,* whereby the simpler and more lowly forms always precede the higher and more specialized forms.

The earth speaks not of a succession of distinct creations but of a continuous ascent, in which, as the millions of years roll by, increasing perfection of structure and beauty of form are found; out of the water-breathing fish arises the air-breathing amphibian; out of the land-living amphibian arises the land-living, air-breathing reptile, these two kinds of creeping things resembling each other closely. The earth speaks loudly and clearly of the ascent of the bird from one kind of reptile and of the mammal from another kind of reptile.

This is not perhaps the way Bryan would have made the animals, but this is the way God made them!

After the long travail of at least a million centuries there appear among the mammals the remote and humble ancestors of that great race which we ourselves have honored with the name of *Primates* because all the members of this race, like ourselves, live upon their wits, relying upon their cleverness and even intelligence in the eternal struggle for existence. In clarion tones, not with uncertain sound, the earth tells us in both the form and the functions of our bodies and of our minds, in every nerve, in every gland, in every muscle which the nerves control, in the lower and higher centres of the brain as the royal seat of our primacy, in the bones which compose our framework, especially in the bones of the skull and jaws and of the foot and hand, that we too have ascended from lowlier ancestors not wholly dissimilar but never identical with other *Primates* to which we feel ourselves proudly superior. Let us regard them as "poor relations" if we will, they are none the less of the same handiwork as ourselves.

In Darwin's day the earth had hardly begun to speak of this relationship of ours to the other *Primates,* but Darwin's was the prophet's ear, close to the earth, which truly interpreted its feeble tones. Today the earth speaks with resonance and clearness, and every ear in every civilized country of the world is attuned to its wonderful message of the creative evolution of man, except the ear of William Jennings Bryan; he alone remains stone-deaf, he alone by his own resounding voice drowns the eternal speech of Nature.

How can I as the author of these essays, a naturalist, a professor of zoology, "a tall professor coming down out of the trees," as he calls me, contend with the resounding voice of Bryan when the voices of Nature are powerless to do so? At once I confess that I cannot contend with him, nor can I still his voice, and this has always been my attitude since February, 1922, when in reply to his article in the New York Times entitled "Evolution of Man," I hastily wrote the first of my rejoinders, "Evolution and Religion," and thus entered the arena of Religion and Science in which the Great Commoner and myself have met at intervals during the past three years. My advice to my opponent is invariably and consistently the same; namely, to drop the methods of the lawyer, of the politician, of the statesman, even of the theologian and of the scientist, and to adopt the simple methods of the naturalist, to observe and hear for himself the great truths which the earth so clearly proclaims.

I do not enter into the well-known details of the wonderful processes of evolution as they have been conscientiously observed in plants and animals for a century and a half. I refer inquirers after truth to the published and readily accessible works of a long line of observers, from Leonardo da Vinci in the fifteenth century to the writers of the eleventh edition of the Encyclopædia Britannica.

As for the creative evolution of man, passing by the early speculative writings of such men as Haeckel, we now have more than a dozen substantial volumes based not upon guesswork or speculation but upon the testimony yielded in the superficial layers of the earth and in caves, embracing hundreds of specimens of the fossilized remains of man, more or less ancient, more or less complete, but invariably, without a single exception, testifying to the gradual *ascent of man* from a lower to a higher state, gradually dropping one primitive bit of anatomy after another until the high, intelligent, fully human aspect is attained.

Again with clarion voice these irrefutable witnesses of our past positively demonstrate two new and somewhat unexpected truths: first, that *man has not descended from any known kind of monkey or ape,* fossil or recent; with this truth, established not by Bryan but by the testimony of the earth, one of the chief sentimental objections to the creative evolution of man disappears forever. Second, *man has a long, independent, superior line of ascent of his own,* with a relatively erect posture, with hands free to grasp and use tools, with the thumb and forefinger capable of handling flint implements such as the graving tools and brush of the artist and, finally, the reed, pen, or crayon, with which to set down his thoughts. Challenge as we may the less perfect fossil discoveries in the Trinil sands, in the Piltdown gravels, in the Heidelberg riverbeds, no man can challenge the convincing testimony to the creative evolution of man afforded by the several complete skeletons of the race of the Neanderthal who lived 100,000 years ago, nor the perfectly preserved fossil remains of the artistic race of the Crô-Magnons who lived 30,000 years ago.

The Neanderthal hunters of 100,000 years ago and the Crô-Magnon artists of 30,000 years ago are not guesswork or the fabric of scientific imagination; they are realities, men like ourselves, the older one a much lower race—a veritable missing link—the other a higher race with all powers equal to our own.

At the time these fossilized artists of the higher Crô-Magnon race lived along the river borders of France all of northern Europe was sinking under the burden of the titanic glacier which covered Belgium and northern France and which drove southward great herds of the reindeer, the woolly rhinoceros, the Arctic hares and lemmings. These artists painted and modelled in clay and rock the fossilized mammoths, and no circumstantial evidence produced in court at any time in the whole history of law has ever been stronger than this evidence that these artists, these reindeer, and these mammoths lived together in the subarctic climate of southern France and northern Spain.

The low-browed Neanderthal hunting race is of far greater antiquity, a fact also established by circumstantial evidence equally strong and equally convincing. When these men hunted the woolly rhinoceros in the half-frozen rivers of southern France the titanic glaciers of the northern hemisphere reached their arms southward from the Scandinavian peaks and from the central and eastern (Laurentian) highlands of Canada, attaining such height and massiveness as to completely bury the entire State of New York, finally reaching their melting-point near the western extremity of Long Island and the centre of the State of New Jersey. This fossilized hunting race of the Neanderthals, low-browed, small-statured, ungainly, hideous of aspect, with retreating chin, broad nostrils, beetling eyebrows, is nevertheless human, beyond challenge. They had tender sentiments, they revered their dead, they believed in the future existence of the hunter in "happy hunting-grounds," as evidenced in their inclusion of the finest flint implements in the burial of their dead.

To sum up the testimony of the rocks, the evidence as regards the creative evolution of man is as unanswerable as that of the creative evolution of the entire plant and animal world. Man is no exception to the universal law that God did use evolution as His plan.

Chapter 5

Chapter 5 starts with the article written by Sewall Wright in which he introduced his metaphor of an adaptive landscape. This paper was given at an international congress of genetics in 1932. Dobzhansky saw it and at once realized that he had here the theory that he needed on which he could hang all of his own empirical research. He said, "I fell in love with Wright," and he was only half joking. Do not be worried if the article seems a bit technical and mathematical. Rest assured that none of this can be really important and that you can skip it. None of the leading synthetic theorists (with the exception of Simpson) could follow any mathematics whatsoever, so the details cannot have been essential! What matters are the pictures—the landscape and then the way in which populations are supposed to drift down from peaks into valleys, quite against the power of selection, and then shoot up the other sides under the influence of selection. I should say that this theory, which has been very influential—although there are now some evolutionists starting to doubt its truth entirely—has absolutely no basis in Darwin's Origin. *However, as I said in the text, it is just the sort of thing that you would expect from a neo-Spencerian.*

The second and third documents are letters. First I give you a letter written by Ernst Mayr, the ornithologist and systematist, when he was the founding editor of the journal Evolution. *See how Mayr is trying to articulate the kind of science to which he thinks evolutionists should aspire. It should be experimental, factual, quantitative, and so forth. In short, all of the things that one finds in the best kinds of science, like physics. Above all else, one must avoid "philosophy and speculation"! One must get into "factors and causes." The second letter is by Dobzhansky, sent to an eminent historian of science, John Greene. Both men were practicing Christians and were trying to work out the relationship between science and religion, in particular between evolution and its ideological underpinnings (progress specifically).*

Dobzhansky was an enthusiast for progress—remember that he was a Teilhard de Chardin supporter—and claimed that evolution itself supports his belief. Progress was not something written in but rather something that could be read out. What this letter shows is that although the synthetic theorists may have worked hard (and successfully) at professionalizing their subject, their motives and interests still overlapped strongly with those of earlier evolutionists in the Thomas Henry Huxley mould. Say what you like, evolutionary theory was and remains more than just a theory.

The Roles of Mutation, Inbreeding, Crossbreeding and Selection in Evolution

Sewall Wright

The enormous importance of biparental reproduction as a factor in evolution was brought out a good many years ago by EAST. The observed properties of gene mutation—fortuitous in origin, infrequent in occurrence and deleterious when not negligible in effect—seem about as unfavorable as possible for an evolutionary process. Under biparental reproduction, however, a limited number of mutations which are not too injurious to be carried by the species furnish an almost infinite field of possible variations through which the species may work its way under natural selection.

Estimates of the total number of genes in the cells of higher organisms range from 1000 up. Some 400 loci have been reported as having mutated in Drosophila during a laboratory experience which is certainly very limited compared with the history of the species in nature. Presumably, allelomorphs of all type genes are present at all times in any reasonably numerous species. Judging from the frequency of multiple allelomorphs in those organisms which have been studied most, it is reasonably certain that many different allelomorphs of each gene are in existence at all times. With 10 allelomorphs in each of 1000 loci, the number of possible combinations is 10^{1000} which is a very large number. It has been estimated that the total number of electrons and protons in the whole visible universe is much less than 10^{100}.

However, not all of this field is easily available in an interbreeding population. Suppose that each type gene is manifested in 99 percent of the individuals, and that most of the remaining 1 percent have the most favorable of the other allelomorphs, which in general means one with only a slight differential effect. The average individual will show the effects of 1 percent of the 1000, or 10 deviations from the type, and since this average has a standard deviation of $\sqrt{10}$ only a small proportion will exhibit more than 20 deviations from type where 1000 are possible. The population is thus confined to an infinitesimal portion of the field of possible gene combinations, yet this portion includes some 10^{40} homozygous combinations, on the above extremely conservative basis, enough so that there is no reasonable chance that any two individuals have exactly the same genetic constitution in a species of millions of millions of individuals persisting over millions of generations. There is no difficulty in accounting for the probable genetic uniqueness of each individual human being or other organism which is the product of biparental reproduction.

If the entire field of possible gene combinations be graded with respect to adaptive value under a particular set of conditions, what would be its nature? Figure 1 shows the combinations in the cases of 2 to 5 paired allelomorphs. In the last case, each of the 32 homozygous combinations is at one remove from 5 others, at two removes from 10, etc. It would require 5 dimensions to represent these relations symmetrically; a sixth dimension is needed to represent level of adaptive value. The 32 combinations here compare with 10^{1000} in a species with 1000 loci each represented by 10 allelomorphs, and the 5 dimensions required for adequate representation compare with 9000. The two dimensions of figure 2 are a very inadequate representation of such a field. The contour lines are intended to represent the scale of adaptive value.

One possibility is that a particular combination gives maximum adaptation and that the adaptiveness of the other combinations falls off more or less regularly according to the number of removes. A species whose individuals are clustered about some combination other than the highest would move up the steepest gradient toward the peak, having reached which it would remain unchanged except for the rare occurrence of new favorable mutations.

But even in the two factor case (figure 1) it is possible that there may be two peaks, and the chance that this may be the case greatly increases with each additional locus.

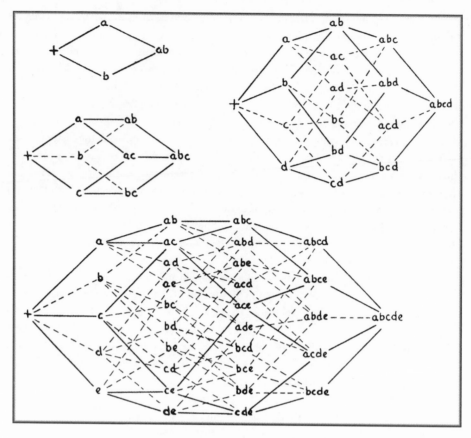

Figure 1: The combination of from 2 to 5 paired allelomorphs

With something like 10^{1000} possibilities (figure 2) it may be taken as certain that there will be an enormous number of widely separated harmonious combinations. The chance that a random combination is as adaptive as those characteristic of the species may be as low as 10^{-100} and still leave room for 10^{800} separate peaks, each surrounded by 10^{100} more or less similar combinations. In a rugged field of this character, selection will easily carry the species to the nearest peak, but there may be innumerable other peaks which are higher but which are separated by "valleys." The problem of evolution as I see it is that of a mechanism by which the species may continually find its way from lower to higher peaks in such a field. In order that this may occur, there must be some trial and error mechanism on a grand scale by which the species may explore the region surrounding the small portion of the field which it occupies. To evolve, the species must not be under strict control of natural selection. Is there such a trial and error mechanism?

At this point let us consider briefly the situation with respect to a single locus. In each graph in figure 3 the abscissas represent a scale of gene frequency, 0 percent of the type genes to the left, 100 percent to the right. The elementary evolutionary process is, of course, change of gene frequency, a practically continuous process. Owing to the symmetry of the Mendelian mechanism, any gene frequency tends to remain constant in the absence of disturbing factors. If the type gene mutates at a certain rate, its frequency tends to move to the left, but at a continually decreasing rate. The type gene would ultimately be lost from the population if there were no opposing factor. But the type gene is in general favored by selection. Under selection, its frequency tends to move to the right. The rate is greatest at some point near the middle of the range. At a certain gene

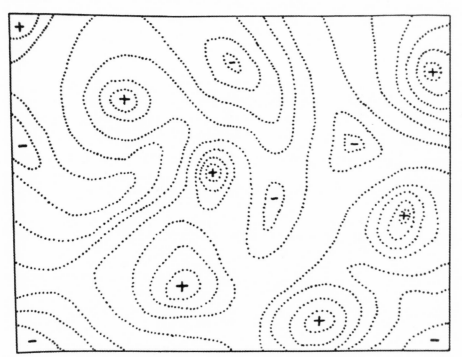

frequency the opposing pressures are equal and opposite, and at this point there is consequently equilibrium. There are other mechanisms of equilibrium among evolutionary factors which need not be discussed here. Note that we have here a theory of the stability of species in spite of continuing mutation pressure, a continuing field of variability so extensive that no two individuals are ever genetically the same, and continuing selection.

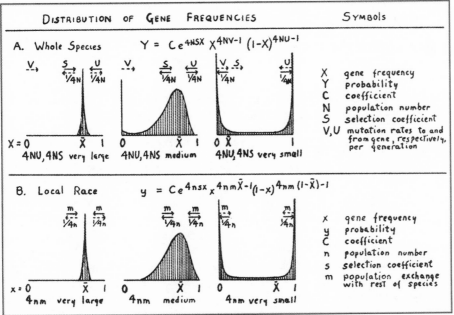

If the population is not indefinitely large, another factor must be taken into account: the effects of accidents of sampling among those that survive and become parents in each generation and among the germ cells of these, in other words, the effects of inbreeding. Gene frequency in a given generation is in general a little different one way or the other from that in the preceding, merely by chance. In time, gene frequency may wander a long way from the position of equilibrium, although the farther it wanders the greater the pressure toward return. The result is a frequency distribution within which gene frequency moves at random. There is considerable spread even with very slight inbreeding and the form of distribution becomes U-shaped with close inbreeding. The rate of movement of gene frequency is very slow in the former case but is rapid in the latter (among unfixed genes). In this case, however, the tendency toward complete fixation of genes, practically irrespective of selection, leads in the end to extinction.

In a local race, subject to a small amount of crossbreeding with the rest of the species (figure 3, lower half), the tendency toward random fixation is balanced by immigration pressure instead of by mutation and selection. In a small sufficiently isolated group all gene frequencies can drift irregular back and forth about their mean values at a rapid rate, in terms of geologic time, without reaching fixation and giving the effects of close inbreeding. The resultant differentiation of races is of course increased by any local differences in the conditions of selection.

Let us return to the field of gene combinations (figure 4). In an indefinitely large but freely interbreeding species living under constant conditions, each gene will reach ultimately a certain equilibrium. The species will occupy a certain field of variation about a peak in our diagram (heavy broken contour in upper left of each figure). The field occupied remains constant although no two individuals are ever identical. Under the above conditions further evolution can occur only by the appearance of wholly new (instead of recurrent) mutations, and ones which happen to be favorable from the first.

Figure 4: Field of gene combinations occupied by a population within the general field of possible combinations. Type of history under specified conditions indicated by relation to initial field (heavy broken contour) and arrow.

A. Increased Mutation or reduced Selection 4NU, 4NS very large

B. Increased Selection or reduced Mutation 4NU, 4NS very large

C. Qualitative Change of Environment 4NU, 4NS very large

D. Close Inbreeding 4NU, 4NS very small

E. Slight Inbreeding 4NU, 4NS medium

F. Division into local Races 4nm medium

Such mutations would change the character of the field itself, increasing the elevation of the peak occupied by the species. Evolutionary progress through this mechanism is excessively slow since the chance of occurrence of such mutations is very small and, after occurrence, the time required for attainment of sufficient frequency to be subject to selection to an appreciable extent is enormous.

The general rate of mutation may conceivably increase for some reason. For example, certain authors have suggested an increased incidence of cosmic rays in this connection. The effect (figure 4A) will be as a rule a spreading of the field occupied by the species until a new equilibrium is reached. There will be an average lowering of the adaptive level of the species. On the other hand, there will be a speeding up of the process discussed above, elevation of the peak itself through appearance of novel favorable mutations. Another possibility of evolutionary advance is that the spreading of the field occupied may go so far as to include another and higher peak, in which case the species will move over and occupy the region about this. These mechanisms do not appear adequate to explain evolution to an important extent.

The effects of reduced mutation rate (figure 4B) are of course the opposite: a rise in average level, but reduced variability, less chance of novel favorable mutation, and less chance of capture of a neighboring peak.

The effect of increased severity of selection (also 4B) is, of course, to increase the average level of adaptation until a new equilibrium is reached. But again this is at the expense of the field of variation of the species and reduces the chance of capture of another adaptive peak. The only basis for continuing advance is the appearance of novel favorable mutations which are relatively rapidly utilized in this case. But at best the rate is extremely slow even in terms of geologic time, judging from the observed rates of mutation.

Relaxation of selection has of course the opposite effects and thus effects somewhat like those of increased mutation rate (figure 4A).

The environment, living and non-living, of any species is actually in continual change. In terms of our diagram this means that certain of the high places are gradually being depressed and certain of the low places are becoming higher (figure 4C). A species occupying a small field under influence of severe selection is likely to be left in a pit and become extinct, the victim of extreme specialization to conditions which have ceased, but if under sufficiently moderate selection to occupy a wide field, it will merely be kept continually on the move. Here we undoubtedly have an important evolutionary process and one which has been generally recognized. It consists largely of change without advance in adaptation. The mechanism is, however, one which shuffles the species about in the general field. Since the species will be shuffled out of low peaks more easily than high ones, it should gradually find its way to the higher general regions of the field as a whole.

Figure 4D illustrates the effect of reduction in size of population below a certain relation to the rate of mutation and severity of selection. There is fixation of one or another allelomorph in nearly every locus, largely irrespective of the direction favored by selection. The species moves down from its peak in an erratic fashion and comes to occupy a much smaller field. In other words there is the deterioration and homogeneity of a closely inbred population. After equilibrium has been reached in variability, movement becomes excessively slow, and, such as there is, is nonadaptive. The end can only be extinction. Extreme inbreeding is not a factor which is likely to give evolutionary advance.

With an intermediate relation between size of population and mutation rate, gene frequencies drift at random without reaching the complete fixation of close inbreeding (figure 4E). The species moves down from the extreme peak but continually wanders in the vicinity. There is some chance that it may encounter a gradient leading to another peak and shift its allegiance to this. Since it will escape relatively easily from low peaks as compared with high ones, there is here a trial and error mechanism by which in time the species may work its way to the highest peaks in the general field. The rate of progress, however, is extremely slow since change of gene frequency is of the order of the reciprocal of the effective population size and this reciprocal must be of the order of the mutation rate in order to meet the conditions for this case.

Finally (figure 4F), let us consider the case of a large species which is subdivided into many small local races, each breeding largely within itself but occasionally crossbreeding. The field of gene combinations occupied by each of these local races shifts continually in a nonadaptive fashion (except in so far as there are local differences in the conditions of selection). The rate of movement may be enormously greater than in the preceding case since the condition for such movement is that the reciprocal of the population number be of the order of the proportion of crossbreeding instead of the mutation rate. With many local races, each spreading over a considerable field and moving relatively rapidly in the more general field about the controlling peak, the chances are good that one at least will come under the influence of another peak. If a higher peak, this race will expand in numbers and by crossbreeding with the others will pull the whole species toward the new position. The average adaptiveness of the species thus advances under intergroup selection, an enormously more effective process than intragroup selection. The conclusion is that subdivision of a species into local races provides the most effective mechanism for trial and error in the field of gene combinations.

It need scarcely be pointed out that with such a mechanism complete isolation of a portion of a species should result relatively rapidly in specific differentiation, and one that is not necessarily adaptive. The effective intergroup competition leading to adaptive advance may be between species rather than races. Such isolation is doubtless usually geographic in character at the outset but may be clinched by the development of hybrid sterility. The usual difference of the chromosome complements of related species puts the importance of chromosome aberration as an evolutionary process beyond question, but, as I see it, this importance is not in the character differences which they bring (slight in balanced types), but rather in leading to the sterility of hybrids and thus making permanent the isolation of two groups.

How far do the observations of actual species and their subdivisions conform to this picture? This is naturally too large a subject for more than a few suggestions.

That evolution involves nonadaptive differentiation to a large extent at the subspecies and even the species level is indicated by the kinds of differences by which such groups are actually distinguished by systematists. It is only at the subfamily and family levels that clear-cut adaptive differences become the rule (ROBSON, JACOT). The principal evolutionary mechanism in the origin of species must thus be an essentially nonadaptive one.

That natural species often are subdivided into numerous local races is indicated by many studies. The case of the human species is most familiar. Aside from the familiar racial differences recent studies indicate a distribution of frequencies relative to an apparently nonadaptive series of allelomorphs, that determining blood groups, of just the sort discussed above. I scarcely need to labor the point that changes in the average of mankind in the historic period have come about more by expansion of some types and decrease and absorption of others than by uniform evolutionary advance. During the recent period, no doubt, the phases of intergroup competition and crossbreeding have tended to overbalance the process of local differentiation, but it is probable that in the hundreds of thousands of years of prehistory, human evolution was determined by a balance between these factors.

Subdivision into numerous local races whose differences are largely nonadaptive has been recorded in other organisms wherever a sufficiently detailed study has been made. Among the land snails of the Hawaiian Islands, GULICK (sixty years ago) found that each mountain valley, often each grove of trees, had its own characteristic type, differing from others in "nonutilitarian" respects. GULICK attributed this differentiation to inbreeding. More recently CRAMPTON has found a similar situation in the land snails of Tahiti and has followed over a period of years evolutionary changes which seem to be of the type here discussed. I may also refer to the studies of fishes by DAVID STARR JORDAN, garter snakes by RUTHVEN, bird lice by KELLOGG, deer mice by OSGOOD, and gall wasps by KINSEY as others which indicate the role of local isolation as a differentiating factor. Many other cases are discussed by OSBORN and especially by RENSCH in recent summaries. Many of these authors insist on the nonadaptive character of most of the differences among local races. Others attribute all differences to the environment, but this seems to be more an expression of faith than a view based on tangible evidence.

An even more minute local differentiation has been revealed when the methods of statistical analysis have been applied. SCHMIDT demonstrated the existence of persistent mean differences at each collecting station in certain species of marine fish of the fjords of Denmark, and these differences were not related in any close way to the environment. That the differences were in part genetic was demonstrated in the laboratory. DAVID THOMPSON has found a correlation between water distance and degree of differentiation within certain fresh water species of fish of the streams of Illinois. SUMNER'S extensive studies of subspecies of Peromyscus (deer mice) reveal genetic differentiations, often apparently nonadaptive, among local populations and demonstrate the genetic heterogeneity of each such group.

The modern breeds of livestock have come from selection among the products of local inbreeding and of crossbreeding between these, followed by renewed inbreeding, rather than from mass selection of species. The recent studies of the geographical distribution of particular genes in livestock and cultivated plants by SEREBROVSKY, PHILIPTSCHENKO and others are especially instructive with respect to the composition of such species.

The paleontologists present a picture which has been interpreted by some as irreconcilable with the Mendelian mechanism, but this seems to be due more to a failure to appreciate statistical consequences of this mechanism than to anything in the data. The horse has been the standard example of an orthogenetic evolutionary sequence preserved for us with an abundance of material. Yet MATHEW'S interpretation as one in which evolution has proceeded by extensive differentiation of local races, intergroup selection, and crossbreeding is as close as possible to that required under the Mendelian theory.

Summing up: I have attempted to form a judgment as to the conditions for evolution based on the statistical consequences of Mendelian heredity. The most general conclusion is that evolution depends on a certain balance among its factors. There must be gene mutation, but an excessive rate gives an array of freaks, not evolution; there must be selection, but too severe a process destroys the field of variability, and thus the basis for further advance; prevalence of local inbreeding within a species has extremely important evolutionary consequences, but too close inbreeding leads merely to extinction. A certain amount of crossbreeding is favorable but not too much. In this dependence on balance the species is like a living organism. At all levels of organization life depends on the maintenance of a certain balance among its factors.

More specifically, under biparental reproduction a very low rate of mutation balanced by moderate selection is enough to maintain a practically infinite field of possible gene combinations within the species. The field actually occupied is relatively small though sufficiently extensive that no two individuals have the same genetic constitution. The course of evolution through the general field is not controlled by direction of mutation and not directly by selection, except as conditions change, but by a trial and error mechanism consisting of a largely nonadaptive differentiation of local races (due to inbreeding balanced by occasional crossbreeding) and a determination of long time trend by intergroup selection. The splitting of species depends on the effects of more complete isolation, often made permanent by the accumulation of chromosome aberrations, usually of the balanced type. Studies of natural species indicate that the conditions for such an evolutionary process are often present.

Literature Cited

CRAMPTON, H. E., 1925. Contemporaneous organic differentiation in the species of Partula living in Moorea, Society Islands. Amer. Nat. 59:5–35.

EAST, E. M., 1918. The role of reproduction in evolution. Amer. Nat. 52 :273–289.

GULICK, J. T., 1905. Evolution, racial and habitudinal. Pub. Carnegie Instn. 25:1–269.

JACOT, A. P., 1932. The status of the species and the genus. Amer. Nat. 66:346–364.

JORDAN, D. S., 1908. The law of geminate species. Amer. Nat. 42:73–80.

KELLOGG, V. L., 1908. Darwinism, today. 403 pp. New York: Henry Holt and Co.

KINSEY, A. C., 1930. The gall wasp genus Cynips. Indiana Univ. Studies. 84–86:1–577.

MATHEW, W. D., 1926. The evolution of the horse. A record and its interpretation. Quart. Rev. Biol. 1:139–185.

OSBORN, H. F., 1927. The origin of species. V. Speciation and mutation. Amer. Nat. 49:193–239.

OSGOOD, W. H., 1909. Revision of the mice of the genus Peromyscus. North American Fauna 28:1–285.

PHILIPTSCHENKO, J., 1927. Variabilität and Variation. 101 pp. Berlin.

RENSCH, B., 1929. Das Prinzip geographischer Rassenkreise und das Problem der Artbildung. 206 pp. Berlin: Gebrüder Borntraeger.

ROBSON, G. C., 1928. The species problem. 283 pp. Edinburgh and London: Oliver and Boyd.

RUTHVEN, A. G., 1908. Variation and genetic relationships of the garter snakes. U. S. Nat. Mus. Bull. 61:1–301.

SCHMIDT, J., 1917. Statistical investigations with *Zoarces viviparus*. L. J. Genet. 7:105–118.

SEREBROVSKY, A. S., 1929. Beitrag zur geographischen Genetic des Haushahns in Sowjet-Russland. Arch. f. Geflügelkunde Jahrgang 3: 161–169.

SUMNER, F. B., 1932. Genetic, distributional, and evolutionary studies of the subspecies of deer mice (Peromyscus). Bibl. genet. 9:1–106.

THOMPSON, D. H., 1931. Variation in fishes as a function of distance. Trans. Ill. State Acad. of Sci. 23:276–281.

WRIGHT, S., 1931. Evolution in Mendelian populations. Genetics 16:97–159.

Letter to G. F. Ferris

E. Mayr

March 29, 1948
Dr. G. F. Ferris
Natural History Museum
Stanford University
California

Dear Dr. Ferris,

I must apologize for not writing you earlier but there was some unexpected delay in getting your manuscript back from one of the readers.

Let me say right at the beginning that I read your paper with the greatest of pleasure. I think it does just what you intend, namely, stimulate the reader to think about a lot of things he normally takes for granted. Your paper is a fine example of a scientific stop-look-and-listen attitude in research and one of the unfortunately so rare cases of formulation of the problem before initiation of a research program (see later parts of this letter for a discussion of some of the points which you brought up). It pains me therefore to have to report to you that both readers of the Editorial Board say that the paper is not suitable for EVOLUTION in its present form. Reluctantly I have come to the conclusion that they may be right that the present paper is more a discussion of scientific methodology than one of the causes and factors of evolution. Much of your discussion appears to be particularly suitable as an introductory chapter of a book rather than as an article in a periodical.

However, in view of the obvious evolutionary importance of your studies on the morphology of the Annulata and the desirability to make them available to the readers of EVOLUTION, I was wondering whether it would not be possible for you to publish elsewhere that part of your manuscript that discusses scientific methods and to utilize the evolutionary working hypotheses of pages 15 and 16 as the introductory pages of one or two factual papers dealing with evolution in the Annulata. It has so far been the editorial policy of EVOLUTION to present concrete facts in every paper followed by the conclusions to be drawn from these facts. This policy was adopted deliberately because the prestige of evolutionary research has suffered in the past because of too much philosophy and speculation. A number of subjects from the domain of your present research have occurred to me which would be in harmony with our editorial policy. For example, it would be extremely interesting to have a paper on the single subject, The Origin of New Structures Among the Annulata. As you say with so much justification in Postulate Number 4, new structures originate extremely rarely. This, of course, is true in the vertebrates also. A paper on stable structures in an otherwise variable environment would also be very fascinating. This subject, incidentally, might permit a tie-in with certain of the problems of genetics. A third paper might be written on evolutionary trends in the morphology of the Annulata. A fourth one might be on parallelism, etc.

Your work appears so novel and important that I hope sincerely that the report of the readers of the Editorial Board does not discourage you because I am very much looking forward to receiving a manuscript from you along one of the above-mentioned lines.

Before closing, I would like to discuss one or two thoughts with you that came to my mind while reading your paper. I think you might say that there are four major evolutionary problems:

1. The fact of evolution
2. The material of evolution
3. The origin of discontinuities
4. The course of evolution.

As far as (1) is concerned, it requires no integration with genetics and is the proper field of phylogenetic paleontology, comparative embryology, and comparative anatomy. It is (2) that is the proper domain of genetics. It is (3) that is the principal domain of taxonomy since the origin of species is what you really meant by the origin of discontinuities. As far as (4) is concerned, it deals with the factors and causes of evolutionary change and it is here where the various fields concur and where the field of ecology enters the picture. Since there is so much agreement nowadays on (1), (2), and (3), the trend of evolutionary research has been very strongly toward (4). This is realized by the geneticists themselves, as most strongly expressed by Muller in some recent statements.

I was wondering as to the propriety of using the word "principle" as you have done. As Northrop says correctly, the first step of research is the formulation of a problem. The second step is the establishment of a working hypothesis while the final, ultimate answer is a Principle. I have a feeling that in a preliminary investigation one should not start with a set of principles. Rather, these principles should be the final conclusion of the investigation. However, this matter being one of terminology, no doubt everybody has his own subjective interpretation.

Let me conclude by saying how much I enjoyed reading your manuscript and I express the hope that you will be able to submit a paper along the lines suggested by me.

Very sincerely yours,
Ernst Mayr

Letter to John Greene

Theodosius Dobzhansky

New York, November 23, 1961
Dear Greene:

Let me assure you that I am grateful to you for your detailed reply, and that I also enjoy an opportunity to discuss matters of such vital concern to us with a man of your intellectual stature and your intellectual background. Even if we find ourselves in rather basic disagreement, let us drop the titles of "professors," and let us discuss things, even though this is very difficult in letters and we have not an opportunity to do so in any other way.

You say you do not understand where I stand. Let me remove all doubts about this. I am a Christian, hence I stand with my good friend Birch, and you, and Teilhard, and certainly not with Huxley, although his is very much a majority opinion among at least the natural scientists. I am not a Whiteheadian, and so differ from Birch who likes White-head, which to me is a rather meaningless philosophy.

It is hard to go much beyond these sweeping statements, but let me try. You and I will agree that the world is not a "devil's vaudeville" (Dostoyevsky's words), but is mean-ingful. Evolution (cosmic + biological + human) is going towards something, we hope some City of God. This belief is not imposed on us by our scientific discoveries, but if we wish (but not if we do not wish) we may see in nature manifestations of the Omega, or your creative ground (you spell this not with capitals?), or simply of God.

But let us face the problem of how the creative ground creates. I think it is here that your view is weak. You yourself say that if God was trying to make an Equus out of an Eohippus, then "He must be branded as a slow, clumsy, inefficient workman." Well, this is the crux of the whole matter. Nothing makes sense if you suppose that God makes special interventions to direct evolution. I refuse to believe in "direction" in any other sense than that the Alpha and the Omega[1] of evolution are simultaneously present in God's eyes (like to Laplace's universal intelligence). To direct evolution in any other sense, God must induce mutations, shuffle the nucleotides in DNA, and give from time to time little pushes to natural selection at critical moments. All this makes no sense to me (nor to Birch, Teilhard is unclear on this). I cannot believe that God becomes from time to time a particularly powerful enzyme.

I see no escape from thinking that God acts not in fits of miraculous interventions, but in all significant and insignificant, spectacular and humdrum events. Pantheism, you may say? I do not think so, but if so then there is this much truth in pantheism. The really tough point is, of course, in what sense can God's action be seen in all that happens. I am not foolish enough to think that I can solve this. Perhaps Teilhard had a hint, very ob-scurely expressed. You are one of the few who could perhaps get on in trying to solve this. But look, if everything that happens is in some sense "trial and error," then it is merely redundant to say that "someone" is trying to accomplish something in particular in one place but not in another. But I refuse to abstain from talking about progress, im-provement, and creativity. Why should I? Some extreme "scientists" would eliminate ex-pressions such as that the eye is built so that the animal can see things. Perhaps extremes converge, and you would join these mechanistic purists? In evolution some organisms progressed and improved and stayed alive, others failed to do so and became extinct. Some adaptations are better than others—for the organisms having them; they are better for survival rather than for death. Yes, life is a value and a success, death is valueless and a failure. So, some evolutionary changes are better than others. Yes, life is trying hard to hang on and to produce more life. I see no need *at this point* to say that some creative ground is trying to get better adaptations and better values and more living substance. For one thing, this creative ground is only very slowly learning how to do it, and is even now clumsy and inefficient about it as you yourself admit! Your surprise at the fact that a process devoid of intelligence and will has produced beings which have intelligence and will is for me hard to understand, I cannot follow you into saying that this *proves* that this

---

[1] These are Teilhard de Chardin's concepts.

process itself is (and was?) *not* devoid of these things. Evolution, biological and especially human, is a process which generates novelties. Very remarkable indeed; but not a proof of the action of a creative ground shuffling genes. I feel really shocked—arguing with you I have to sound like a Huxley, something I do not like at all.

"The Death of Adam" showed clearly that to you evolution is something unwelcome though unavoidable. Your new book and your letter underscore this. To me, like to Birch and to Teilhard, evolution is a bright light. But it does not follow that evolution is a source of natural theology and a "proof" of the existence of God. I am driven to the view of such a conservative as [Karl] Barth which you quote on page 57 of your recent book (although otherwise I have no stomach for the neo-orthodoxy).[2] I am groping for a tolerable self-consistent Weltanschauung but do not claim having found one. Teilhard seemingly did, but perhaps not unexpectedly his book only hints at it and is unable to spell it out. Let me try to say this—evolution should be eventually understood *sub specie religionis* and religion *sub specie evolutionis*. But neither is deducible from the other. Both have to be somehow integrated in one's philosophy of "ultimate concern." I hope this is what you are struggling to do, but I submit that your attempt to view evolution as being actually propelled by a "creative ground" is no more satisfactory than all finalistic theories of evolution (popularized or rather vulgarized by Lecomte du Noüy [1947]).

I fear the above sentences may be misunderstood—I do not doubt that at some level evolution, like everything in the world, is a manifestation of God's activity. All that I say is that *as a scientist* I do not observe anything that would prove this. In short, as scientists Laplace and myself "have no need of this hypothesis," but as a human being I do need this hypothesis! But I cannot follow your advice and put these things in water-tight compartments, and see only "change" and no "progress," only "change" and no "trial and error." For as a scientist I observe that evolution is on the whole progressive, its "creativeness" is increasing, and these findings I find fitting nicely into my general thinking, in which your "creative ground" is perfectly acceptable.

Well, this is a very long letter, and I fear still not succeeding to explain what I would like to explain, doubtless because my thinking is too hazy and unsatisfactory. But this is the best I can do in a letter. And I still hope that some day in a future not too remote we may find an opportunity to spend a good long evening or maybe a part of a night over glasses of wine if you use it or over cups of coffee if you don't, discussing these things and helping each other to see holes in our respective ideas.

With warmest personal regards,

Sincerely yours,

Th. Dobzhansky

P.S. I trust you would not object against sending your letter and mine to Charles Birch in Australia.

---

[2] "... it is not the existence of the world in its manifoldness, from which we are to read off the fact that God is its Creator. The world ... gives us no information about God as the Creator. ... But when God has been known and then known again in the world..., that is because He is to be sought and found by us in Jesus Christ" (Barth 1959, 51. quoted in Greene 1961a, 57–58).

Chapter 6

Now we come to the origin of life. The document for this section is that paper in which J. B. S. Haldane puts forward his suggestion about life's origins. Haldane, I should explain, was the son of an eminent physiologist who was also something of a vitalist, J. S. Haldane. The son—Jack or J. B. S—defined himself with and against the father, on the one hand having imbibed science from birth but on the other hand being as outrageously reductionistic or materialistic (or whatever, as the mood suited him) as possible. After spending World War I happily killing Germans (typically, Haldane ended his life as a pacifistic, vegetarian Hindu), Haldane became not only a biochemist but also a brilliant population geneticist. At the same time he began writing for the general public (equally typically, the upper-class Haldane could be ferociously rude to underlings), and some of his most stimulating ideas appeared in more popular venues. This piece appeared in The Rationalist Annual, *a place for nontechnical material by and for atheists and others.*

It is clear from his piece, "The Origin of Life," that it was the virus that stimulated Haldane's imagination. He saw the virus as something halfway between being fully living and being totally inert. On the one hand, in the right conditions it can multiply at a terrific rate. On the other hand, the right conditions seem to be real, living organisms as its hosts and sources of food and energy generally. The virus therefore is the kind of bridge that Haldane thought must have come into existence at the beginning when life was first developing. Then, thanks to a lack of oxygen, primitive forms were able to survive and even reproduce, eventually leading to more sophisticated forms until one had full-blown organisms. The key therefore is a gradual development rather than a one-step spontaneous generation. At the end of his piece, Haldane could not forbear speculating on the origins of mind—a matter of great interest to his father. He comes to the somewhat double-edged conclusion that the biochemist's "ignorance disqualifies him no more than the historian or the geologist from attempting to solve a historical problem." I wonder what the older Haldane thought of that.

The Origin of Life

J. B. S. Haldane

Until about 150 years ago it was generally believed that living beings were constantly arising out of dead matter. Maggots were supposed to be generated spontaneously in decaying meat. In 1668 Redi showed that this did not happen provided insects were carefully excluded. And in 1860 Pasteur extended the proof to the bacteria which he had shown were the cause of putrefaction. It seemed fairly clear that all the living beings known to us originate from other living beings. At the same time Darwin gave a new emotional interest to the problem. It had appeared unimportant that a few worms should originate from mud. But if man was descended from worms such spontaneous generation acquired a new significance. The origin of life on the Earth would have been as casual an affair as the evolution of monkeys into man. Even if the latter stages of man's history were due to natural causes, pride clung to a supernatural, or at least surprising, mode of origin for his ultimate ancestors. So it was with a sigh of relief that a good many men, whom Darwin's arguments had convinced, accepted the conclusion of Pasteur that life can originate only from life. It was possible either to suppose that life had been supernaturally created on Earth some millions of years ago, or that it had been brought to Earth by a meteorite or by micro-organisms floating through interstellar space. But a large number, perhaps the majority, of biologists, believed, in spite of Pasteur, that at some time in the remote past life had originated on Earth from dead matter as the result of natural processes.

The more ardent materialists tried to fill in the details of this process, but without complete success. Oddly enough, the few scientific men who professed idealism agreed with them. For if one can find evidences of mind (in religious terminology the finger of God) in the most ordinary events, even those which go on in the chemical laboratory, one can without much difficulty believe in the origin of life from such processes. Pasteur's work therefore appealed most strongly to those who desired to stress the contrast between mind and matter. For a variety of obscure historical reasons, the Christian Churches have taken this latter point of view. But it should never be forgotten that the early Christians held many views which are now regarded as materialistic. They believed in the resurrection of the body, not the immortality of the soul. St Paul seems to have attributed consciousness and will to the body. He used a phrase translated in the revised version as 'the mind of the flesh,' and credited the flesh with a capacity for hatred, wrath and other mental functions. Many modern physiologists hold similar beliefs. But, perhaps unfortunately for Christianity, the Church was captured by a group of very inferior Greek philosophers in the third and fourth centuries AD. Since that date views as to the relation between mind and body which St Paul, at least, did not hold, have been regarded as part of Christianity, and have retarded the progress of science.

It is hard to believe that any lapse of time will dim the glory of Pasteur's positive achievements. He published singularly few experimental results. It has even been suggested by a cynic that his entire work would not gain a Doctorate of Philosophy today! But every experiment was final. I have never heard of any one who has repeated any experiment of Pasteur's with a result different from that of the master. Yet his deductions from these experiments were sometimes too sweeping. It is perhaps not quite irrelevant that he worked in his latter years with half a brain. His right cerebral hemisphere had been extensively wrecked by the bursting of an artery when he was only forty-five years old; and the united brain-power of the microbiologists who succeeded him has barely compensated for that accident. Even during his lifetime some of the conclusions which he had drawn from his experimental work were disproved. He had said that alcoholic fermentation was impossible without life. Buchner obtained it with a cell-free and dead extract of yeast. And since his death the gap between life and matter has been greatly narrowed.

When Darwin deduced the animal origin of man, a search began for a 'missing link' between ourselves and the apes. When Dubois found the bones of Pithecanthropus some comparative anatomists at once proclaimed that they were of animal origin, while others were equally convinced that they were parts of a human skeleton. It is now generally recognized that either party was right, according to the definition of humanity adopted. Pithecanthropus was a creature which might legitimately be described either as a man or an ape, and its existence showed that the distinction between the two was not absolute.

Now the recent study of ultramicroscopic beings has brought up at least one parallel case, that of the bacteriophage, discovered by d'Herelle, who had been to some extent anticipated by Twort. This is the case of a disease, or, at any rate, abnormality of bacteria. Before the size of the atom was known there was no reason to doubt that

Big fleas have little fleas
Upon their backs to bite 'em;
The little ones have lesser ones,
And so ad infinitum.

But we now know that this is impossible. Roughly speaking, from the point of view of size, the bacillus is the flea's flea, the bacteriophage the bacillus' flea; but the bacteriophage's flea would be of the dimensions of an atom, and atoms do not behave like fleas. In other words, there are only about as many atoms in a cell as cells in a man. The link between living and dead matter is therefore somewhere between a cell and an atom.

D'Herelle found that certain cultures of bacteria began to swell up and burst until all had disappeared. If such cultures were passed through a filter fine enough to keep out all bacteria, the filtrate could infect fresh bacteria and so on indefinitely. Though the infective agents cannot be seen with a microscope, they can be counted as follows. If an active filtrate containing bacteriophage be poured over a colony of bacteria on a jelly, the bacteria will all, or almost all, disappear. If it be diluted many thousand times, a few islands of living bacteria survive for some time. If it be diluted about ten million fold, the bacteria are destroyed round only a few isolated spots, each representing a single particle of bacteriophage.

Since the bacteriophage multiplies, d'Herelle believes it to be a living organism. Bordet and others have taken an opposite view. It will survive heating and other insults which kill the large majority of organisms, and will multiply only in presence of living bacteria, though it can break up dead ones. Except perhaps in presence of bacteria, it does not use oxygen or display any other signs of life. Bordet and his school therefore regard it as a ferment which breaks up bacteria as our own digestive ferments break up our food, at the same time inducing the disintegrating bacteria to produce more of the same ferment. This is not as fantastic as it sounds, for most cells while dying liberate or activate ferments which digest themselves. But these ferments are certainly feeble when compared with the bacteriophage.

Clearly we are in doubt as to the proper criterion of life. D'Herelle says that the bacteriophage is alive, because, like the flea or the tiger, it can multiply indefinitely at the cost of living beings. His opponents say that it can multiply only as long as its food is alive, whereas the tiger certainly, and the flea probably, can live on dead products of life. They suggest that the bacteriophage is like a book or a work of art, which is constantly being copied by living beings, and is therefore only metaphorically alive, its real life being in its copiers.

The American geneticist Muller has, however, suggested an intermediate view. He compares the bacteriophage to a gene—that is to say, one of the units concerned in heredity. A fully coloured and a spotted dog differ because the latter has in each of its cells one or two of a certain gene, which we know is too small for the microscope to see. Before a cell of a dog divides this gene divides also, so that each of the daughter-cells has one, two, or none according with the number in the parent cell. The ordinary spotted dog is healthy, but a gene common among German dogs causes a roan colour when one is present, while two make the dog nearly white, wall-eyed and generally deaf, blind or both. Most of such dogs die young, and the analogy to the bacteriophage is fairly close. The main difference between such a lethal gene, of which many are known, and the bacteriophage, is that the one is only known inside the cell, the other outside. In the present state of our ignorance we may regard the gene either as a tiny organism which can divide in the environment provided by the rest of the cell; or as a bit of machinery which the 'living' cell copies at each division. The truth is probably somewhere in between these two hypotheses.

Unless a living creature is a piece of dead matter plus a soul (a view which finds little support in modern biology) something of the following kind must be true. A simple organism must consist of parts A, B, C, D and so on, each of which can multiply only in presence of all, or almost all, of the others. Among these parts are genes, and the bacteriophage is such a part which has got loose. This hypothesis becomes more plausible if we believe in the work of Hauduroy, who finds that the ultramicroscopic particles into which the bacteria have been broken up, and which pass through filters that can stop the bacteria, occasionally grow up again into bacteria after a lapse of several months. He brings evidence to show that such fragments of bacteria may cause disease, and d'Herelle and Peyre claim to have found the ultramicroscopic form of a common staphylococcus, along with bacteriophage, in cancers, and suspects that this combination may be the cause of that disease.

On this view the bacteriophage is a cog, as it were, in the wheel of a life-cycle of many bacteria. The same bacteriophage can act on different species and is thus, so to say, a spare part which can be fitted into a number of different machines, just as a human diabetic can remain in health when provided with insulin manufactured by a pig. A great many kinds of molecule have been got from cells, and many of them are very efficient when removed from it. One can separate from yeast one of the many tools which it uses in alcoholic fermentation, an enzyme called invertase, and this will break up six times its weight of cane-sugar per second for an indefinite time without wearing out. As it does not form alcohol from the sugar, but only a sticky mixture of other sugars, its use is permitted in the United States in the manufacture of confectionery and cake-icing. But such fragments do not reproduce themselves, though they take part in the assimilation of food by the living cell. No one supposes that they are alive. The bacteriophage is a step beyond the enzyme on the road to life, but it is perhaps an exaggeration to call it fully alive. At about the same stage on the road are the viruses which cause such diseases as smallpox, herpes, and hydrophobia. They can multiply only in living tissue, and pass through filters which stop bacteria.

With these facts in mind we may, I think, legitimately speculate on the origin of life on this planet. Within a few thousand years from its origin it probably cooled down so far as to develop a fairly permanent solid crust. For a long time, however, this crust must have been above the boiling-point of water, which condensed only gradually. The primitive atmosphere probably contained little or no oxygen, for our present supply of that gas is only about enough to burn all the coal and other organic remains found below and on the Earth's surface. On the other hand, almost all the carbon of these organic substances, and much of the carbon now combined in chalk, limestone, and dolomite, were in the atmosphere as carbon dioxide. Probably a good deal of the nitrogen now in the air was combined with metals as nitride in the Earth's crust, so that ammonia was constantly being formed by the action of water. The Sun was perhaps slightly brighter than it is now, and as there was no oxygen in the atmosphere the chemically active ultra-violet rays from the Sun were not, as they now are, mainly stopped by ozone (a modified form of oxygen) in the upper atmosphere, and oxygen itself lower down. They penetrated to the surface of the land and sea, or at least to the clouds.

Now, when ultra-violet light acts on a mixture of water, carbon dioxide, and ammonia, a vast variety of organic substances are made, including sugars and apparently some of the materials from which proteins are built up. This fact has been demonstrated in the laboratory by Baly of Liverpool and his colleagues. In this present world, such substances, if left about, decay—that is to say, they are destroyed by micro-organisms. But before the origin of life they must have accumulated till the primitive oceans reached the consistency of hot dilute soup. Today an organism must trust to luck, skill, or strength to obtain its food. The first precursors of life found food available in considerable quantities, and had no competitors in the struggle for existence. As the primitive atmosphere contained little or no oxygen, they must have obtained the energy which they needed for growth by some other process than oxidation—in fact, by fermentation. For, as Pasteur put it, fermentation is life without oxygen. If this was so, we should expect that high organisms like ourselves would start life as anaerobic beings, just as we start as single cells. This is the case. Embryo chicks for the first two or three days after fertilization use very little oxygen, but obtain the energy which they need for growth by fermenting sugar into lactic acid, like the bacteria which turns milk sour. So do various embryo mammals, and in all probability you and I lived mainly by fermentation during the first week of our pre-natal life. The cancer cell behaves in the same way. Warburg has shown that with its embryonic habit of unrestricted growth there goes an embryonic habit of fermentation.

The first living or half-living things were probably large molecules synthesized under the influence of the Sun's radiation, and only capable of reproduction in the particularly favourable medium in which they originated. Each presumably required a variety of highly specialized molecules before it could reproduce itself, and it depended on chance for a supply of them. This is the case today with most viruses, including the bacteriophage, which can grow only in presence of the complicated assortment of molecules found in a living cell.

The unicellular organisms, including bacteria, which were the simplest living things known a generation ago, are far more complicated. They are organisms—that is to say, systems whose parts co-operate. Each part is specialized to a particular chemical function, and prepares chemical molecules suitable for growth of the other parts. In consequence, the cell as a whole can usually subsist on a few types of molecule, which are transformed within it into the more complex substances needed for the growth of the parts.

The cell consists of numerous half-living chemical molecules suspended in water and enclosed in an oily film. When the whole sea was a vast chemical laboratory the conditions for the formation of such films must have been relatively favourable; but for all that life may have remained in the virus stage for many millions of years before a suitable assemblage of elementary units was brought together in the first cell. There must have been many failures, but the first successful cell had plenty of food, and an immense advantage over its competitors.

It is probable that all organisms now alive are descended from one ancestor, for the following reason. Most of our structural molecules are asymmetrical, as shown by the fact that they rotate the plane of polarized light, and often form asymmetrical crystals. But of the two possible types of any such molecule, related to one another like a right and left boot, only one is found throughout living nature. The apparent exceptions to this rule are all small molecules which are not used in the building of the large structures which display the phenomena of life. There is nothing, so far as we can see, in the nature of things to prevent the existence of looking-glass organisms built from molecules which are, so to say, the mirror-images of those in our own bodies. Many of the requisite molecules have already been made in the laboratory. If life had originated independently on several occasions, such organisms would probably exist. As they do not, this event probably occurred only once, or, more probably, the descendants of the first living organism rapidly evolved far enough to overwhelm any later competitors when these arrived on the scene.

As the primitive organisms used up the foodstuffs available in the sea some of them began to perform in their own bodies the synthesis formerly performed haphazardly by the sunlight, thus ensuring a liberal supply of food. The first plants thus came into existence, living near the surface of the ocean, and making food with the aid of sunlight as do their descendants today. It is thought by many biologists that we animals are descended from them. Among the molecules in our own bodies are a number whose structure resembles that of chlorophyll, the green pigment with which the plants have harnessed the sunlight to their needs. We use them for other purposes than the plants—for example, for carrying oxygen—and we do not, of course, know whether they are, so to speak, descendants of chlorophyll or merely cousins. But since the oxygen liberated by the first plants must have killed off most of the other organisms, the former view is the more plausible.

The above conclusions are speculative. They will remain so until living creatures have been synthesized in the biochemical laboratory. We are a long way from that goal. It was only this year that Pictet for the first time made cane-sugar artificially. It is doubtful whether any enzyme has been obtained quite pure. Nevertheless I hope to live to see one made artificially. I do not think I shall behold the synthesis of anything so nearly alive as a bacteriophage or a virus, and I do not suppose that a self-contained organism will be made for centuries. Until that is done the origin of life will remain a subject for speculation. But such speculation is not idle, because it is susceptible of experimental proof or disproof.

Some people will consider it a sufficient refutation of the above theories to say that they are materialistic, and that materialism can be refuted on philosophical grounds. They are no doubt compatible with materialism, but also with other philosophical tenets. The facts are, after all, fairly plain. Just as we know of sight only in connection with a particular kind of material system called the eye, so we know only of life in connection with certain arrangements of matter, of which the biochemist can give a good, but far from complete, account. The question at issue is: 'How did the first such system on this planet originate?' This is a historical problem to which I have given a very tentative answer on the not unreasonable hypothesis that a thousand million years ago matter obeyed the same laws that it does today.

This answer is compatible, for example, with the view that pre-existent mind or spirit can associate itself with certain kinds of matter. If so, we are left with the mystery as to why mind has so marked a preference for a particular type of colloidal organic substances. Personally I regard all attempts to describe the relation of mind to matter as rather clumsy metaphors. The biochemist knows no more, and no less, about this question than anyone else. His ignorance disqualifies him no more than the historian or the geologist from attempting to solve a historical problem.

Chapter 7

Chapter 7 takes up the question of human evolution. The first document for this section is from Darwin and includes three extracts. We begin with the short passage in a private notebook that Darwin was keeping in the fall of 1838 (the exact date of the entry is November 27), where we have the first unambiguous reference to the mechanism he was to call "natural selection." Note that the mechanism is applied to humans, and most particularly to our brain and our instincts. Darwin was practically unique in that he never hesitated about the application of his evolutionary ideas to our species. The next reference is the full (and only) paragraph on humans that is given in the Origin. *As I explain in the text, Darwin wanted to get the main ideas on evolution out before he turned to our own species. This he did in 1871 in the* Descent of Man, *and here I give the summary from the end of the book. Note that Darwin thinks that we humans are subject to the same laws as other animals and that sexual selection as well as natural selection were important in our origins. (Darwin also thought that Lamarckism, the inheritance of acquired characteristics, was important. Even the greatest genius makes mistakes.) He also makes it clear that he does not think the Adam and Eve story is compatible with evolutionism and that human intelligence and language, although very important, are part and parcel of the same naturalistic evolutionary package.*

For the second document we jump almost exactly a hundred years. You have already been introduced to "Lucy," the wonderfully preserved specimen of Australopithecus afarensis. *She was discovered by the American paleoanthropologist Donald Johanson, and in the extract reproduced here he tells of his discovery. In the text of this book (in Chapter 8), I quoted the passage from Edward O. Wilson's autobiography in which he remembers the excitement of reading and appreciating William Hamilton's model of kin selection as applied to hymenopteran sociality. I wanted there to give you a sense of the sheer thrill of ideas—ideas that simply sandbag you with their brilliance. Here, in parallel as it were, I want to give you a sense of the thrill of empirical discovery. Johanson is probably one of the most self-confident (and self-regarding) men in the whole of science—ego problems he does not have. But that is the way you have got to be if you are to spend days, weeks, months, traipsing through the hottest, driest, most hell-like places on earth, looking for those elusive bits of bone that point back to our evolutionary history. You have got to have faith in yourself and a brute determination to find what you seek. You must demand the answers of nature and refuse to quit until they are given. And what a reward when you succeed! Science—life—does not get any better than this.*

Charles Darwin's Notebooks

An habitual action must some way affect the brain in a manner which can be transmitted.—this is analogous to a blacksmith having children with strong arms.—The other principle of those children. which *chance?* produced with strong arms, outliving the weaker ones, may be applicable to the formation of instincts, independently of habits.—the limit of these two actions either on form or brain very hard to define.—Consider the acquirement of instinct by dogs, would show habit.—

The Origin of Species

In the distant future I see open fields for far more important researches. Psychology will be based on a new foundation, that of the necessary acquirement of each mental power and capacity by gradation. Light will be thrown on the origin of man and his history.

The Descent of Man

The main conclusion arrived at in this work, and now held by many naturalists who are well competent to form a sound judgment, is that man is descended from some less highly organised form. The grounds upon which this conclusion rests will never be shaken, for the close similarity between man and the lower animals in embryonic development, as well as in innumerable points of structure and constitution, both of high and of the most trifling importance,—the rudiments which he retains, and the abnormal reversions to which he is occasionally liable,—are facts which cannot be disputed. They have long been known, but until recently they told us nothing with respect to the origin of man. Now when viewed by the light of our knowledge of the whole organic world, their meaning is unmistakable. The great principle of evolution stands up clear and firm, when these groups of facts are considered in connection with others, such as the mutual affinities of the members of the same group, their geographical distribution in past and present times and their geological succession. It is incredible that all these facts should speak falsely. He who is not content to look, like a savage, at the phenomena of nature as disconnected, cannot any longer believe that man is the work of a separate act of creation. He will be forced to admit that the close resemblance of the embryo of man to that, for instance, of a dog—the construction of his skull, limbs, and whole frame, independently of the uses to which the parts may be put, on the same plan with that of other mammals—the occasional reappearance of various structures, for instance of several distinct muscles, which man does not normally possess, but which are common to the Quadrumana—and a crowd of analogous facts—all point in the plainest manner to the conclusion that man is the co-descendant with other mammals of a common progenitor.

We have seen that man incessantly presents individual differences in all parts of his body and in his mental faculties. These differences or variations seem to be induced by the same general causes, and to obey the same laws as with the lower animals. In both cases similar laws of inheritance prevail. Man tends to increase at a greater rate than his means of subsistence; consequently he is occasionally subjected to a severe struggle for existence, and natural selection will have affected whatever lies within its scope. A succession of strongly-marked variations of a similar nature are by no means requisite; slight fluctuating differences in the individual suffice for the work of natural selection. We may feel assured that the inherited effects of the long-continued use or disuse of parts will have done much in the same direction with natural selection. Modifications formerly of importance, though no longer of any special use, will be long inherited. When one part is modified, other parts will change through the principle of correlation, of which we have instances in many curious cases of correlated monstrosities. Something may be attributed to the direct and definite action of the surrounding conditions of life, such as abundant food, heat, or moisture; and lastly, many characters of slight physiological importance, some indeed of considerable importance, have been gained through sexual selection.

No doubt man, as well as every other animal, presents structures, which as far as we can judge with our little knowledge, are not now of any service to him, nor have been so during any former period of his existence, either in relation to his general conditions of life or of one sex to the other. Such structures cannot be accounted for by any form of selection, or by the inherited effects of the use and disuse of parts. We know, however, that many strange and strongly-marked peculiarities of structure occasionally appear in our domesticated productions, and if the unknown causes which produce them were to act more uniformly, they would probably become common to all the individuals of the species. We may hope hereafter to understand something about the causes of such occasional modifications, especially through the study of monstrosities; hence the labours of experimentalists, such as those of M. Camille Dareste, are full of promise for the future. In the greater number of cases we can only say that the cause of each slight variation and of each monstrosity lies much more in the nature or constitution of the organism, than in the nature of the surrounding conditions; though new and changed conditions certainly play an important part in exciting organic changes of all kinds.

Through the means just specified, aided perhaps by others as yet undiscovered, man has been raised to his present state. But since he attained to the rank of manhood, he has diverged into distinct races, or as they may be more appropriately called sub-species. Some of these, for instance the Negro and European, are so distinct that, if specimens had been brought to a naturalist without any further information, they would undoubtedly have been considered by him as good and true species. Nevertheless all the races agree in so many unimportant details of structure and in so many mental peculiarities, that these can be accounted for only through inheritance from a common progenitor; and a progenitor thus characterised would probably have deserved to rank as man.

It must not be supposed that the divergence of each race from the other races, and of all the races from a common stock, can be traced back to any one pair of progenitors. On the contrary, at every stage in the process of modification, all the individuals which were in any way best fitted for their conditions of life, though in different degrees, would have survived in greater numbers than the less well fitted. The process would have been like that followed by man, when he does not intentionally select particular individuals, but breeds from all the superior and neglects all the inferior individuals. He thus slowly but surely modifies his stock, and unconsciously forms a new strain. So with respect to modifications, acquired independently of selection, and due to variations arising from the nature of the organism and the action of the surrounding conditions, or from changed habits of life, no single pair will have been modified in a much greater degree than the other pairs which inhabit the same country, for all will have been continually blended through free intercrossing.

By considering the embryological structure of man,—the homologies which he presents with the lower animals,—the rudiments which he retains,—and the perversions to which he is liable, we can partly recall in imagination the former condition of our early progenitors; and can approximately place them in their proper position in the zoological series. We thus learn that man is descended from a hairy quadruped, furnished with a tail and pointed ears, probably arboreal in its habits, and an inhabitant of the Old World. This creature, if its whole structure had been examined by a naturalist, would have been classed amongst the Quadrumana, as surely as would the common and still more ancient progenitor of the Old and New World monkeys. The Quadrumana and all the higher mammals are probably derived from an ancient marsupial animal, and this through a long line of diversified forms, either from some reptile-like or some amphibian-like creature, and this again from some fish-like animal. In the dim obscurity of the past we can see that the early progenitor of all the Vertebrata must have been an aquatic animal, provided with branchiæ, with the two sexes united in the same individual, and with the most important organs of the body (such as the brain and heart) imperfectly developed. This animal seems to have been more like the larvae of our existing marine Ascidians than any other known form.

The greatest difficulty which presents itself, when we are driven to the above conclusion on the origin of man, is the high standard of intellectual power and of moral disposition which he has attained. But every one who admits the general principle of evolution, must see that the mental powers of the higher animals, which are the same in kind with those of mankind, though so different in degree, are capable of advancement. Thus the interval between the mental powers of one of the higher apes and of a fish, or between those of an ant and scale-insect, is immense. The development of these powers in animals does not offer any special difficulty; for with our domesticated animals, the mental faculties are certainly variable, and the variations are inherited. No one doubts that these faculties are of the utmost importance to animals in a state of nature. Therefore the conditions are favourable for their development through natural selection. The same conclusion may be extended to man; the intellect must have been all-important to him, even at a very remote period, enabling him to use language to invent and make weapons, tools, traps, &c.; by which means, in combination with his social habits, he long ago became the most dominant of all living creatures.

A great stride in the development of the intellect will have followed, as soon as, through a previous considerable advance, the half-art and half-instinct of language came into use; for the continued use of language will have reacted on the brain, and produced an inherited effect; and this again will have reacted on the improvement of language. The large size of the brain in man, in comparison with that of the lower animals, relatively to the size of their bodies, may be attributed in chief part, as Mr. Chauncey Wright has well remarked,[1] to the early use of some simple form of language,—that wonderful engine which affixes signs to all sorts of objects and qualities, and excites trains of thought which would never arise from the mere impression of the senses, and if they did arise could not be followed out. The higher intellectual powers of man, such as those of ratiocination, abstraction, self consciousness, &c., will have followed from the continued improvement of other mental faculties; but without considerable culture of the mind, both in the race and in the individual, it is doubtful whether these high powers would be exercised, and thus fully attained.

The development of the moral qualities is a more interesting and difficult problem. Their foundation lies in the social instincts, including in this term the family ties. These instincts are of a highly complex nature, and in the case of the lower animals give special tendencies towards certain definite actions; but the more important elements for us are love, and the distinct emotion of sympathy. Animals endowed with the social instincts take pleasure in each other's company, warn each other of danger, defend and aid each other in many ways. These instincts are not extended to all the individuals of the species, but only to those of the same community. As they are highly beneficial to the species, they have in all probability been acquired through natural selection.

A moral being is one who is capable of comparing his past and future actions and motives,—of approving of some and disapproving of others; and the fact that man is the one being who with certainty can be thus designated makes the greatest of all distinctions between him and the lower animals. But in our third chapter I have endeavoured to shew that the moral sense follows, firstly, from the enduring and always present nature of the social instincts, in which respect man agrees with the lower animals; and secondly,

---

[1] On the "Limits of Natural Selection," in the 'North American Review,' Oct. 1870, p.295.

from his mental faculties being highly active and his impressions of past events extremely vivid, in which respects he differs from the lower animals. Owing to this condition of mind, man cannot avoid looking backwards and comparing the impressions of past events and actions. He also continually looks forward. Hence, after some temporary desire or passion has mastered his social instincts, he will reflect and compare the now weakened impression of such past impulses, with the ever present social instinct; and he will then feel that sense of dissatisfaction which all unsatisfied instincts leave behind them. Consequently he resolves to act differently for the future—and this is conscience. Any instinct which is permanently stronger or more enduring than another gives rise to a feeling which we express by saying that it ought to be obeyed. A pointer dog, if able to reflect on his past conduct, would say to himself, I ought (as indeed we say of him) to have pointed at that hare and not have yielded to the passing temptation of hunting it.

Social animals are partly impelled by a wish to aid the members of the same community in a general manner, but more commonly to perform certain definite actions. Man is impelled by the same general wish to aid his fellows, but has few or no special instincts. He differs also from the lower animals in being able to express his desires by words, which thus become the guide to the aid required and bestowed. The motive to give aid is likewise somewhat modified in man: it no longer consists solely of a blind instinctive impulse, but is largely influenced by the praise or blame of his fellow men. Both the appreciation and the bestowal of praise and blame rest on sympathy; and this emotion, as we have seen, is one of the most important elements of the social instincts. Sympathy, though gained as an instinct, is also much strengthened by exercise or habit. As all men desire their own happiness, praise or blame is bestowed on actions and motives, according as they lead to this end; and as happiness is an essential part of the general good, the greatest-happiness principle indirectly serves as a nearly safe standard of right and wrong. As the reasoning powers advance and experience is gained, the more remote effects of certain lines of conduct on the character of the individual, and on the general good, are perceived; and then the self-regarding virtues, from coming within the scope of public opinion, receive praise, and their opposites receive blame. But with the less civilised nations reason often errs, and many bad customs and base superstitions come within the same scope, and consequently are esteemed as high virtues, and their breach as heavy crimes.

The moral faculties are generally esteemed, and with justice, as of higher value than the intellectual powers. But we should always bear in mind that the activity of the mind in vividly recalling past impressions is one of the fundamental though secondary bases of conscience. This fact affords the strongest argument for educating and stimulating in all possible ways the intellectual faculties of every human being. No doubt a man with a torpid mind, if his social affections and sympathies are well developed, will be led to good actions, and may have a fairly sensitive conscience. But whatever renders the imagination of men more vivid and strengthens the habit of recalling and comparing past impressions, will make the conscience more sensitive, and may even compensate to a certain extent for weak social affections and sympathies.

The moral nature of man has reached the highest standard as yet attained, partly through the advancement of the reasoning powers and consequently of a just public opinion, but especially through the sympathies being rendered more tender and widely diffused through the effects of habit, example, instruction, and reflection. It is not improbable that virtuous tendencies may through long practice be inherited. With the more civilised races, the conviction of the existence of an all-seeing Deity has had a potent influence on the advancement of morality. Ultimately man no longer accepts the praise or blame of his fellows as his chief guide, though few escape this influence, but his habitual convictions controlled by reason afford him the safest rule. His conscience then becomes his supreme judge and monitor. Nevertheless the first foundation or origin of the moral sense lies in the social instincts, including sympathy; and these instincts no doubt were primarily gained, as in the case of the lower animals, through natural selection.

The belief in God has often been advanced as not only the greatest, but the most complete of all the distinctions between man and the lower animals. It is however impossible, as we have seen, to maintain that this belief is innate or instinctive in man. On the other hand a belief in all-pervading spiritual agencies seems to be universal; and apparently follows from a considerable advance in the reasoning powers of man, and from a still greater advance in his faculties of imagination, curiosity and wonder. I am aware that the assumed instinctive belief in God has been used by many persons as an argument for His existence. But this is a rash argument, as we should thus be compelled to believe in the existence of many cruel and malignant spirits, possessing only a little more power than man; for the belief in them is far more general than of a beneficent deity. The idea of a universal and beneficent Creator of the universe does not seem to arise in the mind of man, until be has been elevated by long-continued culture.

He who believes in the advancement of man from some lowly-organized form, will naturally ask how does this bear on the belief in the immortality of the soul? The barbarous races of man, as Sir J. Lubbock has shewn, possess no clear belief of this kind; but arguments derived from the primeval beliefs of savages are, as we have just seen, of little or no avail. Few persons feel any anxiety from the impossibility of determining at what precise period in the development of the individual, from the first trace of the minute germinal vesicle to the child either before or after birth, man becomes an immortal being; and there is no greater cause for anxiety because the period in the gradually ascending organic scale cannot possibly be determined.[2]

I am aware that the conclusions arrived at in this work will be denounced by some as highly irreligious; that he who thus denounces them is bound to shew why it is more irreligious to explain the origin of man as a distinct species by descent from some lower form, through the laws of variation and natural selection, than to explain the birth of the individual through the laws of ordinary reproduction. The birth both of the species and of the individual are equally parts of that grand sequence of events, which our minds refuse to accept as the result of blind chance. The understanding revolts at such a conclusion, whether or not we are able to believe that every slight variation of structure,—the union of each pair in marriage,—the dissemination of each seed,—and other such events, have all been ordained for some special purpose.

Lucy: The Beginnings of Humankind

D. Johanson and M. Edey

On the morning of November 30, 1974, I woke, as I usually do on a field expedition, at daybreak. I was in Ethiopia, camped on the edge of a small muddy river, the Awash, at a

---

[2] The Rev. J. A. Picton gives a discussion to this effect in his 'New Theories and the Old Faith,' 1870.

place called Hadar, about a hundred miles northeast of Addis Ababa. I had been there for several weeks, acting as coleader of a group of scientists looking for fossils.

For a few minutes I lay in my tent, looking up at the canvas above me, black at first but quickly turning to green as the sun shot straight up beyond the rim of hills off to the east. Close to the Equator the sun does that; there is no long dawn as there is at home in the United States. It was still relatively cool, not more than 80 degrees. The air had the unmistakable crystalline smell of early morning on the desert, faintly touched with the smoke of cooking fires. Some of the Afar tribesmen who worked for the expedition had brought their families with them, and there was a small compound of dome-shaped huts made of sticks and grass mats about two hundred yards from the main camp. The Afar women had been up before daylight, tending their camels and goats, talking quietly.

For most of the Americans in camp this was the best part of the day. The rocks and boulders that littered the landscape had bled away most of their heat during the night and no longer felt like stoves when you stood next to one of them. I stepped out of the tent and took a look at the sky. Another cloudless day; another flawless morning on the desert that would turn to a crisper later on. I washed my face and got a cup of coffee from the camp cook, Kabete. Mornings are not my favorite time. I am a slow starter and much prefer evenings and nights. At Hadar I feel best just as the sun is going down. I like to walk up one of the exposed ridges near the camp, feel the first stirrings of evening air and watch the hills turn purple. There I can sit alone for a while, think about the work of the day just ended, plan the next, and ponder the larger questions that have brought me to Ethiopia. Dry silent places are intensifiers of thought, and have been known to be since early Christian anchorites went out into the desert to face God and their own souls.

Tom Gray joined me for coffee. Tom was an American graduate student who had come out to Hadar to study the fossil animals and plants of the region, to reconstruct as accurately as possible the kinds and frequencies and relationships of what had lived there at various times in the remote past and what the climate had been like. My own target—the reason for our expedition—was hominid fossils: the bones of extinct human ancestors and their close relatives. I was interested in the evidence for human evolution. But to understand that, to interpret any hominid fossils we might find, we had to have the supporting work of other specialists like Tom.

"So, what's up for today?" I asked.

Tom said he was busy marking fossil sites on a map.

"When are you going to mark in Locality 162?"

"I'm not sure where 162 is," he said.

"Then I guess I'll have to show you." I wasn't eager to go out with Gray that morning. I had a tremendous amount of work to catch up on. We had had a number of visitors to the camp recently. Richard and Mary Leakey, two well-known experts on hominid fossils from Kenya, had left only the day before. During their stay I had not done any paperwork, any cataloguing. I had not written any letters or done detailed descriptions of any fossils. I *should* have stayed in camp that morning—but I didn't. I felt a strong subconscious urge to go with Tom, and I obeyed it. I wrote a note to myself in my daily diary: *Nov. 30, 1974. To Locality 162 with Gray in AM. Feel good.*

As a paleoanthropologist—one who studies the fossils of human ancestors—I am superstitious. Many of us are, because the work we do depends a great deal on luck. The fossils we study are extremely rare, and quite a few distinguished paleoanthropologists have gone a lifetime without finding a single one. I am one of the more fortunate. This was only my third year in the field at Hadar, and I had already found several. I know I am lucky, and I don't try to hide it. That is why I wrote "feel good" in my diary. When I got up that morning I felt it was one of those days when you should press your luck. One of those days when something terrific might happen.

Throughout most of that morning, nothing did. Gray and I got into one of the expedition's four Land Rovers and slowly bounced our way to Locality 162. This was one of several hundred sites that were in the process of being plotted on a master map of the Hadar area, with detailed information about geology and fossils being entered on it as fast as it was obtained. Although the spot we were headed for was only about four miles from camp, it took us half an hour to get there because of the rough terrain. When we arrived it was already beginning to get hot.

At Hadar, which is a wasteland of bare rock, gravel and sand, the fossils that one finds are almost all exposed on the surface of the ground. Hadar is in the center of the Afar desert, an ancient lake bed now dry and filled with sediments that record the history of past geological events. You can trace volcanic ash falls there, deposits of mud and silt washed down from distant mountains, episodes of volcanic dust, more mud, and so on. Those events reveal themselves like layers in a slice of cake in the gullies of new young rivers that recently have cut through the lake bed here and there. It seldom rains at Hadar, but when it does it comes in an overpowering gush—six months' worth overnight. The soil, which is bare of vegetation, cannot hold all that water. It roars down the gullies, cutting back their sides and bringing more fossils into view.

Gray and I parked the Land Rover on the slope of one of those gullies. We were careful to face it in such a way that the canvas water bag that was hanging from the side mirror was in the shade. Gray plotted the locality on the map. Then we got out and began doing what most members of the expedition spent a great deal of their time doing: we began surveying, walking slowly about, looking for exposed fossils.

Some people are good at finding fossils. Others are hopelessly bad at it. It's a matter of practice, of training your eye to see what you need to see. I will never be as good as some of the Afar people. They spend all their time wandering around in the rocks and sand. They have to be sharp-eyed; their lives depend on it. Anything the least bit unusual they notice. One quick educated look at all those stones and pebbles, and they'll spot a couple of things a person not acquainted with the desert would miss.

Tom and I surveyed for a couple of hours. It was now close to noon, and the temperature was approaching 110. We hadn't found much: a few teeth of the small extinct horse *Hipparion;* part of the skull of an extinct pig; some antelope molars; a bit of a monkey jaw. We had large collections of all these things already, but Tom insisted on taking these also as added pieces in the overall jigsaw puzzle of what went where.

"I've had it," said Tom. "When do we head back to camp?"

"Right now, But let's go back this way and survey the bottom of that little gully over there."

The gully in question was just over the crest of the rise where we had been working all morning. It had been thoroughly checked out at least twice before by other workers, who had found nothing interesting. Nevertheless, conscious of the "lucky" feeling that had been with me since I woke, I decided to make that small final detour. There was virtually no bone in the gully. But as we turned to leave, I noticed something lying on the ground partway up the slope.

"That's a bit of a hominid arm," I said.

"Can't be. It's too small. Has to be a monkey of some kind."

We knelt to examine it.

"Much too small," said Gray again.

I shook my head. "Hominid."

"What makes you so sure?" he said.

"That piece right next to your hand. That's hominid too."

"Jesus Christ," said Gray. He picked it up. It was the back of a small skull. A few feet away was part of a femur: a thighbone. "Jesus Christ," he said again. We stood up, and began to see other bits of bone on the slope: a couple of vertebrae, part of a pelvis— all of them hominid. An unbelievable, impermissible thought flickered through my mind. Suppose all these fitted together? Could they be parts of a single, extremely primitive skeleton? No such skeleton had ever been found—anywhere.

"Look at that," said Gray. "Ribs."

A single individual?

"I can't believe it," I said. "I just can't believe it."

"By God, you'd better believe it!" shouted Gray. "Here it is. Right here!" His voice went up into a howl. I joined him. In that 110-degree heat we began jumping up and down. With nobody to share our feelings, we hugged each other, sweaty and smelly, howling and hugging in the heat-shimmering gravel, the small brown remains of what now seemed almost certain to be parts of a single hominid skeleton lying all around us.

"We've got to stop jumping around," I finally said. "We may step on something. Also, we've got to make sure."

"Aren't you sure, for Christ's sake?"

"I mean, suppose we find two left legs. There may be several individuals here, all mixed up. Let's play it cool until we can come back and make absolutely sure that it all fits together."

We collected a couple of pieces of jaw, marked the spot exactly and got into the blistering Land-Rover for the run back to camp. On the way we picked up two expedition geologists who were loaded down with rock samples they had been gathering.

"Something big," Gray kept saying to them. "Something big. Something *big.*"

"Cool it," I said.

But about a quarter of a mile from camp, Gray could not cool it. He pressed his thumb on the Land Rover's horn, and the long blast brought a scurry of scientists who had been bathing in the river. "We've got it," he yelled. "Oh, Jesus, we've got it. We've got The Whole Thing!"

That afternoon everyone in camp was at the gully, sectioning off the site and preparing for a massive collecting job that ultimately took three weeks. When it was done, we had recovered several hundred pieces of bone (many of them fragments) representing about forty percent of the skeleton of a single individual. Tom's and my original hunch had been right. There was no bone duplication.

But a single individual of what? On preliminary examination it was very hard to say, for nothing quite like it had ever been discovered. The camp was rocking with excitement. That first night we never went to bed at all. We talked and talked. We drank beer after beer. There was a tape recorder in the camp, and a tape of the Beatles song "Lucy in the Sky with Diamonds" went belting out into the night sky, and was played at full volume over and over again out of sheer exuberance. At some point during that unforgettable evening—I no longer remember exactly when—the new fossil picked up the name of Lucy, and has been so known ever since, although its proper name—its acquisition number in the Hadar collection—is AL 288-1.

Chapter 8

Chapter 8 takes us into the sociobiology controversy. There are three documents here. First we have a statement about the sociobiological program by Edward O. Wilson and a critique by a group of radical scientists who went under the somewhat ungainly name of the "Science for the People Sociobiology Study Group." (The group included many eminent scientists, among them Richard Lewontin and Stephen Jay Gould.) Of course, part of this debate is as scientific as you like. Wilson argues that we have now reached the point where Darwinian evolutionary theory can be extended out to social behavior and that this includes human social behavior. The critics beg to differ. Part of the debate takes us beyond science. Wilson, as we see, and as we see even more in the final chapter of the text, is trying to give us a vision of life for the modern post-Christian age. His critics think that he is simply giving us a fancy version of a bourgeois, middle-class, white American's view of life, dressed up with pseudoscientific language to sound like the objective truth.

Debates like this are furious, fun (let us not be hypocrites; we all like a good fight), and frustrating. Really and ultimately, the proof is in the pudding. Less metaphorically, the human sociobiology program is going to succeed, no matter what its critics say, if it starts churning out results and leading to new findings. The human sociobiology program is going to fail, no matter what its supporters say, if it gives no new results and findings. Good graduate students do not risk their careers on ideas going nowhere. In fact, as I suggest in the text, things have already settled down since the initial battles. Animal sociobiology is a thriving enterprise, and the sociobiologists are trying—at least sometimes successfully—to get some new insights into human nature. I am a philosopher, so what really interests me is whether we can use sociobiology to help understand questions about epistemology (what can I know?) and ethics (what should I do?). In the final document of this section, I join forces with Edward O. Wilson looking to see if and how Darwinian biology throws light on ethical inquiry. I cannot say that the philosophical world has rushed to endorse our thinking, but we have made some progress. No longer are we ignored. Now teachers use our ideas, for their students, as paradigms of what one should not do when philosophizing! Before I die, I fully expect to find that our ideas are part of the accepted canon.

Sociobiology—Another Biological Determinism

Sociobiology Study Group of Science for the People

Biological determinism represents the claim that the present states of human societies are the specific result of biological forces and the biological "nature" of the human species. Determinist theories all describe a particular model of society which corresponds to the socioeconomic prejudices of the writer. It is then asserted that this pattern has arisen out of human biology and that present human social arrangements are either unchangeable or if altered will demand continued conscious social control because these changed conditions will be "unnatural." Moreover, such determinism provides a direct justification for the status quo as "natural," although some determinists dissociate themselves from some of the consequences of their arguments. The issue, however, is not the motivation of individual creators of determinist theories, but the way these theories operate as powerful forms of legitimation of past and present social institutions such as aggression, competition, domination of women by men, defense of national territory, individualism, and the appearance of a status and wealth hierarchy.

The earlier forms of determinism in the current wave have now been pretty well discredited. The claims that there is a high heritability of IQ, which implies both the unchangeability of IQ and a genetic difference between races or between social classes, have now been thoroughly debunked.

The simplistic forms of the human nature argument given by Lorenz, Ardrey, Tiger and Fox, and others have no scientific credit and have been scorned as works of "advocacy" by E. O. Wilson, whose own book, *Sociobiology: The New Synthesis,* is the manifesto of a new, more complex, version of biological determinism, no less a work of "advocacy" than its rejected predecessors. This book, whose first chapter is on "The Morality of the Gene," is intended to establish sociology as a branch of evolutionary biology, encompassing all human societies, past and present. Wilson believes that "sociology and the other social sciences, as well as the humanities, are the last branches of biology waiting to be included in the Modern Synthesis" (p. 4).

This is no mere academic exercise. For more than a century the idea that human social behavior is determined by evolutionary imperatives operating on inherited dispositions has been seized upon and widely entertained not so much for its alleged correspondence with reality as for its more obvious political value. Among the better known examples are Herbert Spencer's argument in *Social Statics* (1851) that poverty and starvation were natural agents cleansing society of the unfit, and Konrad Lorenz's call in 1940 in Germany for "the extermination of elements of the population loaded with dregs," based upon his ethological theories.

In order to make their case, determinists construct a selective picture of human history, ethnography, and social relations. They misuse the basic concepts and facts of genetics and evolutionary theory, asserting things to be true that are totally unknown, ignoring whole aspects of the evolutionary process, asserting that conclusions follow from premises when they do not. Finally, they invent ad hoc hypotheses to take care of the contradictions and carry on a form of "scientific reasoning" that is untestable and leads to unfalsifiable hypotheses. What follows is a general examination of these elements in sociobiological theory, especially as elaborated in E. O. Wilson's *Sociobiology.*

A Version of Human Nature

For the sociobiologist the first task is to delineate a model of human nature that is to be explained. Among Wilson's universal aspects of human nature are:

1. territoriality and tribalism (pp. 564–565);
2. indoctrinability—"Human beings are absurdly easy to indoctrinate—they *seek* it" (p. 562);
3. spite and family chauvinism—"True spite is commonplace in human societies, undoubtedly because human beings are keenly aware of their own blood lines and have the intelligence to plot intrigue" (p. 119);
4. reciprocal altruism (as opposed to true unselfishness)—"Human behavior abounds with reciprocal altruism," as for example, "aggressively

moralistic behavior," "self-righteousness, gratitude and sympathy" (p. 120);

5. blind faith—"Men would rather believe than know" (p. 561);
6. warfare (p. 572) and genocide (p. 573)—"the most distinctive human qualities" emerged during the "autocatalytic phase of social evolution" which occurred through intertribal warfare, "genocide" and "genosorption."

The list is not exhaustive and is meant only to show how the outlines of human nature are viewed myopically, through the lens of modern Euro-American culture.

To construct such a view of human nature, Wilson must abstract himself totally from any historical or ethnographic perspective. His discussion of the economy of scarcity is an excellent example. An economy of relative scarcity and unequal distribution of rewards is stated to be an aspect of human nature:

> The members of human society sometimes cooperate closely in *insectan* fashion [our emphasis], but more frequently they compete for the limited resources allocated to their role sector. The best and the most entrepreneurial of the role-actors usually gain a disproportionate share of the rewards (p. 554).

There is a great deal of ethnographic and historical description entirely contradicting this conception of social organization. It ignores, for example, the present and historical existence of societies not differentiated in any significant way by "role sectors"; without scarcities differentially induced by social institutions for different subpopulations of the society; not differentiated by lower and higher ranks and strata (Birket-Smith 1959; Fried 1967; Harris 1968; Krader 1968).

Realizing that history and ethnography do not support the universality of their description of human nature, sociobiologists claim that the exceptions are "temporary aberrations" or deviations. Thus, although genocidal warfare is (assertedly) universal, "it is to be expected that some isolated cultures will escape the process for generations at a time, in effect reverting temporarily to what ethnographers classify as a pacific state" (p. 574).

Another related ploy is the claim that ethnographers and historians have been too narrow in their definitions and have not realized that apparently contradictory evidence is really confirmatory.

> Anthropologists often discount territorial behavior as a general human attribute. This happens when the narrowest concept of the phenomenon is borrowed from zoology ... it is necessary to define territory more broadly animals respond to their neighbors in a highly variable manner the scale may run from open hostility ... to oblique forms of advertisement or *no territorial behavior at all* [our emphasis].
>
> If these qualifications are accepted it is reasonable to conclude that territoriality is a general trait of hunter-gatherer societies. (pp. 564–565)

Wilson's views of aggression and warfare are subject to this ploy of all-embracing definition on the one hand and erroneous historical-ethnographic data on the other. "Primitive" warfare is rarely lethal to more than one or at most a few individuals in an episode of warfare, virtually without significance genetically or demographically (Livingstone 1968). Genocide was virtually unknown until state-organized societies appeared in history (as far as can be made out from the archeological and documentary records).

We have given only examples of the general advocacy method employed by sociobiologists in a procedure involving definitions which exclude nothing and the laying of Western conceptual categories onto "primitive" societies.

Humans as Animals—the Meaning of Similarity

To support a biologistic explanation of human institutions it is useful to claim an evolutionary relationship between the nature of human social institutions and "social" behavior in other animals. Obviously sociobiologists would prefer to claim evolutionary homology, rather than simple analogy, as the basis for the similarity in behavior between humans and other animals; then they would have a prima facie case for genetic determination. In some sections of *Sociobiology,* Wilson attempts to do this by listing "universal" features of behavior in higher primates including humans. But claimed external similarity between humans and our closest relatives (which are by no means very close to us) does not imply genetic continuity. A behavior that may be genetically coded in a higher primate may be purely learned and widely spread among human cultures as a consequence of the enormous flexibility of our brain.

More often Wilson argues from evolutionary analogy. Such arguments operate on shaky grounds. They can never be used to assert genetic similarity, but they can serve as a plausibility argument for natural selection of human behavior by assuming that natural selection has operated on different genes in the two species but has produced convergent responses as independent adaptations to similar environments. The argument is not even worth considering unless the similarity is so precise that identical function cannot be reasonably denied, as in the classic case of evolutionary convergence—the eyes of vertebrates and octopuses. Here Wilson fails badly, for his favorite analogies arise by a twisted process of imposing human institutions on animals by metaphor, and then rederiving the human institutions as special cases of the more general phenomenon "discovered" in nature. In this way human institutions suddenly become "natural" and can be viewed as a product of evolution.

A classic example, long antedating *Sociobiology,* is "slavery" in ants. "Slavemaking" species capture the immature stages of "slave" species and bring them back to their own nests. When the captured workers hatch, they perform housekeeping tasks with no compulsion as if they were members of the captor species. Why is this "slavemaking" instead of "domestication"? Human slavery involves members of one's own species under continued compulsion. It is an economic institution in societies producing an economic surplus, with both slave and products as commodities in exchange. It has nothing to do with ants except by weak and meaningless analogy. Wilson expands the realm of these weak analogies (chapter 27) to find barter, division of labor, role playing, culture, ritual, religion, magic, esthetics, and tribalism among nonhumans. But if we insist upon seeing animals in the mirror of our own social arrangements, we cannot fail to find any human institutions we want among them.

Genetic Bases of Behavior

We can dispense with the direct evidence for a genetic basis of various human social forms in a single word, "None." The genetics of normal human behavior is in a rudimentary state because of the impossibility of reproducing particular human genotypes over and over, or of experimentally manipulating the environments of individuals or groups. There is no evidence that meets the elementary requirements of experimental design, that such traits as xenophobia, religion, ethics, social dominance, hierarchy formation, slavemaking, etc., are in any way coded specifically in the genes of human beings.

And indeed, Wilson offers no such evidence. Instead, he makes confused and contradictory statements about what is an essential element in the argument. If there are no genes for parent-offspring conflict, then there is no sense in talking about natural selection for this phenomenon. Thus, he speaks of "genetically programmed sexual and parent-offspring conflict" (p. 563), yet there is the "considerable technical problem of distinguishing behavioral elements and combinations that emerge ... independently of learning and those that are shaped at least to some extent by learning" (p. 159). In fact, it cannot be done.

Elsewhere, the *capacity* to learn is stated to be genetic in the species, so that "it does not matter whether aggression is wholly innate or acquired partly or wholly by learning" (p. 255). But it does matter. If all that is genetically programmed into people is that "genes promoting flexibility in social behavior are strongly selected" (p. 549) and if "genes have given away most of their sovereignty" (p. 550), then biology and evolution give no insight into the human condition except the most trivial one, that the *possibility* of social behavior is part of human biology. However, in the next phrase Wilson reasserts the sovereignty of the genes because they "maintain a certain amount of influence in at least the behavioral qualities that underly the variations between cultures." It is stated as *fact* that genetical differences underlie variations between cultures, when no evidence at all exists for this assertion and there is some considerable evidence against it.

Since sociobiologists can adduce no facts to support the genetic basis for human social behavior, they try two tacks. First, the suggestion of evolutionary homology between behavior in the human species and other animals, if correct, would imply a genetic basis in us. But the evidence for homology as opposed to analogy is very weak. Second, they postulate genes right and left and then go on to argue as if the genes were demonstrated facts. There are hypothetical altruist genes, conformer genes, spite genes, learning genes, homosexuality genes, and so on. An instance of the technique is on pages 554–555 of Wilson's book: "Dahlberg showed that *if* a single gene appears that is responsible for success and upward shift in status. ..." and "Furthermore, *there are many* Dahlberg genes ..." (our emphases throughout). Or on page 562: "*If we assume* for argument that indoctrinability evolves ..." and "Societies containing higher frequencies of conformer genes replace those that disappear ..."(our emphasis). Or consult nearly any page of Trivers (1971) for many more examples.

Geneticists long ago abandoned the naive notion that there are genes for toes, genes for ankles, genes for the lower leg, genes for the kneecap, or the like. Yet sociobiologists break the totality of human social phenomena into arbitrary units, which they reify as "organs of behavior," postulating particular genes for each.

Everything Is Adaptive

The next step in the sociobiological argument is to try to show that the hypothetical, genetically programmed behavior organs have evolved by natural selection. The assertion that all human behavior is or has been adaptive is an outdated expression of Darwinian evolutionary theory, characteristic of Darwin's 19th century defenders who felt it necessary to prove everything adaptive. It is a deeply conservative politics, not an understanding of modern evolutionary theory, that leads one to see the wonderful operation of adaptation in every feature of human social organization.

There is no hint in *Sociobiology* that at this very moment the scientific community of evolutionary geneticists is deeply split on the question of how important adaptive as opposed to random processes are in manifest evolution. More important, there is a strain in modern evolutionary thought, going back to Julian Huxley, that avoids much of the tortured logic required by extreme selectionism, by emphasizing allometry. Organs, not themselves under direct natural selection, may change because of their developmental links to other features that are under selection. Many aspects of human social organization, if not all, may be simply the consequence of increased plasticity of neurological response and cognitive capacity.

The major assertion of sociobiologists that human social structures exist because of their superior adaptive value is only an assumption for which no tests have even been proposed. The entire theory is so constructed that *no tests are possible*. The mode of explanation involves three postulated levels of the operation of natural selection: (1) classical individual selection to account for obviously self-serving behaviors; (2) kin selection to account for altruistic behaviors or submissive acts toward relatives; (3) reciprocal altruism to account for altruistic behaviors toward unrelated persons. All that remains is to make up a "just-so" story of adaptation with the appropriate form of selection acting. For some traits it is easy to invent a story. The "genes" for social dominance, aggression, entrepreneurship, successful deception, and so on will "obviously" be advantageous at the individual level. For example, evidence is presented (p. 288) that dominant males impregnate a disproportionate share of females in mice, baboons, and Yanamamo Indians. In fact, in the ethnographic literature there are numerous examples of groups whose political "leaders" do not have greater access to mates. In general it is hard to demonstrate a correlation of any of the sociobiologists' "adaptive" social behaviors with actual differential reproduction.

Other traits require more ingenuity. Homosexuality would seem to be at a reproductive disadvantage since "of course, homosexual men marry much less frequently and have far fewer children" (Dr. Kinsey disagreed, and what about homosexual women?). But a little ingenuity solves the problem: "The homosexual members of primitive societies may have functioned as helpers ... [operated] with special efficiency in assisting close relatives" (p. 555). Kin selection saves the day when one's imagination for individual selection fails.

Only one more imaginative mechanism is needed to rationalize such phenomena as friendship, morality, patriotism, and submissiveness, even when the bonds do not involve relatives. The theory of reciprocal altruism (Trivers 1971) proposes that selection has operated such that risk taking and acts of kindness can be recognized and reciprocated so that the net fitness of both participants is increased.

The trouble with the whole system is that nothing is explained because everything is explained. If individuals are selfish, that is explained by simple individual selection. If, on the contrary, they are altruistic, it is kin selection or reciprocal altruism. If sexual identities are unambiguously heterosexual, individual fertility is increased. If, however, homosexuality is common, it is a result of kin selection. Sociobiologists give us no example that might conceivably contradict their scheme of perfect adaptation.

Variations of Cultures in Time and Space

There does exist one possibility of tests of sociobiological hypotheses when they make specific *quantitative* predictions about rates of change of characters in time and about the degree of differentiation between populations of a species. Population genetics makes specific predictions about rates of change, and there are hard data on the degree of genetic differentiation between human populations for biochemical traits. Both the theoretical rates of *genetic* change in time and the observed *genetic* differentiation between populations are too small to agree with the very rapid changes that have occurred in human *cultures* historically and the very large *cultural* differences observed among contemporaneous populations. So, for example, the rise of Islam after the 7th century to supreme cultural and political power in the West, to its subsequent rapid decline after the 13th century (a cycle occupying fewer than 30 generations) was too rapid by orders of magnitude for any large change by natural selection. The same problem arises for the immense cultural differences between contemporary groups, since we know from the study of enzyme-specifying genes that there is very little genetic differentiation between nations and races.

Wilson acknowledges and deals with both of these dilemmas by a bold stroke: He invents a new phenomenon. It is the "multiplier effect" (pp. 11–13, 569–572), which postulates that very small differences in the frequency of hypothetical genes for altruism, conformity, indoctrinability, etc., could move a whole society from one cultural pattern to another. The only evidence offered for this "multiplier effect" is a description of differences in behavior between closely related species of insects and of baboons. There is, however, no evidence about the amount of *genetic* difference between these closely related species nor how many tens or hundreds of thousands of generations separate the members of these species pairs since their divergence. The multiplier effect, by which any arbitrary but unknown genetic difference can be converted to any cultural difference you please, is a pure invention of convenience without any evidence to support it. It has been created out of whole cloth to seal off the last aperture through which the theory might have been tested against the real world.

An Alternative View

It is often stated by biological determinists that those who oppose them are "environmental determinists," who believe that the behavior of individuals is precisely determined by some sequence of environmental events in childhood. Such an assertion reveals the essential narrowness of viewpoint in determinist ideologies. First, they see the individual as the basic elements of determination and behavior, whereas society is simply the sum of all the individuals in it. But the truth is that the individual's social activity is to be understood only by first understanding social institutions. We cannot understand what it is to be a slave or a slave owner without first understanding the institution of slavery, which defines and creates both slave and owner.

Second, determinists assert that the evolution of societies is the result of changes in the frequencies of different sorts of individuals within them. But this confuses cause and effect. Societies evolve because social and economic activity alter the physical and social conditions in which these activities occur. Unique historical events, actions of some individuals, and the altering of consciousness of masses of people interact with social and economic forces to influence the timing, form, and even the possibility of particular changes; individuals are not totally autonomous units whose individual qualities determine the direction of social evolution. Feudal society did not pass away because some autonomous force increased the frequency of entrepreneurs. On the contary, the economic activity of Western feudal society itself resulted in a change in economic relations which made serfs into peasants and then into landless industrial workers with all the immense changes in social institutions that were the result.

Finally, determinists assert that the possibility of change in social institutions is limited by the biological constraints on individuals. But we know of no relevant constraints placed on social processes by human biology. There is no evidence from ethnography, archaeology, or history that would enable us to circumscribe the limits of possible human social organization. What history and ethnography do provide us with are the materials for building a theory that will itself be an instrument of social change.

References

Birket-Smith, K. (1959). *The Eskimos,* 2nd ed. (London: Methuen).

Fried, M. (1967). *The Evolution of Political Society* (New York: Random House).

Harris, M. (1968). Law and Order in Egalitarian Societies. Pages 369–391 in *Culture, Man and Nature* (New York: Crowell).

Krader, L. (1968). "Government Without the State." In *Formation of the State* (Englewood Cliffs, N.J.: Prentice Hall) pp. 53–110.

Lorenz, K. (1940). "Durch Domestikation verursachte Störungen arteigenen Verhaltens." *Zeitschrift für angewandte Psychologie und Characterkunde* 59, pp. 56–75. (As quoted in Cloud, W., 1973, "Winners and Sinners." *The Sciences* 13, pp. 16–21.)

Livingstone, F. (1968). "The Effects of Warfare on the Biology of the Human Species." In M. Fried, M. Harris, and R. Murphy, eds. *War: The Anthropology of Armed Conflict and Aggression* (Garden City: Natural History Press) pp. 3–15.

Spencer, H. (1851). *Social Statics* (London: Chapman).

Trivers, R. (1971). "The Evolution of Reciprocal Altruism." *Q. Rev. Biol.* 46, pp. 35–57.

Wilson, E. O. (1975). *Sociobiology: The New Synthesis* (Cambridge, Mass.: Harvard University Press).

Academic Vigilantism and the Political Significance of Sociobiology

Edward O. Wilson

The best response to a political attack of the kind exemplified by the preceding article, "Sociobiology—Another Biological Determinism," is perhaps no response at all. Some of my colleagues have offered that advice. But the problem is larger than the personal distress that this and earlier activities of the Science for the People group have caused me. The issue at hand, I submit, is vigilantism: the judgment of a work of science according to whether it conforms to the political convictions of the judges, who are self-appointed. The sentence for scientists found guilty is to be given a label and to be associated with past deeds that all decent persons will find repellent.

Thus, in a statement published earlier in *The New York Review of Books* (Allen et al. 1975), the Science for the People group characterized my book *Sociobiology: The New*

Synthesis (Wilson 1975a) as the latest attempt to reinvigorate theories that in the past "provided an important basis for the enactment of sterilization laws and restrictive immigration laws by the United States between 1910 and 1930 and also for the eugenics policies which led to the establishment of gas chambers in Nazi Germany." To this malicious charge they added, "Wilson joins the long parade of biological determinists whose work has served to buttress the institutions of their society by exonerating them from responsibility for social problems." The tone of the present *BioScience* article is muted, but the innuendo is clear and remains the same.

This tactic, which has been employed by members of Science for the People against other scientists, throws the person criticized into the role of defendant and renders his ideas easier to discredit. Free and open discussion becomes difficult, as the critics continue to press their campaign, and the target struggles to clear his name. The problem is increased by difficulties in knowing with whom one is dealing. The statements are often published over long lists of names, shifts in committee membership occur through time, and the authors' names are withheld from some of the documents. (All have occurred during the present controversy.)

Despite the protean physical form taken by the Sociobiology Study Group of Science for the People, the belief system they promote is clear-cut and rigid. They postulate that human beings need only decide on the kind of society they wish, and then find the way to bring it into being. Such a vision can be justified if human social behavior proves to be infinitely malleable. In their earlier *New York Review* statement (Allen et al. 1975) the group therefore maintained that although eating, excreting, and sleeping may be genetically determined, social behavior is entirely learned; this belief has been developed further in the *BioScience* article. In contrast, and regardless of all they have said, I am ideologically indifferent to the degree of determinism in human behavior. If human beings proved infinitely malleable, as they hope, then one could justify any social or economic arrangement according to his personal value system. If on the other hand, human beings proved completely fixed, then the status quo could be justified as unavoidable.

Few reasonable persons take the first extreme position and none the second. On the basis of objective evidence the truth appears to lie somewhere in between, closer to the environmentalist than to the genetic pole. That was my wholly empirical conclusion in *Sociobiology: The New Synthesis* and continues to be in later writings. There is no reasonable way that this generalization can be construed as a support of the status quo and continued injustice, as the Science for the People group have now, on four painful occasions, claimed. I have personally argued the opposite conclusion, most fully and explicitly in my *New York Times Magazine* article of 12 October 1975 (Wilson 1975b). The Science for the People group have not found it convenient to mention this part of my writings.

With the exception of the Science for the People group, all of the many biologists and social scientists whose reviews of *Sociobiology: The New Synthesis* I have seen understood the book correctly. None has read a reactionary political message into it, even though the reviewers represent a variety of personal political persuasions; and none has found my assessment of the degree of determinism in human social behavior out of line with the empirical evidence. The Science for the People group have utterly misrepresented the spirit and content of the portions of *Sociobiology* devoted to human beings. They have done so, it would seem, in order to have a conspicuous straw man against which their views can be favorably pitted, and to obscure the valid points in *Sociobiology* which do indeed threaten their own extreme position. Let me document this interpretation with responses to the specific criticisms made by the 35 cosigners.

Response to Criticisms

First, it should be noted that *Sociobiology: The New Synthesis* is a large book, within which only chapter 27 and scattered paragraphs in earlier chapters refer to man. The main theses of sociobiology are based on studies of a myriad of animal species conducted by hundreds of investigators in various biological disciplines. It has been possible to derive propositions by the traditional postulational and deductive methods of theoretical science, and to test many of them rigorously by quantitative studies. One can cite the work on kin selection in social Hymenoptera, the elaboration of caste systems in social insects, the economic functions of vertebrate territories, the ecological causes of ungulate social behavior, the repertory size and transmission characteristics of communication systems, and others. These ideas and data provide the main thrust of general sociobiology.

In my book human sociobiology was approached tentatively and in a taxonomic rather than a political spirit. The final chapter opens with the following passage: "Let us now consider man in the free spirit of natural history, as though we were zoologists from another planet completing a catalog of social species on Earth. In this macroscopic view the humanities and social sciences shrink to specialized branches of biology; history, biography, and fiction are the research protocols of human ethology; and anthropology and sociology together constitute the sociobiology of a single primate species."

It is the intellectually viable contention of the final chapter that the sociobiological methods which have proved effective in the study of animals can be extended to human beings, even though our vastly more complex, flexible behavior will make the application technically more difficult. The degree of success cannot yet be predicted. Chapter 27 was intended to be a beginning rather than a conclusion, and other reviewers have so interpreted it. In it I have characterized the distinctive human traits as best I could from the literature of the social sciences, and I have offered a set of hypotheses about the evolution of the traits stated in a way that seemed to make them most susceptible to analysis by sociobiological methods.

The Science for the People group ignore this main thrust of the book. They cite piece by piece incorrectly, or out of context, and then add their own commentary to furnish me with a political attitude I do not have and the book with a general conclusion that is not there. The following examples cover nearly all of their points.

Roles

The 35 cosigners have me saying that role sectors, and thus certain forms of economic role behavior associated with role sectors, are universal in man. On pages 552 and 554, the reader will find that I did not include role sectors among the widespread or universal traits. What I said was that when role sectors occur, certain economic features are associated with them.

Territory

It is now well known that animal territories commonly vary in size and quality of defense according to habitat, season, and population density. Under some circumstances many species show no territorial behavior, but it is necessary for them to display the behavior under other, specified circumstances in order to be called territorial—an obvious condition. This is the reason I have called the human species territorial. No contradiction in definitions exists; the cosigners have made it appear to exist by simply deleting three key pieces from the quoted statement. Most human societies are territorial most of the time.

Warfare

In *Sociobiology* I presented widespread lethal warfare in early human groups as a working hypothesis, not as a fact, contrary to what the cosigners suggest. And it is a hypothesis wholly consistent with the evidence: military activity and territorial expansion have been concomitants throughout history and at all levels of social organization (Otterbein 1970), and they can hardly fail to have had significant demographic and genetic consequences.

Slavery and Other Terms

The cosigners state that I claim to have found barter, religion, magic, and tribalism among nonhumans. I have made no such claim. The cosigners do not like to see terms such as slavery, division of labor, and ritual used in both zoology and the social sciences. Do they wish also to expunge communication, dominance, monogamy, and parental care from the vocabulary of zoology?

Genetic Bases of Behavior

The cosigners claim that no evidence exists for the genetic basis of particular forms of social behavior. Their statement indicates that they do not use the same criteria as other biologists. To postulate the existence of genes for the diagnostic human traits is not to imply that there exists one gene for spite, another for homosexuality, and so on, as one might envision the inheritance of flower color or seed texture in garden peas. The tendency to develop such behaviors, in a distinctively human form, is part of an immensely complex social repertory which is undoubtedly dependent on large numbers of genes.

My emphasis in *Sociobiology* was on the most widespread, distinctive qualities of human behavior—"human nature" if you wish—and the possible reasons why the underlying genes are different from those affecting social behavior in other species. Certain forms of human social behavior, such as the facial expressions used to convey the basic emotions, are relatively inflexible and transcultural. Human expressions, in fact, are so similar to those of the higher cercopithecoid primates as to suggest the possible existence of true homology (*Sociobiology*, pp. 227–228). Other kinds of response, including those under the categories of aggression, sexuality, and conformity, are of course subject to great variation through differences in experience. But as plastic as these latter behaviors might seem to us, they still form only a small subset of the many versions found in social species as a whole. It seems inconceivable that human beings could be socialized into the distinctive patterns of, say, ring-tailed lemurs, hamadryas baboons, or gibbons, or vice versa. This is the ordinary criterion on which the expression "genetic control of human social behavior" in sociobiology is based. The main idea conveyed by the final chapter of my book is that such a comparison with other social species will place human behavior in a clearer evolutionary perspective.

With reference to genetic variation between human populations, there is no firm evidence. As usual, the cosigners misrepresent what I said. Here is their claim: "It is stated as a *fact* that genetical differences underlie variations between cultures, when no evidence at all exists for this assertion and there is some considerable evidence against it" (emphasis theirs). Here is what I really said, in the very sentences to which they allude (p. 550): "Even a small portion of this [genetic] variance invested in population differences *might* predispose societies toward cultural differences. At the very least, we should try to measure this amount. It is not valid to point to the absence of a behavioral trait in one or a few societies as conclusive evidence that the trait is environmentally induced and has no genetic disposition in man. The very opposite *could* be true" (italics newly added).

Adaptation versus Non-adaptation
The Science for the People group state that I believe all social behavior to be adaptive and hence "normal." This is so patently false that I am surprised the cosigners could bring themselves to say it. I have on the contrary discussed circumstances under which certain forms of animal social behavior become maladaptive, with examples and ways in which the deviations can be analyzed (pp. 33–34). With reference to human social behavior I have said (Wilson 1975b, an article well known to the cosigners): "When any genetic bias is demonstrated, it cannot be used to justify a continuing practice in present and future societies. Since most of us live in a radically new environment of our own making, the pursuit of such a practice would be bad biology; and like all bad biology, it would invite disaster." I then cited examples of maladaptive behavior in human beings. Furthermore, both R. L. Trivers and I have provided varieties of adaptation hypotheses that compete with each other and against the non-adaptation hypothesis, contrary to the assertion of the Science for the People group (see, e.g., *Sociobiology*: pp. 123–124, 309–311, 326–327, 416–418).

Cultural Evolution
The cosigners propose that "sociobiological hypotheses" can be tested by seeing whether certain short-term episodes in history, such as the rise and decline of Islam, occurred too rapidly to be due to genetic change. They conclude that the theory of population genetics excludes that possibility. I agree, and that is why neither I nor any other sociobiologist of my acquaintance has ever proposed such hypotheses. The examples I used in Sociobiology to make the same point are the origin of the slave society of Jamaica, the decline of the Ik in Uganda, the alteration of Irish society following the potato famine, and the shift in the Japanese authority structure following World War II (pp. 548–550). I see no reason why the subject was even brought up. (A fuller discussion of the rates of cultural evolution and the complementarity of cultural to genetic evolution can be found in pages 168–175 and 555–562 of *Sociobiology*.)

Comments on the Debate

I now invite readers to check each of the pronouncements in the article by the 35 cosigners against the actual statements in my book, in the true context in which the statements were made. I suggest that they will encounter very little correspondence, and I am confident that they will be left with no doubt as to my true meaning.

How is it possible for the Science for the People group to misrepresent so consistently the content of a book, in contrast to all of the many other reviewers among their scientific colleagues? There is first the circumstance of the size and composition of the group. It has grown from 16, when it called itself The Genetic Engineering Group of Scientists and Engineers for Social and Political Action (in the magazine *Science for the People,* November 1975), to the present 35 now identified as the Sociobiology Study Group of Science for the People. The membership is heterogeneous: from the best count I can make there are eight professors in several fields of science in the Boston area; other members include at least one psychiatrist, a secondary school teacher, students and research assistants. Furthermore, in conformity with their political convictions the group really does believe in collective decision making and writing, so perhaps the result is not all that surprising. (In the issue of *Science for the People* just mentioned, the two main targets of criticism were myself, for biological determinism, and the Soviet Union, for revisionism.)

But the other, more important cause of the problem, and the reason I have not been able to find the matter as humorous as have some of my colleagues, is the remorseless zeal of the cosigners. By their own testimony they worked for months on the project. They appear to have been alarmed by the impact a critical success of the book might have on the acceptability of their own political views. One of the faculty members, in a *Harvard Crimson* interview on 3 November 1975, stated that the group was formed of persons who became interested "in breaking down the screen of approval" around the book. Clamorous denunciations followed during a closely packed series of lectures, work sessions, and release of printed statements. In October 1975 a second professorial member of the group drafted a 5,000-word position paper for *The New York Times* which characterized me as an ideologue and a privileged member of modern Western industrial society whose book attempts to preserve the status quo *(The New York Times,* 9 November 1975). Later the same person (who shares the identical privileges at Harvard) startled me even more by declaring that "Sociobiology is not a racist doctrine" but "any kind of genetic determinism can and does feed other kinds, including the belief that some races are superior to others" *(Harvard Crimson,* 3 December 1975).

The latter argument is identical to that advanced simultaneously by student members of the Harvard-Radcliffe Committee against Racism, who, citing the Science for the People statement for authority, did not hesitate to label the book "dangerously racist" in leaflets distributed through the Boston area. Both the logic and the accusation were false and hurtful, and at this point the matter was close to getting out of hand.

On various occasions and with only limited success the Harvard faculty has attempted to protect itself from activities of this kind. During an earlier, similar episode 100 of its members published a statement that "In an academic community the substitution of personal harassment for reasoned inquiry is intolerable. The openminded search for truth cannot proceed in an atmosphere of political intimidation." This is the melancholy principle which has been confirmed by the exchange now extended to *BioScience.* In the Boston area at the present time it has become difficult to conduct an open forum on human sociobiology, or even general sociobiology, without falling into the role of either prosecutor or defendant.

The Political Significance of Sociobiology

Finally and briefly, let me express what I consider to be the real significance of human sociobiology for political and social thought. The question that science is now in a position to approach is the very origin and meaning of human values, from which all ethical pronouncements and much of political practice flow. Philosophers themselves have not explored the problem; traditional ethical philosophy begins with premises that are examined with reference to their consequences but not their origins. Thus, John Rawls opens his celebrated *A Theory of Justice* (1971) with a proposition he regards as beyond dispute: "In a just society the liberties of equal citizenship are taken as settled; the rights secured by justice are not subject to political bargaining or to the calculus of social interests." Robert Nozick launches his equally celebrated *Anarchy, State, and Utopia* (1974) with a similarly firm proposition: "Individuals have rights, and there are things no person or group may do to them (without violating their rights). So strong and far-reaching are these rights that they raise the question of what, if anything, the state and its officials may do."

These two premises are somewhat different in content, and they lead to radically different prescriptions. Rawls would allow rigid social control to secure as close an approach as possible to the equal distribution of society's rewards. Nozick sees the ideal society as one governed by a minimal state, empowered only to protect its citizens from force and fraud, and with unequal distribution of rewards wholly permissible. Rawls rejects the meritocracy; Nozick accepts it as desirable except in those cases where local communities voluntarily decide to experiment with egalitarianism.

Whether in conflict or agreement, where do such fundamental premises come from? What lies behind the intuition on which they are based? Contemporary philosophers have progressed no further than Sophocles' Antigone, who said of moral imperatives, "They were not born today or yesterday; they die not, and none knoweth whence they sprung."

At this point the 35 members of the Science for the People group also come to a halt. At the close of their essay they imply the central issue to be a decision about the kind of the society we want to live in; humanity can then find the way to bring this society into being. But which persons are the "we" who will decide, and whose moral precepts must thereby be validated? The group believe that all social behavior is learned and transmitted by culture. But if this is true, the value system by which "we" will decide social policy is created by the culture in which the most powerful decision makers were reared and hence must inevitably validate the status quo, the very condition which the Science for the People group reject. The solution to the conundrum must be that their premise of complete environmentalism is wrong.

The evidence that human nature is to some extent genetically influenced is in my opinion decisive. In the present space I can only suggest that the reader consider the facts presented in *Sociobiology* and in the very extensive primary literature on the subject, some of which is cited in this work. It follows that value systems are probably influenced, again to an unknown extent, by emotional responses programmed in the limbic system of the brain. The qualities that comprise human nature in the Maring of New Guinea as recognizably as they did in the Greeks at Troy are surely due in part to constraints within the unique human genotype. The challenge of human sociobiology, shared with the social sciences, is to measure the degree of these constraints and to infer their significance through the reconstruction of the evolutionary history of the mind. The enterprise is the logical complement to the continued study of cultural evolution.

Even if that formidable challenge is successfully met, however, it will still leave the ethical question: To what extent should the censors and motivators in the emotive centers of the brain be obeyed? Given that these controls deeply and unconsciously affect our moral decisions, how faithfully must they be consulted once they have been defined and assayed as a biological process? The answer must confront what appears to me to be the true human dilemma. We cannot follow the suggestions of the censors and motivators blindly. Although they are the source of our deepest and most compelling feelings, their genetic constraints evolved during the millions of years of prehistory, under conditions that to a large extent no longer exist. At some time in the future it will be necessary to decide how human we wish to remain, in this the ultimate biological sense, and to pick and choose consciously among the emotional guides we have inherited.

This dilemma should engender a sense of reserve about proposals for radical social change based on utopian intuition. To the extent that the biological interpretation noted here proves correct, men have rights that are innate, rooted in the ineradicable drives for survival and self-esteem, and these rights do not require the validation of ad hoc theoretical constructions produced by society. If culture is all that created human rights, as the extreme environmentalist position holds, then culture can equally well validate their removal. Even some philosophers of the radical left see this flaw in the position taken by Science for the People. Noam Chomsky, whose own linguistic research has provided evidence for the existence of genetic influence, considers extreme environmentalism to be a belief susceptible to dictatorships of both the left and the right:

> One can easily see why reformers and revolutionaries should become radical environmentalists, and there is no doubt that concepts of immutable human nature can be and have been employed to erect barriers against social change and to defend established privilege. But a deeper look will show that the concept of the "empty organism," plastic and unstructured, apart from being false, also serves naturally as the support for the most reactionary social doctrines. If people are, in fact, malleable and plastic beings with no essential psychological nature, then why should they not be controlled and coerced by those who claim authority, special knowledge, and a unique insight into what is best for those less enlightened? ... The principle that human nature, in its psychological aspects, is nothing more than a product of history and given social relations removes all barriers to coercion and manipulation by the powerful. This too, I think, may be a reason for its appeal to intellectual ideologists, of whatever political persuasion (Chomsky 1975, p. 132).

Chomsky and I, not to mention Herbert Marcuse (who has a similar belief in the biological conservatism of human nature), can scarcely be accused of having linked arms to preserve the status quo, and yet that would seem to follow from the strange logic employed by the Science for the People group.

In their corybantic attentions to sociobiology, the Science for the People group have committed what can be usefully termed the Fallacy of the Political Consequent. This is the assumption that political belief systems can be mapped one-on-one onto biological or psychological generalizations. Another particularly ironic example is the response to B. F. Skinner's writings. Skinner is a radical environmentalist, whose conclusions about human behavior are essentially indistinguishable from those of the Science for the People group. Yet the particular political conclusions he has drawn are anathema to the radical left, who reject them as elitist, reactionary, and so forth. The cause of the Fallacy of the Political Consequent is the failure to appreciate adequately that scientific theories and political ideas are both complex and tenuously linked, and that political ideas are shaped in good part by personal judgments lying outside the domain of scientific evaluation.

All political proposals, radical and otherwise, should be seriously received and debated. But whatever direction we choose to take in the future, social progress can only be enhanced, not impeded, by the deeper investigation of the genetic constraints of human nature, which will steadily replace rumor and folklore with testable knowledge. Nothing is to be gained by a dogmatic denial of the existence of the constraints or attempts to discourage public discussion of them. Knowledge humanely acquired and widely shared, related to human needs but kept free of political censorship, is the real science for the people.

References

Allen, E. et al. (1975). "Against 'Sociobiology.'" *The New York Review of Books,* November 13.

Chomsky, N. (1975). *Reflections on Language* (New York: Pantheon Books, Random House).

Nozick, R. (1974). *Anarchy, State and Utopia* (New York: Basic Books).

Otterbein, K. F. (1970). *The Evolution of War* (New Haven, Conn.: Human Relations Area Files Press).

Rawls, J. (1971). *A Theory of Justice* (Cambridge, Mass.: Harvard University Press).

Wilson, E. O. (1975a). *Sociobiology: The New Synthesis* (Cambridge, Mass.: Harvard University Press).

———. (1975b). "Human Decency Is Animal." *The New York Times Magazine,* October 12, pp. 38–50.

The Evolution of Ethics

Michael Ruse and Edward O. Wilson

Attempts to link evolution and ethics first sprang up in the middle of the last century, as people turned to alternative foundations in response to what they perceived as the collapse of Christianity. If God does not stand behind the Sermon on the Mount, then what does? Such attempts at evolutionary ethicising became known collectively as "social Darwinism," although they owed less to Charles Darwin and more to that quintessentially Victorian man of ideas, Herbert Spencer. Finding worth in what he perceived to be the upward progress of evolution from amoeba to human, from savage to *Homo britannicus,* Spencer argued that right conduct lies in the cherishing of the evolutionary process, in order that the best or fittest be able to survive and the inadequate be rigorously eliminated.

While Spencer's ideas attracted strong support in some quarters, for example the North American barons of industry, evolutionary ethics in this mode never really caught fire. On the one hand, social Darwinism seems so immoral! Right conduct surely cannot entail stamping on widows and babies. And no amount of tinkering by revisionists, such as Prince Peter Kropotkin in the last century and Sir Julian Huxley and C. H. Waddington in this, changes the fact. On the other hand, the very basis of a Spencerian-type approach is shaky. There is no progress to evolution. In a purely Darwinian sense, an amoeba is as good as a person.

Most people, therefore, have happily agreed with the 18th-century philosopher David Hume that there is an impassible gulf between matters of fact (for example, evolution) and matters of morality (disinterested help of others). To use phrasing made popular in this century by the Cambridge philosopher G. E. Moore, evolutionary ethics commits "the naturalistic fallacy" by trying to translate *is* into *ought*.

It is true that past efforts to create an evolutionary ethics have come to very little. Yet to revert to the opposite conclusion, that evolution and ethics have nothing to say to each other, is altogether too quick. Recent advances in evolutionary theory have cast a new light on the matter, giving substance to the dreams of the old theorisers, although not in the way or for the reasons they thought.

Our starting point is with the science. Two propositions appear to have been established beyond any reasonable doubt. First, the social behaviour of animals is firmly under the control of the genes, and has been shaped into forms that give reproductive advantages. Secondly, humans are animals. Darwin knew that the first claim was true, and a multitude of recent studies, from fruit flies to frogs, have affirmed it repeatedly. Darwin knew also that the second claim is true, and positive evidence continues to pour in from virtually every biological discipline. Genetically, we are a sibling species to the chimpanzee, having evolved with them for more than 3 1/2 billion years, parting a mere 6 million or so years ago.

What do these facts have to do with morality? A chain of reasoning leads us to a distinctly human but still biologically based ethical sense. First, note that we are not just talking about behaviour, but about *social* behaviour. Today's students of this subject, sociobioiogists, know that it is often in an individual's biological self-interest to cooperate with its fellows, rather than (as traditional evolutionary ethicists thought) to fight flat out. After all, a loaf shared is better than a whole loaf, if the latter carries the risk of being killed or seriously hurt.

Secondly, and less obviously, there are ways in which nature can bring about "altruism," in the sense of self-sacrifice for the benefit of others. If those benefited are relatives, the altruist is still favouring genes identical to his own, even if he dies without leaving any direct offspring. Thus we say that the individual is altruistic but his genes are "selfish." Note that such behaviour implies nothing about good intentions or other ways of being "nice." To get altruism you can go the way of the ants. They are genetically hardwired, performing their duties in perfect cooperative harmony. They have no thoughts, at least of a human kind, only actions. Alternatively, you could go to the other extreme, and evolve super-brains, where every possible action is first weighed and assessed, and a policy of rationally assessed self-interest is always followed.

Neither of these options has proved attractive to animals like humans, and we have avoided both. If we had become hardwired in the course of evolution, we could never deviate from our course. Were something untoward to happen, we would be stuck with maladaptive behaviour. Worker ants are relatively cheap to produce, so this rigidity matters relatively little to their colonies. Humans require a great deal of parental investment, and it would be stupid in the literal sense of the word if we were to go wrong at the slightest environmental quiver. Alternatively, if we possessed super-brains, we would require even more resources than we do now; such as parental care stretched over many more years. Additionally, like those chess machines that survey every move, we would be forever making up our minds. Crises would be upon us, and we would still be thinking.

Nature's Moral Imperative

How then has nature made humans "altruistic"? The clue lies in the chess machines we just mentioned. The new breed, those that can beat grandmasters, forgo omnipotence for utility. They follow certain strategies that have proved successful. So with humans. Our minds are not tabulae rasae, but moulded according to certain innate dispositions. These dispositions, known technically as "epigenetic rules," incline us to particular courses of action, such as learning rapidly to fear heights and snakes, although they certainly do not lock us, ant-like, into undeviating behaviour.

The best studied epigenetic rules, such as those affecting fears or the avoidance of incest, appear to have been put into place because of their biological virtues. Although altruism is less well documented (there is some evidence, for example, that varying degrees of its expression have a genetic component), such behaviour is also adaptive—at least when directed in appropriate measure toward kin and allies. We need to be altruistic. Thus, we have rules inclining us to such courses of behaviour. The key question is then: how are these rules expressed in our conscious awareness? We need something to spur us against our usual selfish dispositions. Nature, therefore, has made us (via the rules) believe in a disinterested moral code, according to which we *ought* to help our fellows. Thus, we are inclined to go out and work with our fellows. In short, to make us altruistic in the adaptive, biological sense, our biology makes us altruistic in the more conventionally understood sense of acting on deeply held beliefs about right and wrong.

Such is the modern scientific account of morality; at least the one most consistent with biology. But, what has any of this to do with the concerns of the traditional evolutionary ethicist? Even if the explanation were proved to be entirely true, it does not reveal whether in some ultimate, absolute sense, evolution stands behind morality. Does the sociobiological scenario just sketched justify the same moral code that religionists believe to be decreed by God? Or that some philosophers believe to exist apart from humanity, like a mathematical theorem?

It used to be thought, in the bad old days of social Darwinism when evolution was poorly understood, that life is an uninterrupted struggle "—nature red in tooth and claw." But this is only one side of natural selection. What we have just seen is that the same process also leads to altruism and reciprocity in highly social groups. Thus the human species has evolved genuine sentiments of obligation, of the duty to be loving and kind. In no way does this materialist explanation imply that we are hypocrites consciously trying to further our biological ends and paying lip-service to ethics. We function better because we believe. In this sense, evolution is consistent with conventional views of morality.

On the other hand, the question of ultimate foundations requires a different and more subtle answer. As evolutionists, we see that no justification of the traditional kind is possible. Morality, or more strictly our belief in morality, is merely an adaptation put in place to further our reproductive ends. Hence the basis of ethics does not lie in God's will—or in the metaphorical roots of evolution or any other part of the framework of the Universe. In an important sense, ethics as we understand it is an illusion fobbed off on us by our genes to get us to cooperate. It is without external grounding. Ethics is produced by evolution but not justified by it, because, like Macbeth's dagger, it serves a powerful purpose without existing in substance.

In speaking thus of illusion, we are not saying that ethics is nothing, and should now be thought of as purely dreamlike. Unlike Macbeth's dagger, ethics is a *shared* illusion of the human race. If it were not so, it would not work. The moral ones among us would be outbred by the immoral. For this reason, since all human beings are dependent on the "ethics game," evolutionary reasoning emphatically does not lead to moral relativism. Human minds develop according to epigenetic rules that distinguish between proper moral claims like "Be kind to children" and crazy imperatives like "Treat cabbages with the respect you show your mother."

Ethical codes work because they drive us to go against our selfish day-to-day impulses in favour of long-term group survival and harmony and thus, over our lifetimes, the multiplication of our genes many times. Furthermore, the way our biology enforces its ends is by making us think that there is an objective higher code, to which we are all subject. If we thought ethics to be no more than a question of personal desires, we would tend to ignore it. Why should we base our life's plan on your love of French cuisine? Because we think that ethics is objectively based, we are inclined to obey moral rules. We help small children because it is right even though it is personally inconvenient to us.

If this perception of human evolution is correct, it provides a new basis for moral reasoning. Ethics is seen to have a solid foundation, not in divine guidance or pure moral imperatives, but in the shared qualities of human nature and the desperate need for reciprocity. The key is the deeper, more objective study of human nature, and for this reason we need to turn ethical philosophy into an applied science.

Some philosophers have argued that even if ethics could be explained wholly in such a materialist fashion, this alone would not eliminate the possibility that moral imperatives exist, sitting apart like mathematical truths. Perhaps human evolution is moving toward such celestial perfection, and the apprehension of such truths. There are biological reasons for seeing and hearing the moving train, but it still exists!

Unfortunately, the cases of mathematical principles, material objects and ethics are not parallel. Natural selection is above all opportunistic. Suppose that, instead of evolving from savannah-dwelling primates, we had evolved in a very different way. If, like the termites, we needed to dwell in darkness, eat each other's faeces and cannibalise the dead, our epigenetic rules would be very different from what they are now. Our minds would be strongly prone to extol such acts as beautiful and moral. And we would find it morally disgusting to live in the open air, dispose of body waste and bury the dead. Termite ayatollahs would surely declare such things to be against the will of God. Termite social theorists would surely argue for a stricter caste system.

Ethics does not have the objective foundation our biology leads us to think it has. But this is no negative conclusion. Human beings face incredible social problems, primarily because their biology cannot cope with the effects of their technology. A deeper understanding of this biology is surely a first step towards solving some of these pressing worries. Seeing morality for what it is, a legacy of evolution rather than a reflection of eternal, divinely inspired verities, is part of this understanding.

Chapter 9

Chapter 9 deals with the critics, particularly the ideas of Stephen Jay Gould and others working on problems of macroevolution. My first document is one of his justly celebrated essays, "The Episodic Nature of Evolutionary Change," from his monthly column in Natural History. *Here Gould lays out his theory of punctuated equilibria in the clear prose for which he is properly praised. He argues that evolutionary change as revealed by the fossil record is much more jerky or spasmodic than is believed and claimed by classical Darwinism, arguing also that this is no artifact of the record—incomplete fossilization or whatever—but a reflection of reality. Evolution's history is one of inaction or lack of change, broken by periods when there is a rapid move from one form to another. Although Gould is careful to hedge his bets and certainly does come out foursquare for this philosophy, note the favorable reference to Marxism. As I explain in the text, I am inclined to think that Gould's real philosophical roots lie further back in the German idealism of the early nineteenth century, and I draw your attention to Gould's mention of the philosopher Hegel, who was himself an ardent proponent of* Naturphilosophie.

My second document, countering Gould, is by John Maynard Smith. Asking "Did Darwin Get It Right?" Maynard Smith—a student of J. B. S. Haldane—answers his question with a ringing affirmative. As explained in the text, Maynard Smith feels that Gould misrepresents or caricatures the Darwinian position. No one ever said that everything must be adaptive all of the time. The point rather is that organic features are rooted in and related to adaptation. Having four limbs rather than six may not now be a necessity, but in the past, when vertebrates were aquatic, it was really important. Natural selection was all-important even if now it plays no direct role. You should judge Maynard Smith's arguments on their merits, but you might like to know that he too was a Marxist, breaking from that philosophy in the 1950s because of the treatment of geneticists when Lysenko was in power and because of political events, particularly the Hungarian Revolution in 1956, which was put down brutally by the Soviets. Do we have here at work a philosophical disagreement as much as anything purely scientific?

The Episodic Nature of Evolutionary Change

S. J. Gould

On November 23, 1859, the day before his revolutionary book hit the stands, Charles Darwin received an extraordinary letter from his friend Thomas Henry Huxley. It offered warm support in the coming conflict, even the supreme sacrifice: "I am prepared to go to the stake, if requisite ... I am sharpening up my claws and beak in readiness." But it also contained a warning: "You have loaded yourself with an unnecessary difficulty in adopting *Natura non facit saltum* so unreservedly."

The Latin phrase, usually attributed to Linnaeus, states that "nature does not make leaps." Darwin was a strict adherent to this ancient motto. As a disciple of Charles Lyell, the apostle of gradualism in geology, Darwin portrayed evolution as a stately and orderly process, working at a speed so slow that no person could hope to observe it in a lifetime. Ancestors and descendants, Darwin argued, must be connected by "infinitely numerous transitional links" forming "the finest graduated steps." Only an immense span of time had permitted such a sluggish process to achieve so much.

Huxley felt that Darwin was digging a ditch for his own theory. Natural selection required no postulate about rates; it could operate just as well if evolution proceeded at a rapid pace. The road ahead was rocky enough; why harness the theory of natural selection to an assumption both unnecessary and probably false? The fossil record offered no support for gradual change: whole faunas had been wiped out during disarmingly short intervals. New species almost always appeared suddenly in the fossil record with no intermediate links to ancestors in older rocks of the same region. Evolution, Huxley believed, could proceed so rapidly that the slow and fitful process of sedimentation rarely caught it in the act.

The conflict between adherents of rapid and gradual change had been particularly intense in geological circles during the years of Darwin's apprenticeship in science. I do not know why Darwin chose to follow Lyell and the gradualists so strictly, but I am certain of one thing: preference for one view or the other had nothing to do with superior perception of empirical information. On this question, nature spoke (and continues to speak) in multifarious and muffled voices. Cultural and methodological preferences had as much influence upon any decision as the constraints of data.

On issues so fundamental as a general philosophy of change, science and society usually work hand in hand. The static systems of European monarchies won support from legions of scholars as the embodiment of natural law. Alexander Pope wrote:

> Order is Heaven's first law; and this confessed,
> Some are, and must be, greater than the rest.

As monarchies fell and as the eighteenth century ended in an age of revolution, scientists began to see change as a normal part of universal order, not as aberrant and exceptional. Scholars then transferred to nature the liberal program of slow and orderly change that they advocated for social transformation in human society. To many scientists, natural cataclysm seemed as threatening as the reign of terror that had taken their great colleague Lavoisier.

Yet the geologic record seemed to provide as much evidence for cataclysmic as for gradual change. Therefore, in defending gradualism as a nearly universal tempo, Darwin had to use Lyell's most characteristic method of argument—he had to reject literal appearance and common sense for an underlying "reality." (Contrary to popular myths, Darwin and Lyell were not the heros of true science, defending objectivity against the theological fantasies of such "catastrophists" as Cuvier and Buckland. Catastrophists were as committed to science as any gradualist; in fact, they adopted the more "objective" view that one should believe what one sees and not interpolate missing bits of a gradual record into a literal tale of rapid change.) In short, Darwin argued that the geologic record was exceedingly imperfect—a book with few remaining pages, few lines on each page, and few words on each line. We do not see slow evolutionary change in the fossil record because we study only one step in thousands. Change seems to be abrupt because the intermediate steps are missing.

The extreme rarity of transitional forms in the fossil record persists as the trade secret of paleontology. The evolutionary trees that adorn our textbooks have data only at the tips and nodes of their branches; the rest is inference, however reasonable, not the evidence of fossils. Yet Darwin was so wedded to gradualism that he wagered his entire theory on a denial of this literal record:

> The geological record is extremely imperfect and this fact will to a large extent explain why we do not find interminable varieties, connecting together

all the extinct and existing forms of life by the finest graduated steps. He who rejects these views on the nature of the geological record, will rightly reject my whole theory.

Darwin's argument still persists as the favored escape of most paleontologists from the embarrassment of a record that seems to show so little of evolution directly. In exposing its cultural and methodological roots, I wish in no way to impugn the potential validity of gradualism (for all general views have similar roots). I wish only to point out that it was never "seen" in the rocks.

Paleontologists have paid an exorbitant price for Darwin's argument. We fancy ourselves as the only true students of life's history, yet to preserve our favored account of evolution by natural selection we view our data as so bad that we almost never see the very process we profess to study.

For several years, Niles Eldredge of the American Museum of Natural History and I have been advocating a resolution of this uncomfortable paradox. We believe that Huxley was right in his warning. The modern theory of evolution does not require gradual change. In fact, the operation of Darwinian processes should yield exactly what we see in the fossil record. It is gradualism that we must reject, not Darwinism.

The history of most fossil species includes two features particularly inconsistent with gradualism:

1. *Stasis.* Most species exhibit no directional change during their tenure on earth. They appear in the fossil record looking much the same as when they disappear; morphological change is usually limited and directionless.
2. *Sudden appearance.* In any local area, a species does not arise gradually by the steady transformation of its ancestors; it appears all at once and "fully formed."

Evolution proceeds in two major modes. In the first, phyletic transformation, an entire population changes from one state to another. If all evolutionary change occurred in this mode, life would not persist for long. Phyletic evolution yields no increase in diversity, only a transformation of one thing into another. Since extinction (by extirpation, not by evolution into something else) is so common, a biota with no mechanism for increasing diversity would soon be wiped out. The second mode, speciation, replenishes the earth. New species branch off from a persisting parental stock.

Darwin, to be sure, acknowledged and discussed the process of speciation. But he cast his discussion of evolutionary change almost totally in the mold of phyletic transformation. In this context, the phenomena of stasis and sudden appearance could hardly be attributed to anything but imperfection of the record; for if new species arise by the transformation of entire ancestral populations, and if we almost never see the transformation (because species are essentially static through their range), then our record must be hopelessly incomplete.

Eldredge and I believe that speciation is responsible for almost all evolutionary change. Moreover, the way in which it occurs virtually guarantees that sudden appearance and stasis shall dominate the fossil record.

All major theories of speciation maintain that splitting takes place rapidly in very small populations. The theory of geographic, or allopatric, speciation is preferred by most evolutionists for most situations (allopatric means "in another place").* A new species can arise when a small segment of the ancestral population is isolated at the periphery of the ancestral range. Large, stable central populations exert a strong homogenizing influence. New and favorable mutations are diluted by the sheer bulk of the population through which they must spread. They may build slowly in frequency, but changing environments usually cancel their selective value long before they reach fixation. Thus, phyletic transformation in large populations should be very rare—as the fossil record proclaims.

But small, peripherally isolated groups are cut off from their parental stock. They live as tiny populations in geographic corners of the ancestral range. Selective pressures are usually intense because peripheries mark the edge of ecological tolerance for ancestral forms. Favorable variations spread quickly. Small, peripheral isolates are a laboratory of evolutionary change.

What should the fossil record include if most evolution occurs by speciation in peripheral isolates? Species should be static through their range because our fossils are the remains of large central populations. In any local area inhabited by ancestors, a descendent species should appear suddenly by migration from the peripheral region in which it evolved. In the peripheral region itself, we might find direct evidence of speciation, but such good fortune would be rare indeed because the event occurs so rapidly in such a small population. Thus, the fossil record is a faithful rendering of what evolutionary theory predicts, not a pitiful vestige of a once bountiful tale.

Eldredge and I refer to this scheme as the model *of punctuated equilibria*. Lineages change little during most of their history, but events of rapid speciation occasionally punctuate this tranquillity. Evolution is the differential survival and deployment of these punctuations. (In describing the speciation of peripheral isolates as very rapid, I speak as a geologist. The process may take hundreds, even thousands of years; you might see nothing if you stared at speciating bees on a tree for your entire lifetime. But a thousand years is a tiny fraction of one percent of the average duration for most fossil invertebrate species—5 to 10 million years. Geologists can rarely resolve so short an interval at all; we tend to treat it as a moment.)

If gradualism is more a product of Western thought than a fact of nature, then we should consider alternate philosophies of change to enlarge our realm of constraining prejudices. In the Soviet Union, for example, scientists are trained with a very different philosophy of change—the so-called dialectical laws, reformulated by Engels from Hegel's philosophy. The dialectical laws are explicitly punctuational. They speak, for example, of the "transformation of quantity into quality." This may sound like mumbo

---

* I wrote this essay in 1977. Since then, a major shift of opinion has been sweeping through evolutionary biology. The allopatric orthodoxy has been breaking down and several mechanisms of sympatric speciation have been gaining both legitimacy and examples. (In sympatric speciation, new forms arise within the geographic range of their ancestors.) These sympatric mechanisms are united in their insistence upon the two conditions that Eldredge and I require for our model of the fossil record—*rapid* origin in a *small* population. In fact, they generally advocate smaller groups and more rapid change than conventional allopatry envisages (primarily because groups in potential contact with their forebears must move quickly towards reproductive isolation, lest their favorable variants be diluted by breeding with the more numerous parental forms). See White (1978) for a thorough discussion of these sympatric models.

jumbo, but it suggests that change occurs in large leaps following a slow accumulation of stresses that a system resists until it reaches the breaking point. Heat water and it eventually boils. Oppress the workers more and more and bring on the revolution. Eldredge and I were fascinated to learn that many Russian paleontologists support a model similar to our punctuated equilibria.

I emphatically do not assert the general "truth" of this philosophy of punctuational change. Any attempt to support the exclusive validity of such a grandiose notion would border on the nonsensical. Gradualism sometimes works well. (I often fly over the folded Appalachians and marvel at the striking parallel ridges left standing by gradual erosion of the softer rocks surrounding them.) I make a simple plea for pluralism in guiding philosophies, and for the recognition that such philosophies, however hidden and unarticulated, constrain all our thought. The dialectical laws express an ideology quite openly; our Western preference for gradualism does the same thing more subtly.

Nonetheless, I will confess to a personal belief that a punctuational view may prove to map tempos of biological and geologic change more accurately and more often than any of its competitors—if only because complex systems in steady state are both common and highly resistant to change. As my colleague British geologist Derek V. Ager writes in supporting a punctuational view of geologic change: "The history of any one part of the earth, like the life of a soldier, consists of long periods of boredom and short periods of terror."

Did Darwin Get It Right?

John Maynard Smith

> I think I can see what is breaking down in evolutionary theory—the strict construction of the modern synthesis with its belief in pervasive adaptation, gradualism and extrapolation by smooth continuity from causes of change in local populations to major trends and transitions in the history of life.

> A new and general evolutionary theory will embody this notion of hierachy and stress a variety of themes either ignored or explicitly rejected by the modern synthesis.

These quotations come from a recent paper in *Palaeobiology* by Stephen Jay Gould. What is the new theory? Is it indeed likely to replace the currently orthodox "neo-Darwinian" view? Proponents of the new view make a minimum and a maximum claim. The minimum claim is an empirical one concerning the nature of the fossil record. It is that species, once they come into existence, persist with little or no change, often for millions of years ("stasis"), and that evolutionary change is concentrated into relatively brief periods ("punctuation"), these punctuational changes occurring at the moment when a single species splits into two. The maximal claim is a deduction from this, together with arguments drawn from the study of development: it is that evolutionary change, when it does occur, is not caused by natural selection operating on the genetic differences between members of populations, as Darwin argued and as most contemporary evolutionists would agree, but by some other process. I will discuss these claims in turn; as will be apparent, it would be possible to accept the first without being driven to accept the second.

The claim of stasis and punctuation will ultimately be settled by a study of the fossil record. I am not a palaeontologist, and it might therefore be wiser if I were to say merely that some palaeontologists assert that it is true, and others are vehemently denying it. There is something, however, that an outsider can say. It is that the matter can be settled only by a statistical analysis of measurements of fossil populations from different levels in the rocks, and not by an analysis of the lengths of time for which particular named species or genera persist in the fossil record. The trouble with the latter method is that one does not know whether one is studying the rates of evolution of real organisms, or merely the habits of the taxonomists who gave the names to the fossils. Suppose that in some lineage evolutionary change took place at a more or less steady rate, to such an extent that the earliest and latest forms are sufficiently different to warrant their being placed in different species. If there is at some point a gap in the record, because suitable deposits were not being laid down or have since been eroded, then there will be a gap in the sequence of forms, and taxonomists will give fossils before the gap one name and after it another. It follows that an analysis of named forms tells us little: measurements of populations, on the other hand, would reveal whether change was or was not occurring before and after the gap.

My reason for making this rather obvious point is that the only extended presentation of the punctuationist view—Stanley's book, *Macroevolution*—rests almost entirely on an analysis of the durations of named species and genera. When he does present population measurements, they tend to support the view that changes are gradual rather than sudden. I think that at least some of the changes he presents as examples of sudden change will turn out on analysis to point the other way. I was unable to find any evidence in the book which supported, let alone established, the punctuationist view.

Of course, that is not to say that the punctuationist view is not correct. One study, based on a proper statistical analysis, which does support the minimal claim, but not the maximal one, is Williamson's study of the freshwater molluscs (snails and bivalves) of the Lake Turkana region of Africa over the last five million years. Of the 21 species studied, most showed no substantial evolutionary change during the whole period: "stasis" was a reality. The remaining six species were more interesting. They also showed little change for most of the period. There was, however, a time when the water table fell and the lake was isolated from the rest of the rift valley. When this occurred, these six species changed rather rapidly. Through a depth of deposit of about one meter, corresponding roughly to 50,000 years, successive populations show changes of shape great enough to justify placing the later forms in different species. Later, when the lake was again connected to the rest of the rift valley, these new forms disappear suddenly, and are replaced by the original forms, which presumably re-entered the lake from outside, where they had persisted unchanged.

This is a clear example of stasis and punctuation. However, it offers no support for the view that changes, when they do occur, are not the result of selection acting within populations. Williamson does have intermediate populations, so we know that the change did not depend on the occurrence of a "hopeful monster" (see below), or on the existence of an isolated population small enough to permit random changes to outweigh natural selection. The example is also interesting in showing how we may be misled if we study the fossil record only in one place. Suppose that, when the water table rose again, the new form had replaced the original one in the rest of the rift valley, instead of the other way round. Then, if we had examined the fossil record anywhere else but in Lake

Turkana, we would have concluded, wrongly, that an effectively instantaneous evolutionary change had occurred.

Williamson's study suggests an easy resolution of the debate. Both sides are right, and the disagreement is purely semantic. A change taking 50,000 years is sudden to a palaeontologist but gradual to a population geneticist. My own guess is that there is not much more to the argument than that. However, the debate shows no signs of going away.

One question that arises is how far the new ideas are actually new. Much less so, I think, than their proponents would have us believe. They speak and write as if the orthodox view is that evolution occurs at a rate which is not only "gradual" but uniform. Yet George Gaylord Simpson, one of the main architects of the "modern synthesis" now under attack, wrote a book, *Tempo and Mode in Evolution,* devoted to emphasizing the great variability of evolutionary rates. It has never been part of the modern synthesis that evolutionary rates are uniform.

Yet there is a real point at issue. If it turns out to be the case that all, or most, evolutionary change is concentrated into brief periods, and associated with the splitting of lineages, that would require some serious rethinking. Oddly enough, it is not so much the sudden changes which would raise difficulties, but the intervening stasis. Why should a species remain unchanged for millions of years? The explanation favored by most punctuationists is that there are "developmental constraints" which must be overcome before a species can change. The suggestion is that the members of a given species share a developmental pathway which can be modified so as to produce some kinds of change in adult structure rather easily, and other kinds of change only with great difficulty, or not at all. I do not doubt that this is true: indeed, in my book *The Theory of Evolution,* published in 1958 and intended as a popular account of the modern synthesis, I spent some time emphasizing that "the pattern of development of a given species is such that there are only a limited number of ways in which it can be altered without causing complete breakdown." Neo-Darwinists have never supposed that genetic mutation is equally likely to produce changes in adult structure in any direction: all that is assumed is that mutations do not, as a general rule, adapt organisms to withstand the agents which caused them. What is at issue, then, is not whether there are developmental constraints, because clearly there are, but whether such constraints can account for stasis in evolution.

I find it hard to accept such an explanation for stasis, for two reasons. The first is that artificial selection can and does produce dramatic morphological change: one has only to look at the breeds of dogs to appreciate that. The second is that species are not uniform in space. Most species with a wide geographical range show differences between regions. Often these differences are so great that one does not know whether the extreme forms would behave as a single species if they met. Occasionally we know that they would not. This requires that a ring of forms should arise, with the terminal links overlapping. The Herring Gull and Lesser Black-Backed Gull afford a familiar example. In Britain and Scandinavia they behave as distinct species, without hybridizing, but they are linked by a series of forms encircling the Arctic.

Stasis in time is, therefore, a puzzle, since it seems not to occur in space. The simplest explanation is that species remain constant in time if their environments remain constant. It is also worth remembering that the hard parts of marine invertebrates, on which most arguments for stasis are based, tell us relatively little about the animals within. There are on our beaches two species of periwinkle whose shells are indistinguishable, but which do not interbreed and of which one lays eggs and the other bears live young.

The question of stasis and punctuation will be settled by a statistical analysis of the fossil record. But what of the wider issues? Is mutation plus natural selection within populations sufficient to explain evolution on a large scale, or must new mechanisms be proposed?

It is helpful to start by asking why Darwin himself was a believer in gradual change. The reason lies, I believe, in the nature of the problem he was trying to solve. For Darwin, the outstanding characteristic of living organisms which called for an explanation was the detailed way in which they are adapted to their forms of life. He knew that "sports"—structural novelties of large extent—did arise from time to time, but felt that fine adaptation could not be explained by large changes of this kind: it would be like trying to perform a surgical operation with a mechanically-controlled scalpel which could only be moved a foot at a time. Gruber has suggested that Darwin's equating of gradual with natural and of sudden with supernatural was a permanent feature of this thinking, which predated his evolutionary views and his loss of religious faith. It may have originated with Archbishop Sumner's argument (on which Darwin made notes when a student at Cambridge) that Christ must have been a divine rather than a human teacher because of the suddenness with which his teachings were accepted. Darwin seems to have retained the conviction that sudden changes are supernatural long after he had rejected Sumner's application of the idea.

Whatever the source of Darwin's conviction, I think he was correct both in his emphasis on detailed adaptation as the phenomenon to be explained, and in his conviction that to achieve such adaptation requires large numbers of selective events. It does not, however, follow that all the steps had to be small. I have always had a soft spot for "hopeful monsters": new types arising by genetic mutation, strikingly different in some respects from their parents, and taking a first step in the direction of some new adaptation, which could then be perfected by further smaller changes. We know that mutations of large effect occur: our only problem is whether they are ever incorporated during evolution, or are always eliminated by selection. I see no *a priori* reason why such large steps should not occasionally happen in evolution. What genetic evidence we have points the other way, however. On the relatively few occasions when related species differing in some morphological feature have been analyzed genetically, it has turned out, as Darwin would have expected had he known of the possibility, that the difference is caused by a number of genes, each of small effect.

As I see it, a hopeful monster would still stand or fall by the test of natural selection. There is nothing here to call for radical rethinking. Perhaps the greatest weakness of the punctuationists is their failure to suggest a plausible alternative mechanism. The nearest they have come is the hypothesis of "species selection." The idea is that when a new species arises, it differs from its ancestral species in ways which are random relative to any long-term evolutionary trends. Species will differ, however, in their likelihood of going extinct, and of splitting again to form new species. Thus selection will operate between species, favoring those characteristics which make extinction unlikely and splitting likely. In "species selection," as compared to classical individual selection, the species replaces the individual organism, extinction replaces death, the splitting of species into two replaces birth, and mutation is replaced by punctuational changes at the time of splitting.

Some such process must take place. I have argued elsewhere that it may have been a relevant force in maintaining sexual reproduction in higher animals. It is, however, a weak force compared to typical Darwinian between-individual selection, basically because the origin and extinction of species are rare events compared to the birth and death of individuals. Some critics of Darwinism have argued that the perfection of adaptation is too great to be accounted for by the selection of random mutations. I think, on quantitative grounds, that they are mistaken. If however, they were to use the same argument to refute species selection as the major cause of evolutionary trends, they might well be right. For punctuationists, one way out of the difficulty would be to argue that adaptation is in fact less precise than biologists have supposed. Gould has recently tried this road. As it happens, I think he is right to complain of some of the more fanciful adaptive explanations that have been offered, but I also think that he will find that the residue of genuine adaptive fit between structure and function is orders of magnitude too great to be explained by species selection.

One other extension of the punctuationist argument is worth discussing. As explained above, stasis has been explained by developmental constraints. This amounts to saying that the developmental processes are such that only certain kinds of animals are possible and viable. The extension is to apply the same idea to explain the existence of the major patterns of organization, or "bauplans," observable in the natural world. The existence of such bauplans is not at issue. For example, all vertebrates, whether swimming, flying, creeping or burrowing, have the same basic pattern of an internal jointed backbone with a hollow nerve cord above it and segmented body muscles on either side of it, and the vast majority have two pairs of fins, or of legs which are derived from fins (although a few have lost one or both pairs of appendages). Why should this be so?

Darwin's opinion is worth quoting. In *The Origin of Species,* he wrote:

> It is generally acknowledged that all organic beings have been formed on two laws—Unity of Type, and the Conditions of Existence. By unity of type is meant that fundamental agreement in structure which we see in organic beings of the same class, and which is quite independent of their habits of life. On my theory, unity of type is explained by unity of descent. The expression of conditions of existence, so often insisted on by the illustrious Cuvier, is fully embraced by the principle of natural selection. For natural selection acts by either now adapting the varying parts of each being to its organic and inorganic conditions of life; or by having adapted them during the long-past periods of time ... Hence, in fact, the law of Conditions of Existence is the higher law; as it includes, through the inheritance of former adaptations, that of Unity of Type.

That is, we have two pairs of limbs because our remote ancestors had two pairs of fins, and they had two pairs of fins because that is an efficient number for a swimming animal to have.

I fully share Darwin's opinion. The basic vertebrate pattern arose in the first place as an adaptation for sinusoidal swimming. Early fish have two pairs of fins for the same reason that most early aeroplanes had wings and tailplane: two pairs of fins is the smallest number that can produce an upward or downward force through any point in the body. In the same vein, insects (which are descended from animals with many legs) have six legs because that is the smallest number which permits an insect to take half its legs off the ground and not fall over.

The alternative view would be that there are (as yet unknown) laws of form or development which permit only certain kinds of organisms to exist—for example, organisms with internal skeletons, dorsal nerve cords and four legs, or with external skeletons, ventral nerve cords and six legs—and which forbid all others, in the same way that the laws of physics permit only elliptical planetary orbits, or the laws of chemistry permit only certain compounds. This view is a manifestation of the "physics envy" which still infects some biologists. I believe it to be mistaken. In some cases it is demonstrably false. For example, some of the earliest vertebrates had more than two pairs of fins (just as some early aeroplanes had a noseplane as well as a tailplane). Hence there is no general law forbidding such organisms.

What I have said about bauplans does not rule out the possibility that there may be a limited number of kinds of unit developmental processes which occur, and which are linked together in various ways to produce adult structures. The discovery of such processes would be of profound importance for biology, and would no doubt influence our views about evolution.

One last word needs to be said about bauplans. They may, as Darwin thought, have arisen in the first place as adaptations to particular ways of life, but, once having arisen, they have proved to be far more conservative in evolution than the way of life which gave them birth. Apparently it has been easier for organisms to adapt to new ways of life by modifying existing structures than by scrapping them and starting afresh. It is for this reason that comparative anatomy is a good guide to relationship.

Punctuationist views will, I believe, prove to be a ripple rather than a revolution. Their most positive achievement may be to persuade more people to study populations of fossils with adequate statistical methods. In the meanwhile, those who would like to believe that Darwin is dead, whether because they are creationists, or because they dislike the apparently Thatcherite conclusions which have been drawn from his theory, or find the mathematics of population genetics too hard for them, would be well advised to be cautious: the reports of his death have been exaggerated.

Chapter 10

Chapter 10 takes on the science-religion relationship. The first document is an extract taken from a book by the leader of the New Creationists, Berkeley lawyer Phillip Johnson. In Darwin on Trial, *the old evolutionist is charged, tried, found guilty, and led away in chains. Nothing is left standing. Evolution is false, Darwinism is trivial, and the underlying philosophy of materialism is shown to be shallow and dangerous. Those who think one can sup with the devil of evolutionism had better use a mighty long spoon. In fact, Johnson thinks that no cutlery will do and implores us to turn from the vile, seductive doctrine. In the passage given here, Johnson raises one of the charges frequently made against the Darwinian mechanism of natural selection: it is a truism or a "tautology." It is little more than a necessary redescription, like saying that a bachelor is an unmarried man—and then being told that the definition of an unmarried man is a "bachelor"! Supposedly natural selection is equivalent to the "survival of the fittest," but who are the fittest other than those that survive—which means that natural selection means no more than that those who survive are those that survive! True, but not that informative.*

As you might imagine, Darwinians have counters to this line of argument. It really is silly to pretend that it is just an empty truism to say that there is a struggle for existence—tell that to

Dawkins's antelope. It is no less silly to say that it is just an empty truism that some of those who succeed have variations that help them and that the losers do not have the variations. What is tricky is the Darwinian's underlying assumption that natural selection is a regular sort of thing—it is not just random what helps on one occasion rather than another. Hence you can identify a butterfly, say, with a certain kind of camouflage as being fitter than one without it. And here we do have a labeling, a definition. But this is not to say that we have here a mere tautology. After all it could be false. Indeed, if Sewall Wright's genetic drift has any truth, then it is sometimes false that the fitter survive and reproduce. The very notion of drift is that natural selection does not always work.

The second document of the section is a letter written and disseminated by Pope John Paul II in 1997. Although doctrinally very conservative, this pope has always been friendly toward science, perhaps in part because of his pride in his fellow countryman and (as was he himself) former professor at Krakow University in Poland: Nicholas Copernicus. In his letter John Paul takes on the whole question of evolution, arguing that not only is there now good evidence of evolution but that modern theories of change (natural selection?) are now well supported. Where he does not give an inch is over the question of human souls. These are created and put in place miraculously. On the one hand, I suppose this is no worry for the evolutionist. Souls are clearly not scientific entities and thus are outside the bounds of scientific (including evolutionary) explanation. On the other hand, I think this is a worry for the evolutionist. Miraculous creation and insertion of anything seems to go against the spirit of evolution, Darwinism particularly. If you think that the mind came through evolution, do you need a soul in addition? What price Christianity now? At the least, more work is needed on the science-religion relationship.

Finally, in the third and concluding document, we have Richard Dawkins's response to the pope. It needs little comment here. It is violently antireligion and shows no interest in attempts by believers to forge a friendly relationship between science and religion. "Cowardly flabbiness of the intellect" is a phrase that stays with one long after the details of the critique are forgotten. As you know, this is hardly the attitude that I have taken in this book. Now it is up to you to judge which of us two, Ruse or Dawkins, takes the correct approach to these matters.

Darwin on Trial

P. E. Johnson

Natural Selection as a Tautology

Many of the most prominent neo-Darwinists have written at one time or another that natural selection is a tautology, a way of saying the same thing twice. In this formulation the theory predicts that the fittest organisms will produce the most offspring, and it defines the fittest organisms as the ones which produce the most offspring. It is important to document this point, because many Darwinists have convinced themselves that the tautology idea is a misunderstanding introduced into the literature by creationists and other uncomprehending faultfinders. But here are a few examples collected by Norman Macbeth:

> J. B. S. Haldane (1935): "... the phrase, 'survival of the fittest,' is something of a tautology. So are most mathematical theorems. There is no harm in saying the same truth in two different ways."

Ernst Mayr (1963): "... those individuals that have the most offspring are by definition ... the fittest ones."

George Gaylord Simpson (1964): "Natural selection favors fitness only if you define fitness as leaving more descendants. In fact geneticists do define it that way, which may be confusing to others. To a geneticist fitness has nothing to do with health, strength, good looks, or anything but effectiveness in breeding."

The explanation by Simpson just quoted indicates why it is not easy to formulate the theory of natural selection other than as a tautology. It may seem obvious, for example, that it is advantageous for a wild stallion to be able to run faster, but in the Darwinian sense this will be true only to the extent that a faster stallion sires more offspring. If greater speed leads to more frequent falls, or if faster stallions tend to outdistance the mares and miss opportunities for reproduction, then the improvement may be disadvantageous.

Just about any characteristic can be either advantageous or disadvantageous, depending upon the surrounding environmental conditions. Does it seem that the ability to fly is obviously an advantage? Darwin hypothesized that natural selection might have caused beetles on Madeira to lose the ability to fly, because beetles capable of flight tended to be blown out to sea. The large human brain requires a large skull which causes discomfort and danger to the mother in childbirth. We assume that our brain size is advantageous because civilized humans dominate the planet, but it is far from obvious that the large brain was a net advantage in the circumstances in which it supposedly evolved. Among primates in general, those with the largest brains are not the ones least in danger of extinction.

In all such cases we can presume a characteristic to be advantageous because a species which has it seems to be thriving, but in most cases it is impossible to identify the advantage independently of the outcome. That is why Simpson was so insistent that "advantage" has no inherent meaning other than actual success in reproduction. All we can say is that the individuals which produced the most offspring must have had the qualities required for producing the most offspring.

The famous philosopher of science Karl Popper at one time wrote that Darwinism is not really a scientific theory because natural selection is an all-purpose explanation which can account for anything, and which therefore explains nothing. Popper backed away from this position after he was besieged by indignant Darwinist protests, but he had plenty of justification for taking it. As he wrote in his own defense, "some of the greatest contemporary Darwinists themselves formulate the theory in such a way that it amounts to the tautology that those organisms that leave most offspring leave most offspring," citing Fisher, Haldane, Simpson, "and others." One of the others was C. H. Waddington, whose attempt to make sense of the matter deserves to be preserved for posterity:

> Darwin's major contribution was, of course, the suggestion that evolution can be explained by the natural selection of random variations. Natural selection, which was at first considered as though it were a hypothesis that was in need of experimental or observational confirmation, turns out on closer inspection to be a tautology, a statement of an inevitable but previously unrecognized relation. It states that the fittest individuals in a population (de-

fined as those which leave most offspring) will leave most offspring. This fact in no way reduces the magnitude of Darwin's achievement; only after it was clearly formulated, could biologists realize the enormous power of the principle as a weapon of explanation.

That was not an offhand statement, but a considered judgment published in a paper presented at the great convocation at the University of Chicago in 1959 celebrating the hundredth anniversary of the publication of *The Origin of Species*. Apparently, none of the distinguished authorities present told Waddington that a tautology does not explain anything. When I want to know how a fish can become a man, I am not enlightened by being told that the organisms that leave the most offspring are the ones that leave the most offspring.

It is not difficult to understand how leading Darwinists were led to formulate natural selection as a tautology. The contemporary neo-Darwinian synthesis grew out of population genetics, a field anchored in mathematics and concerned with demonstrating how rapidly very small mutational advantages could spread in a population. The advantages in question were assumptions in a theorem, not qualities observed in nature, and the mathematicians naturally tended to think of them as "whatever it was that caused the organism and its descendants to produce more offspring than other members of the species." This way of thinking spread to the zoologists and paleontologists, who found it convenient to assume that their guiding theory was simply true by definition. As long as outside critics were not paying attention, the absurdity of the tautology formulation was in no danger of exposure.

What happened to change this situation was that Popper's comment received a great deal of publicity, and creationists and other unfriendly critics began citing it to support their contention that Darwinism is not really a scientific theory. The Darwinists themselves became aware of a dangerous situation, and thereafter critics raising the tautology claim were firmly told that they were simply demonstrating their inability to understand Darwinism.

Message to Pontifical Academy of Sciences, October 22, 1996

Pope John Paul II

To the Members of the Pontifical Academy of Sciences meeting in the Plenary Session:
    With great pleasure I address cordial greetings to you, Mr President, and to all of you who constitute the Pontifical Academy of Sciences, on the occasion of your plenary assembly. I offer my best wishes in particular to the new academicians, who have come to take part in your work for the first time. I would also like to remember the academicians who died during the past year, whom I commend to the Lord of life.
    1. In celebrating the 60th anniversary of the Academy's refoundation, I would like to recall the intentions of my predecessor Pius XI, who wished to surround himself with a select group of scholars, relying on them to inform the Holy See in complete freedom about developments in scientific research, and thereby to assist him in his reflections.
    He asked those whom he called the Church's *Senatus scientificus* to serve the truth. I again extend this same invitation to you today, certain that we will all be able to profit

from the fruitfulness of a trustful dialogue between the Church and science (cf. *Address to the Academy of Sciences,* n. 1, 28 October 1986, *L'Osservatore Romano* English edition, 24 November 1986, p. 22).

Science at the Dawn of the Third Millennium

2. I am pleased with the first theme you have chosen, that of the origins of life and evolution, an essential subject which deeply interests the Church, since Revelation, for its part, contains teaching concerning the nature and origins of man. How do the conclusions reached by the various scientific disciplines coincide with those contained in the message of Revelation? And if, at first sight, there are apparent contradictions, in what direction do we look for their solution? We know, in fact, that truth cannot contradict truth (cf. Leo XIII, Encyclical *Providentissimus Deus*). Moreover, to shed greater light on historical truth, your research on the Church's relations with science between the 16th and 18th centuries is of great importance.

During this plenary session, you are undertaking a "reflection on science at the dawn of the third millennium," starting with the identification of the principal problems created by the sciences and which affect humanity's future. With this step you point the way to solutions which will be beneficial to the whole human community. In the domain of inanimate and animate nature, the evolution of science and its applications gives rise to new questions. The better the Church's knowledge is of their essential aspects, the more she will understand their impact. Consequently, in accordance with her specific mission she will be able to offer criteria for discerning the moral conduct required of all human beings in view of their integral salvation.

3. Before offering you several reflections that more specifically concern the subject of the origin of life and its evolution, I would like to remind you that the Magisterium of the Church has already made pronouncements on these matters within the framework of her own competence. I will cite here two interventions.

In his Encyclical *Humani generis* (1950), my predecessor Pius XII had already stated that there was no opposition between evolution and the doctrine of the faith about man and his vocation, on condition that one did not lose sight of several indisputable points (cf. AAS 42 [1950], pp. 575–576).

For my part, when I received those taking part in your Academy's plenary assembly on 31 October 1992, I had the opportunity, with regard to Galileo, to draw attention to the need of a rigorous hermeneutic for the correct interpretation of the inspired word. It is necessary to determine the proper sense of Scripture, while avoiding any unwarranted interpretations that make it say what it does not intend to say. In order to delineate the field of their own study, the exegete and the theologian must keep informed about the results achieved by the natural sciences (cf. AAS 85 [1993], pp. 764–772; *Address to the Pontifical Biblical Commission,* 23 April 1993, announcing the document on *The interpretation of the Bible in the Church:* AAS 86 [1994], pp. 232–243).

Evolution and the Church's Magisterium

4. Taking into account the state of scientific research at the time as well as of the requirements of theology, the Encyclical *Humani generis* considered the doctrine of "evolutionism" a serious hypothesis, worthy of investigation and in-depth study equal to that of the

opposing hypothesis. Pius XII added two methodological conditions: that this opinion should not be adopted as though it were a certain, proven doctrine and as though one could totally prescind from Revelation with regard to the questions it raises. He also spelled out the condition on which this opinion would be compatible with the Christian faith, a point to which I will return.

Today, almost half a century after the publication of the Encyclical, new knowledge has led to the recognition of more than one hypothesis in the theory of evolution. It is indeed remarkable that this theory has been progressively accepted by researchers, following a series of discoveries in various fields of knowledge. The convergence, neither sought nor fabricated, of the results of work that was conducted independently is in itself a significant argument in favour of this theory.

What is the significance of such a theory? To address this question is to enter the field of epistemology. A theory is a metascientific elaboration, distinct from the results of observation but consistent with them. By means of it a series of independent data and facts can be related and interpreted in a unified explanation. A theory's validity depends on whether or not it can be verified; it is constantly tested against the facts; wherever it can no longer explain the latter, it shows its limitations and unsuitability. It must then be rethought.

Furthermore, while the formulation of a theory like that of evolution complies with the need for consistency with the observed data, it borrows certain notions from natural philosophy.

And, to tell the truth, rather than *the* theory of evolution, we should speak of *several* theories of evolution. On the one hand, this plurality has to do with the different explanations advanced for the mechanism of evolution, and on the other, with the various philosophies on which it is based. Hence the existence of materialist, reductionist and spiritualist interpretations. What is to be decided here is the true role of philosophy and, beyond it, of theology.

5. The Church's Magisterium is directly concerned with the question of evolution, for it involves the conception of man: Revelation teaches us that he was created in the image and likeness of God (cf. Gn 1:27–29). The conciliar Constitution *Gaudium et spes* has magnificently explained this doctrine, which is pivotal to Christian thought. It recalled that man is "the only creature on earth that God has wanted for its own sake" (n. 24). In other terms, the human individual cannot be subordinated as a pure means or a pure instrument, either to the species or to society, he has value *per se*. He is a person. With his intellect and his will, he is capable of forming a relationship of communion, solidarity and self-giving with his peers. St Thomas observes that man's likeness to God resides especially in his speculative intellect, for his relationship with the object of his knowledge resembles God's relationship with what he has created (*Summa Theologica*, I-II, q. 3, a. 5, ad 1). But even more, man is called to enter into a relationship of knowledge and love with God himself, a relationship which will find its complete fulfilment beyond time, in eternity. All the depth and grandeur of this vocation are revealed to us in the mystery of the risen Christ (cf. *Gaudium et spes*, n. 22). It is by virtue of his spiritual soul that the whole person possesses such a dignity even in his body. Pius XII stressed this essential point: if the human body takes its origin from pre-existent living matter, the spiritual soul is immediately created by God ("*animas enim a Deo immediate creari catholica fides nos retinere iubet*"; Encyclical *Humani generis*, AAS 42 [1950], p. 575).

Consequently, theories of evolution which, in accordance with the philosophies inspiring them, consider the mind as emerging from the forces of living matter, or as a mere epiphenomenon of this matter, are incompatible with the truth about man. Nor are they able to ground the dignity of the person.

6. With man, then, we find ourselves in the presence of an ontological difference, an ontological leap, one could say. However, does not the posing of such ontological discontinuity run counter to that physical continuity which seems to be the main thread of research into evolution in the field of physics and chemistry? Consideration of the method used in the various branches of knowledge makes it possible to reconcile two points of view which would seem irreconcilable. The sciences of observation describe and measure the multiple manifestations of life with increasing precision and correlate them with the time line. The moment of transition into the spiritual cannot be the object of this kind of observation, which nevertheless can discover at the experimental level a series of very valuable signs indicating what is specific to the human being. But the experience of metaphysical knowledge, of self-awareness and self-reflection, of moral conscience, freedom, or again, of aesthetic and religious experience, falls within the competence of philosophical analysis and reflection, while theology brings out its ultimate meaning according to the Creator's plans.

We Are Called to Enter Eternal Life

7. In conclusion, I would like to call to mind a Gospel truth which can shed a higher light on the horizon of your research into the origins and unfolding of living matter. The Bible in fact bears an extraordinary message of life. It gives us a wise vision of life inasmuch as it describes the loftiest forms of existence. This vision guided me in the Encyclical which I dedicated to respect for human life, and which I called precisely *Evangelium vitae.*

It is significant that in St John's Gospel life refers to the divine light which Christ communicates to us. We are called to enter into eternal life, that is to say, into the eternity of divine beatitude.

To warn us against the serious temptations threatening us, our Lord quotes the great saying of *Deuteronomy:* "Man shall not live by bread alone, but by every word that proceeds from the mouth of God" (Dt 8:3, cf. Mt 4:4).

Even more, "life" is one of the most beautiful titles which the Bible attributes to God. He is the *living God.*

I cordially invoke an abundance of divine blessings upon you and upon all who are close to you.

From the Vatican, 22 October 1996.

Obscurantism to the Rescue

Richard Dawkins

A cowardly flabbiness of the intellect afflicts otherwise rational people confronted with long-established religions (though, significantly, not in the face of younger traditions such as Scientology or the Moonies). S. J. Gould (1997), commenting in his *Natural History* column on the Pope's attitude to evolution, is representative of a dominant strain of conciliatory thought, among believers and non-believers alike:

> Science and religion are not in conflict, for their teachings occupy distinctly different domains (p 16).

I believe, with all my heart, in a respectful, even *loving* concordat ... (p 60; my emphasis).

Well, what are these two distinctly different domains, these "nonoverlapping magisteria" that should snuggle up together in a respectful and loving concordat? Gould again:

The net of science covers the empirical universe: what is it made of (fact) and why does it work this way (theory). The net of religion extends over questions of moral meaning and value (p 60).

Would that it were that tidy. In a moment I'll look at what the Pope actually says about evolution, and then at other claims of his church, to see if they really are so neatly distinct from the domain of science. First though, a brief aside on the claim that religion has some special expertise to offer us on moral questions. This is often blithely accepted even by the non-religious, presumably in the course of a civilized "bending over backwards" to concede the best point your opponent has to offer—however weak that best point may be.

The question, "What is right and what is wrong?" is a genuinely difficult question that science certainly cannot answer. Given a moral premise or a priori moral belief, the important and rigorous discipline of secular moral philosophy can pursue scientific or logical modes of reasoning to point up hidden implications of such beliefs, and hidden inconsistencies between them. But the absolute moral premises themselves must come from elsewhere, presumably from unargued conviction. Or, it might be hoped, from religion—meaning some combination of authority, revelation, tradition and scripture.

Unfortunately, the hope that religion might provide a bedrock, from which our otherwise sand-based morals can be derived, is a forlorn one. In practice no civilized person uses scripture as ultimate authority for moral reasoning. Instead, we pick and choose the nice bits of scripture (like the Sermon on the Mount) and blithely ignore the nasty bits (like the obligation to stone adulteresses, execute apostates and punish the grandchildren of offenders). The God of the Old Testament himself, with his pitilessly vengeful jealousy, his racism, sexism and terrifying bloodlust, will not be adopted as a literal role model by anybody you or I would wish to know. Yes, *of course* it is unfair to judge the customs of an earlier era by the enlightened standards of our own. But that is precisely my *point!* Evidently, we have some alternative source of ultimate moral conviction which overrides scripture when it suits us.

That alternative source seems to be some kind of liberal consensus of decency and natural justice, which changes over historical time, frequently under the influence of secular reformists. Admittedly, that doesn't sound like bedrock. But in practice we, including the religious among us, give it higher priority than scripture. In practice we more or less ignore scripture, quoting it when it supports our liberal consensus, quietly forgetting it when it doesn't. And, wherever that liberal consensus comes from, it is available to all of us, whether we are religious or not.

Similarly, great religious teachers like Jesus or Gautama Buddha may inspire us, by their good example, to adopt their personal moral convictions. But again we pick and choose among religious leaders, avoiding the bad examples of Jim Jones or Charles Manson, and we may choose good secular role models such as Jawaharlal Nehru or Nelson Mandela. Traditions too, however anciently followed, may be good or bad, and we use our secular judgment of decency and natural justice to decide which ones to follow, which to give up.

But that discussion of moral values was a digression. I now turn to my main topic of evolution, and whether the Pope lives up to the ideal of keeping off the scientific grass. His Message on Evolution to the Pontifical Academy of Sciences begins with some casuistical doubletalk designed to reconcile what John Paul is about to say with the previous, more equivocal pronouncements of Pius XII, whose acceptance of evolution was comparatively grudging and reluctant. Then the Pope comes to the hard task of reconciling scientific evidence with "revelation."

> Revelation teaches us that [man] was created in the image and likeness of God ... if the human body takes its origin from pre-existent living matter, the spiritual soul is immediately created by God. ... Consequently, theories of evolution which, in accordance with the philosophies inspiring them, consider the mind as emerging from the forces of living matter, or as a mere epiphenomenon of this matter, are incompatible with the truth about man. ... With man, then, we find ourselves in the presence of an ontological difference, an ontological leap, one could say (John Paul II 1996, this issue p 383).

To do the Pope credit, at this point he recognizes the essential contradiction between the two positions he is attempting to reconcile:

> However, does not the posing of such ontological discontinuity run counter to that physical continuity which seems to be the main thread of research into evolution in the field of physics and chemistry? (John Paul II 1996, this issue p 383).

Never fear. As so often in the past, obscurantism comes to the rescue:

> Consideration of the method used in the various branches of knowledge makes it possible to reconcile two points of view which would seem irreconcilable. The sciences of observation describe and measure the multiple manifestations of life with increasing precision and correlate them with the time line. The moment of transition to the spiritual cannot be the object of this kind of observation, which nevertheless can discover at the experimental level a series of very valuable signs indicating what is specific to the human being (John Paul II 1996, this issue p 383).

In plain language, there came a moment in the evolution of hominids when God intervened and injected a human soul into a previously animal lineage (When? A million years ago? Two million years ago? Between *Homo erectus* and *Homo sapiens*? Between "archaic" *Homo sapiens* and *H. sapiens sapiens*?). The sudden injection is necessary, of course, otherwise there would be no distinction upon which to base Catholic morality, which is speciesist to the core. You can kill adult animals for meat, but abortion and euthanasia are murder because human life is involved.

Catholicism's "net" is not limited to moral considerations, if only because Catholic morals have scientific implications. Catholic morality demands the presence of a great gulf between *Homo sapiens* and the rest of the animal kingdom. Such a gulf is fundamentally antievolutionary. The sudden injection of an immortal soul in the time-line is an antievolutionary intrusion into the domain of science.

More generally, it is completely unrealistic to claim, as Gould and many others do, that religion keeps itself away from science's turf, restricting itself to morals and values. A universe with a supernatural presence would be a fundamentally and qualitatively different kind of universe from one without. The difference is, inescapably, a scientific difference. Religions make existence claims, and this means scientific claims.

The same is true of many of the major doctrines of the Roman Catholic Church. The Virgin Birth, the bodily Assumption of the Blessed Virgin Mary, the Resurrection of Jesus, the survival of our own soul after death: these are all claims of a clearly scientific nature. Either Jesus had a corporeal father or he didn't. This is not a question of "values" or "morals"; it is a question of sober fact. We may not have the evidence to answer it, but it is a scientific question, nevertheless. You may be sure that, if any evidence supporting the claim were discovered, the Vatican would not be reticent in promoting it.

Either Mary's body decayed when she died, or it was physically removed from this planet to Heaven. The official Roman Catholic doctrine of Assumption, promulgated as recently as 1950, implies that Heaven has a physical location and exists in the domain of physical reality—how else could the physical body of a woman go there? I am not, here, saying that the doctrine of the Assumption of the Virgin is necessarily false (although of course I think it is). I am simply rebutting the claim that it is outside the domain of science. On the contrary, the Assumption of the Virgin is transparently a scientific theory. So is the theory that our souls survive bodily death, and so are all stories of angelic visitations, Maryan manifestations and miracles of all types.

There is something dishonestly self-serving in the tactic of claiming that all religious beliefs are outside the domain of science. On the one hand miracle stories and the promise of life after death are used to impress simple people, win converts and swell congregations. It is precisely their scientific power that gives these stories their popular appeal. But at the same time it is considered below the belt to subject the same stories to the ordinary rigors of scientific criticism: these are religious matters and therefore outside the domain of science. But you cannot have it both ways. At least, religious theorists and apologists should not be allowed to get away with having it both ways. Unfortunately all too many of us, including nonreligious people, are unaccountably ready to let them get away with it.

I suppose it is gratifying to have the Pope as an ally in the struggle against fundamentalist creationism. It is certainly amusing to see the rug pulled out from under the feet of Catholic creationists such as Michael Behe. Even so, given a choice between honest to goodness fundamentalism on the one hand, and the obscurantist, disingenuous doublethink of the Roman Catholic Church on the other, I know which I prefer.

References

Gould, S. J. 1997. Nonoverlapping magisteria: science and religion are not in conflict, for their teachings occupy distinctly different domains. *Natural History* 3 March: 16.

John Paul II. 1996. Message to Pontifical Academy of Sciences, 22 October. *L'Osservatore Romano* 30 October: 3, 7.

Illustration and Document Credits and Permissions

Chapter 1

Illustrations

2　Archive Photos.

3　Archive Photos.

6　National Portrait Gallery, London.

7　Wellcome Institute Library, London.

10　From Edward Drinker Cope, *The Vertebrata of the Tertiary Formations of the West* (New York: Arno Press, 1883).

15　From Richard Owen, *Palaeontology* (1860).

17　Wellcome Institute Library, London.

19　From Ramon Lull, *Ladder of Ascent and Descent of the Mind* (1305; first printed edition 1512). Used in Ruse, *Monad to Man* (Harvard, 1996).

22　From M. J. S. Rudwick, *The Meaning of Fossils* (London: Macdonald/New York: Elsevier, 1972).

Chapter 2

Illustrations

30　*Charles Darwin, 1840* by George Richmond. Used by kind permission of the Charles Darwin Museum, Down House. Down House and its contents are now owned and managed by English Heritage, Keysign House, London.

32　Darwin Archive, Cambridge University Library.

34　National Maritime Museum.

35　From Charles Lyell, *Principles of Geology* (New York: D. Appleton, 1872).

38　From David Lack, *Darwin's Finches* (Cambridge: Cambridge University Press, 1947).

39　Archive Photos.

40　From Thomas H. Huxley, *Man's Place in Nature and Other Anthropological Essays*, Collected Essays Vol. 7 (New York: Appleton, 1903).

43　From Charles Darwin, *Origin of Species by Means of Natural Selection, or the Preservation of Favored Races in the Struggle for Life* (Charles Darwin, 1872).

Chapter 3

Illustrations

Chapter 4

Illustrations

111   Bryan College Archives.
112   Bryan College Archives.

Chapter 5

Illustrations

118   Austrian Information Service.
119   Corbis-Bettmann.
120   From W. Bateson, *Mendel's Principles of Heredity* (Cambridge: Cambridge University Press, 1913).
122   Courtesy of T. D. Lysenko.
123   Courtesy of Sewall Wright.
125   Courtesy Columbia University Press.
128   G. G. Simpson, *Tempo and Mode in Evolution* (New York: Columbia University Press, 1944, p. 92).
129   Natural History Museum, London.
141   Camera Press Ltd.
142   From the author's collection.

Documents

Sewall Wright, "The Roles of Mutation, Inbreeding, Crossbreeding and Selection in Evolution," *Proceedings of the Sixth International Congress of Genetics* 1932, 356–366. Reprinted with permission of the Genetics Society of America.

Ernst Mayr to G. F. Ferris, March 29, 1948. Reprinted with the permission of the American Philosophical Society.

Theodosius Dobzhansky to John Greene, November 23, 1961. Reprinted with the permission of the American Philosophical Society.

Chapter 6

Illustrations

146   University of Strasbourg.
149   Wellcome Institute Library, London.
155   From Sir C. Wyville Thomson, *The Depths of the Sea: An Account of the General Results of the Dredging Cruises of H.M.Ss. 'Porcupine' and 'Lightning' during the Summers of 1868, 1869, and 1870, under the Scientific Direction of Dr. Carpenter, F.R.S., J. Gwyn Jeffreys, F.R.S., and Dr. Wyville Thomson, F.R.S.* (New York, London: Macmillan, 1873).
157   From the author's collection.
158   Royal College of Surgeons.
160   From T. Dobzhansky, F. J. Ayala, G. L. Stebbins, and J. W. Valentine, *Evolution* (San Francisco: W. H. Freeman, 1977).
166   From M. Glaessner and M. Wade, "The late precambrian fossils from Ediacara, South Australia." *Paleontology* 9 (1966): 599–628. Used by permission of the authors and the Paleontological Association.

167　From J. William Schopf, "The Evolution of the Earliest Cells." *Scientific American* 239:3 (September 1978): 110–138.

168　From J. W. Valentine, "The evolution of multicellular plants and animals." *Scientific American* 239:3 (September 1978), p.142.

Chapter 7

Illustrations

172　Ira Block.

173　Collection of Roger Lewin.

174　Ira Block.

176　From T. H. Huxley, *Evidence as to Man's Place in Nature* (New York: D. Appleton, 1863).

177　From Ernst Haeckel, *The Last Link: Our Present Knowledge of the Descent of Man,* 2d. ed. (London: Black, 1899).

178　John Reader/Science Photo Library.

179　Corbis-Bettmann.

181　Corbis-Bettmann.

195　Courtesy of Jeffrey Laitman.

Documents

Reprinted from Charles Darwin: *Charles Darwin's Notebooks, 1836–1844: Geology, Transmutation of Species, Metaphysical Enquiries.* Edited by Paul H. Barrett, Peter J. Gautry, Sandra Herbert, David Kohn, and Sydney Smith. Copyright © 1987 by Paul H. Barrett, Peter J. Gautrey, Sandra Herbert, David Kohn, Sydney Smith. Used by permission of the publisher, Cornell University Press.

Donald C. Johanson and Maitland A. Edey: Reprinted with permission of Simon and Schuster from *Lucy: The Beginnings of Humankind* by Donald C. Johanson and Maitland A. Edey. Copyright © 1981 by Donald C. Johanson and Maitland A. Edey.

Chapter 8

Illustrations

204　Harvard University, photo by Jon Chase.

206　From Charles Darwin, *The Descent of Man and Selection in Relation to Sex* (New York: Appleton, 1883).

209　From the author's collection, photo by Jane Clelland.

211　From the author's collection.

212　From the author's collection.

214　From the author's collection.

215　Courtesy of Geoffrey Parker.

219　From the author's collection.

223 From Howells, *Scientific American,* 1960.
224 Courtesy of Sarah Hrdy.
225 From the author's collection.

Documents

Science for the People Sociobiological Study Group, "Sociobiology: Another Biological Determinism," *BioScience* 26,182–186. Copyright © 1976 American Institute of Biological Sciences.

E. O. Wilson, "Academic Vigilantism and the Political Significance of Sociobiology," *BioScience* 26, 183–190. Copyright © 1976 American Institute of Biological Sciences.

M. Ruse and E. O. Wilson, "The Evolution of Ethics," *New Scientist* 1478 (1985): 108–128. Reprinted with the permission of the publisher.

Chapter 9

Illustrations

232 Harvard University, photo by Jon Chase.
233 From Ruse, Michael, *Darwinism Defended* (Addison-Wesley, 1982).
236 From Richard Owen, *On the Nature of Limbs* (1849). Reproduced in F. S. Russell, *Form and Function* (London: John Murray, 1816), and in Owen, *Anatomy of Vertebrates* (London: Longmans, Green, 1866, 3 volumes; reprint New York: AMS, 1973).
237 From S. J. Gould and R. C. Lewontin, "The Spandrels of San Marco and the Panglossian Paradigm: A Critique of the Adaptationist Program," *Proceedings of the Royal Society of London, B: Biological Sciences* 205 (1979): 581–598.
241 From S. J. Gould, "Morphological channeling by structural constraint: convergence in styles of dwarfing and gigantism in *Cerion,* with a description of two new fossil species and a report on the discovery of the largest *Cerion.*" *Paleobiology* 10(2):173.
242 Photograph courtesy of the Smithsonian Institution (Dr. D. Erwin).
243 Photograph courtesy of the Smithsonian Institution (Dr. D. Erwin).
250 From S. J. Gould, *Ontogeny and Phylogeny* (Belknap/Harvard, 1977). Hypermorphosis in the Irish elk. Figure from J. G. Millais, *Mammals of Great Britain and Ireland* (London: Longmans, Green, 1906).
252 From Lynn Margulis and Dorian Sagan, *What Is Life?* (Simon and Schuster, 1995).
256 Courtesy of John Sepkoski.
257 From J. J. Sepkoski, "A Kinetic Model of Phanerozoic Taxonomic Diversity II: Early Paleozoic Families and Mass Extinctions." *Paleobiology* 5 (1979): 222–252.

Documents

"The Episodic Nature of Evolutionary Change," pp. 179–185 of *The Panda's Thumb: More Reflections in Natural History* by Stephen Jay Gould. Copyright © 1980 by Stephen Jay Gould. Used by permission of W. W. Norton and Company, Inc.

J. Maynard Smith, "Did Darwin Get It Right?" *London Review of Books* 3 (11) (1981): 10–11. Reprinted with the permission of the publisher.

Chapter 10

Illustrations

262   Courtesy of Michael Ruse.
266   Courtesy of Philip Johnson.
268   Sally A. Morgan; Ecoscene/Corbis.
278   Fritz Polking; Frank Lane Picture Agency/Corbis.

Documents

P. E. Johnson, *Darwin on Trial*. Washington D.C.: Regnery Gateway. Reprinted with the permission of the Publisher.

John Paul II, "The Pope's Message on Evolution," *The Quarterly Review of Biology* 72 (1997): 377–383. English translation ("Magisterium Is Concerned with Question of Evolution for It Involves Conception of Man") reprinted courtesy L'Osservatore Romano.

R. Dawkins, "Obscurantism to the rescue." *Quarterly Review of Biology* 72 (1997): 397–399. Copyright © The University of Chicago Press. Reprinted courtesy of The University of Chicago Press and the author.

Index

Lucy *(Australopithecus afarensis)*,
181–182, 181(fig.), 200–201,
365(doc.), 370–374(doc.)
Lyell, Charles, 37, 175
Darwin and gradualism,
393–394(doc.)
deism, 34
fossil record, 19
grand theory of climate, 35–36
Huxley on, 330(doc.)
influence on Tennyson, 315(doc.)
Lysenko, Trofim, 122(fig.), 134, 143,
158
The Lysenko Affair (Joravsky), 143

MacArthur, Robert, 256
MacArthur-Wilson hypothesis, 256
Macbeth, Norman, 403(doc.)
Macroevolution (Stanley), 398(doc.)
Malthus, Thomas Robert, 40, 44–45
Marcuse, Herbert, 388(doc.)
Marine fauna, 256–257
Marrella, 243(fig.)
Marsh, Othniel Charles, 99–102,
100(fig.)
Martin, H. N., 66
Marx, Karl, 80, 81(fig.)
Marxism, 158–159, 244–245,
393(doc.)
Mastodon, 22(fig.)
Materialism, 21, 150, 283, 360(doc.),
364(doc.)
Mathews, Shailer, 110–111
Maynard Smith, John, 214, 214(fig.),
231–232, 247, 251,
397–402(doc.)
Mayr, Ernst, 404(doc.)
founder principle of speciation,
234–235
on good scientific methodology,
346(doc.), 355–356(doc.)
gradualism, 127–128
hostility to molecular biology, 140
synthetic theory, 130
McCosh, James, 76
Mead, Margaret, 208
*The Meaning of Evolution: A Study of the
History of Life* (Simpson), 133
Medawar, Peter, 132
Medical profession, 65–66
Mencken, H. L., 90, 113
Mendel, Gregor, 118, 118(fig.)
Mendelian genetics, 118–120
Metabolism, 166–167
Metaphysical naturalism, 269–272
Methodological naturalism, 269–272
Meyranx, Pierre-Stanislas, 1
Microevolutionary processes, 127
Miller, Kenneth, 285
Miller, Stanley, 159–160
Mind, concept of, 368–369(doc.)

Miracles, 271–274
Missing links, 5–6, 68, 175–176,
267–268, 268(fig.)
Mitochondrial Eve hypothesis,
191–192
Mivart, St. George, 73
Modernism, 111
Molecular biology, 140–143
ape-human divergence, 182–183
Creationist perspective, 267–268
mitochondrial Eve hypothesis,
191–192
molecular drift, 240, 248–249
Molecular drift, 240, 248–249
Monistic approach to consciousness,
199–200
Moore, Aubrey, 106
Morals and values
altruism as biological behavior,
390–392(doc.)
cruelty of nature, 277–278
evolution of ethics and morals,
217, 368–369(doc.),
375(doc.), 409–410(doc.)
political aspects of selection,
77–78
predestination, 77
Social Darwinism and, 73–74
Morgan, Thomas Hunt, 118, 119(fig.)
Morris, Henry M., 264–265
Morris, Simon Conway, 259
Moule interieur (internal living force),
148–149
Muller, Herman J.
antimilitary stance, 139
bacteriophages, 361(doc.)
balance-classical dispute, 142
classical hypothesis, 117–118
radiation-incurred variations,
120–121, 137–138
Multiregional theory, of human
evolution, 174, 189–194,
189(fig.), 193
Multivariate analysis, 190–192
Museums, 71, 93, 102–105
Mutations, 267
adaptive landscape theory,
347(doc.), 350–351(doc.)
balanced superior heterozygote
fitness, 136–138
Darwin on, 366–367(doc.)
developmental constraints and,
399(doc.)
good vs. bad mutations, 140
hopeful monsters, 238, 400(doc.)
pseudomutant strategy, 253–254
saltationism and, 121
Mystery of Mysteries (Ruse), 27

National Academy of Sciences, 100,
131

Natural selection, 41, 45, 236–238,
393(doc.)
adaptation of human behavior,
378–379(doc.)
Darwin on, 323–326(doc.),
366–367(doc.)
Darwin's failure to address
heredity, 53–54
genetics and, 121
Gould's disagreement with,
236–239
and gradualism, 393–397(doc.)
Gray's divine view of, 95
group selection, 205–206,
210–212
guided variations, 342–343(doc.)
hierarchical theory of, 239–243,
248–249, 254–255, 258
Huxley's view of, 70–71
internal and external constraints,
241
kin selection, 211–213, 379(doc.)
levels of, 209–210, 248–249
optimization of, 253–254
painful nature of, 277–282
at phenotype level, 249–251
principle of divergence, 46,
326–327(doc.)
racial selection, 56–57, 223(fig.)
sexual selection, 46, 55–56,
205–207, 215–220
social selection, 336–339(doc.)
species selection, 400–401(doc.)
as tautology, 267, 402–405(doc.)
through proper conduct, 389(doc.)
See also Acquired characteristics;
Adaptation; Artificial selection;
Existence, struggle for;
Individual selection
Natural Theology (Paley), 39
Naturalism and the origins of life,
162–164
Naturalist (Wilson), 285
Naturphilosophie, 80, 158, 393(doc.)
influence on Agassiz, 91–93
influence on Gould, 235–236, 245
spontaneous generation, 150–151
Nazism, 79, 81–82
Neanderthal man, 201
controversy surrounding, 172–174,
172(fig.), 187–194
DNA comparison with humans,
193
language development in,
194–197
as missing link, 345–346(doc.)
Newman, John Henry, Cardinal, 106
Newton, Sir Isaac, xii, 148
NOMA principle, 266
Nuclear weapons stockpiling, 139
Numbers, Ronald, 114